Tackling the Concussion Epidemic

Tom A. Schweizer • Andrew J. Baker
Editors

Tackling the Concussion Epidemic

A Bench to Bedside Approach

Springer

Editors
Tom A. Schweizer
Keenan Research Centre
St. Michael's Hospital, Unity Health
Toronto and University of Toronto
(Neurosurgery)
Toronto, ON, Canada

Andrew J. Baker
Departments of Critical Care
and Anesthesia
St. Michael's Hospital, Unity Health
Toronto and University of Toronto
Toronto, ON, Canada

ISBN 978-3-030-93812-3 ISBN 978-3-030-93813-0 (eBook)
https://doi.org/10.1007/978-3-030-93813-0

This Springer imprint is published by the registered company Springer Nature Switzerland AG
The registered company address is: Gewerbestrasse 11, 6330 Cham, Switzerland

To my wife Melanie, I am grateful to share my life with you.

To my children Evan and Averie – watching you grow and excel brings me constant pride and joy.

TAS

For my wife Jane, and for people everywhere with experiences of concussion.

AJB

Foreword

Tackling the Concussion Epidemic: A Bench to Bedside Approach is an impressive collection by an international group of leaders in the field of concussion. This book comes at a time when concussion is firmly in the public consciousness with increasing academic publications and funding to support this work and even a recent Hollywood movie focused on the subject. This is a big shift from a decade ago. Surprisingly, when my group resurveyed Canadian medical schools in 2017, concussion was still absent from the curriculum in some schools, although the number of schools without concussion on the curriculum was shrinking [1]. I joined the Canadian Academy of Sport Medicine (CASM) concussion committee which ultimately published in 2000 one of the most important concussion guidelines ever issued entitled "Guidelines for Assessment and Management of Sports Related Concussion" [2]. It contained two extremely important and original principles of management of sport concussion: (1) A player suspected of a concussion should be removed from the game or practice. (2) Every player suspected of a concussion should be examined by a "medical doctor." These were revolutionary principles at the time, and their value has been recognized ever since. Indeed, they have been enshrined in almost all subsequent iterations of concussion guidelines in Canada, and internationally, including the Ontario Neurotrauma Foundation Guidelines [3], Parachute Canada guidelines [4], and the International Concussion Consensus Conferences organized by the Concussion in Sport Group (CISG), the last one in Berlin in 2016, the proceedings of which were published in 2017 [5].

The Sport Concussion Conferences held by the CISG every 3–4 years for the last 20 years have served well as the basis for recognition and management of sports concussions and concussions sustained in other ways including work, school, motor vehicle crashes, falls at home, and intimate partner violence. Table 1 summarizes the major accomplishments of the last 20 years in the field of concussion, and Table 2 lists the challenges that must be met in the future to provide more informed management and prevention of concussions.

Table 1 Accomplishments in the concussion field since 2000

CASM concussion committee "guidelines for assessment and management of sports related concussion"
Concussion in sport group's international consensus conferences
Establishment of a working definition for sport concussion unencumbered by the GCS
National Guidelines for Management of Sports Concussions (Parachute Canada)
Guidelines for management of school-based concussions (Ontario Ministries of Sport, Education and Health, PPM158)
Gradual acceptance of the sports concussion definition for non-sports concussions
Almost every medical school includes concussion in its curriculum
Acceptance that concussion consequences are a spectrum of disorders
Realization that Second Impact Syndrome (SIS) exists and is preventable
Realization that repetitive concussion can cause brain degeneration
Improvement in return to play guidelines and international adoption
Refinement of recommendations for never return to contact sports or other risk environment.
Improvement in return to school/learn guidelines and types of accommodations
Improvement in return to work guidelines and types of accommodations
Realization that anxiety and depression are due to brain injury and not just reactions because of factors such as inability to return to play
Improved ability to differentiate concussion from whiplash, although they frequently co-exist
Improvement in understanding of the biomechanical forces that cause concussion
Management of concussion Requires a multidisciplinary, individualized team approach
More insurers, lawyers and judges now understand that a brain injury can occur without bruising or swelling of the scalp, without a direct blow to the head, and without a positive CT scan or MRI.
Enhanced likelihood that blood biomarkers will aid in the diagnosis of concussion and the assessment of neurodegeneration following concussion.

Definitions of Concussion and the Consequences of Concussion

Nosological debates have raged on about the definitions of concussion and the consequences of concussion mainly due to our lack of knowledge about the precise underlying biological mechanisms. Classically, it has been acceptable to use symptoms and syndromes to define conditions like concussion and post-concussion syndrome (PCS) when the precise pathophysiology is unknown. Thus, questions like "should we call it concussion or mild traumatic brain injury (mTBI)?" or "should we call it post-concussion syndrome or persisting concussion symptoms?" (also PCS) are often asked by students of concussion. Similarly, does one use the definitions provided by the Diagnostic and Statistical Manual of Mental Health Disorders (currently DSM-5) of the American Psychiatric Association, the American Academy of Neurology or the Physical Medicine and Rehabilitation Society definitions? In my view, the most accurate and useful definitions have been provided by the CISG of which I have been a participant for several years, with the most recent definitions

Table 2 Persisting problems in the concussion field

No proven imaging or blood biomarker
Failure to substantially decrease concussions in professional and amateur collision sports
Failure to provide adequate and timely access to medical doctors for the diagnosis and management of concussion in many parts of developed and less developed countries. therapists are not adequate substitutes.
Failure to fully investigate all sports deaths by autopsy and inquest
Lack of sufficient enforcement of existing concussion management guidelines in sports and non-sports, including the need for all governments to enact concussion legislation
Failure to find more effective treatments for many of the symptoms of concussion including headache, anxiety, depression, PTSD, vertigo, fatigue, photosensitivity and screen intolerance, etc.
Failure to develop a laboratory model of concussion proven to simulate human concussion
Failure to develop proven physical methods for preventing concussion other than abstinence. For example, there is no helmet or other device that is proven to prevent concussions.
More work needs to be done on demonstrating the value of education and rules for prevention of sports concussions.
Failure to develop proven methods of detection of concussions in athletes by sensors.
Failure to define the early clinical manifestations of CTE in time to prevent or treat it.
Lack of a defined clinical entity that is accompanied by the presence of CTE at autopsy.
Lack of consensus concerning the pathological diagnosis of CTE
Insufficient knowledge about the incidence and prevention of concussions and CTE in intimate partner violence

published in 2017 [6]. Also, my view is that mTBI is an oxymoron and potentially implies an under recognition that patients may have lifelong symptoms after a concussion. Also, mTBI is based on the Glasgow Coma Scale (GCS) which although invaluable for more severe brain injuries is useless and irrelevant in concussion. mTBI as a diagnosis is also harmful when patients do not get fully compensated for their concussions and PCS by the insurance and legal systems in many countries because their injuries are "mild" even though the consequences may be lifelong. Figure 1 shows the gradation of severity of brain injuries with concussion the mildest form. Therefore, it is time to fully endorse the term concussion and discard mTBI. The worth of the CISG definitions is proven by their usefulness for non-sport concussions listed above.

With concussions and their consequences, symptoms are still the most important diagnostic criteria because of the lack of reliable biomarkers (see below). Thus, it is surprising that the term "post-concussion syndrome" has been eliminated from DSM-5 presumably because a syndrome cannot be based on symptoms only. No matter, the CISG recommended that children and adults who remain symptomatic from concussion for more than 28 days have "persisting" concussion symptoms [6]. Ironically, we can continue to use the handy acronym PCS, but now it denotes "persisting concussion symptoms." Thus, symptoms are still important in concussion, perhaps more important than in any other field because that is all we have in our present state of knowledge! Fortunately, the human brain is so smart that it has

		Increasing Severity		
GCS (Glasgow Coma Scale)	All 15 = Normal (Therefore, GCS is irrelevent!)	13–15	9–12	3–8
[redacted]				
Clinical neurological deficits	None	There may be some	Always	Always
[redacted]				

Fig. 1 The severity of brain injuries is shown with increasing severity from left to right. Concussion is the least severe brain injury and can be differentiated from mTBI. Concussion is a more homogeneous category than mTBI which includes more severe brain injuries with focal neurological deficits and focal imaging abnormalities

provided us with a multitude of symptoms of concussion, about 100, in fact. Eventually, symptoms may help us with the task of identifying the anatomy and physiology of concussions in the brain's multilocation "network."

What follows is a brief list of some of the important topics covered by this book:

Biomarkers

The continuing lack of concrete neuroanatomical, neurophysiological, and neuropathological characterization of such a common condition as acute concussion is almost unique in the neurological field. Indeed, we still do not have reliable biomarkers of concussion, although there are many candidate biofluid and imaging biomarkers for the concussion spectrum. For example, neurofilament light chain has considerable promise as a biomarker for acute concussion [7, 8], and possibly also for the severity of neurodegeneration after concussions [9]. Imaging has been extensively studied as a biomarker for the concussion spectrum of disorders, and there are some promising leads such as PET scanning with F18 labelled phosphorylated tau ligands for CTE [10]. Magnetic resonance imaging (MRI) in several forms has been examined as biomarkers for the concussion spectrum, and recently MRI-based diagnostic criteria have been proposed for PCS [11].

Persisting Symptoms of Concussion

We are making progress in determining why some people do not recover in the first month after an acute concussion. A large number of pre-existing factors including migraine, heredity, sex, LOC, number of concussions previously sustained, and others may be implicated. It is important to continue the research since their identification may lead to prevention strategies. We have improved the treatment of many of the symptoms of Persisting Symptoms of Concussion (PCS) including headaches, vertigo, photophobia, screen intolerance [12], and especially the mental health sequelae including depression, anxiety, posttraumatic stress disorder, and concomitant injuries, especially cervical spine whiplash. Multidisciplinary clinics have made significant progress in treating PSC with measures like graded, "subthreshold" exercise [13]. Critical components of care in PSC are expert management of return to learn (RTL), return to play (RTP), and return to work (RTW), all of which are facilitated by an individualized multidisciplinary approach. Again, symptoms are important in PCS because treatment is targeted towards relief of specific symptoms because no treatment can be targeted toward the altered pathophysiology of concussion which remains unknown. Unfortunately, this dependence on symptom-based therapy has led to an enormous number of unproven therapies which are recommended to unsuspecting, desperate patients. This is tragic because of the waste of the patients' time and financial resources. Referral to a multidisciplinary clinic led by or at least including a medical doctor is recommended.

Chronic Traumatic Encephalopathy (CTE)

CTE is a type of neurodegenerative disease that is highly controversial with respect to its clinical characteristics, incidence after multiple concussions, and neuropathological features [14, 15]. It is uncertain what percentage of repetitively concussed people will develop this neurodegenerative disorder, and how often it is preceded by a state of chronic PSC. The Canadian Concussion Centre has actively promoted brain donations among athletes and others so that research can resolve some of these uncertainties. We have accumulated about 50 brains mainly from professional athletes who had sustained multiple concussions. Our initial publication in 2013 showed that CTE was present in only one-third of the first six cases, all of whom had sustained multiple concussions in sports [16]. One of the well-known National Hockey League players with CTE in our series was chronicled in a book authored by Ken Dryden. It is likely that other types of repetitive concussions such as in multiple motor vehicle crashes or intimate partner violence can also lead to CTE. Fortunately, the international multidisciplinary team of experts required to elucidate answers about the prevalence and clinical-pathological correlation and possible treatment and prevention of CTE are now working together to do the joint clinical-pathological studies required to further our knowledge of the prevention and possible treatment of this disorder.

Conclusions

We have made considerable gains in the concussion field in the last 20 years or so as shown in Table 1, and although I can't claim a single one to be my own, I have certainly enjoyed joining the chase for improvement. Similarly, I have enjoyed trying to unravel the mysteries of concussion that still remain as shown in Table 2. Indeed, Tables 1 and 2 demonstrate that the concussion spectrum of disorders is a major public health concern, and that further work is required to enhance concussion prevention, recognition, management, and research. It is my privilege and pleasure to have been asked by Drs. Schweizer and Baker to contribute to the solution through this foreword to their book which is aimed at solving some of the mysteries and deficiencies in Table 2 that remain either unsolved or uncorrected.

The present book is very timely. Tom A. Schweizer and Andrew J. Baker are both well-recognized experts in the field of concussion. They have gathered an impressive international team of world leaders in the field to discuss the state-of-the-art research and clinical management of concussion. As discussed above the book covers a very diverse list of topics relevant to the field and represents a significant introduction and summary of the current state. I strongly recommend this book to researchers and healthcare practitioners with an interest in this topic.

References

1. Mathieu F, Ellis MJ, Tator CH. Concussion education in Canadian medical schools: a 5 year follow-up survey. BMC Med Educ. 2018;18(1):316.
2. Canadian Academy of Sport Medicine Concussion Committee. Guidelines for assessment and management of sport-related concussion. Clin J Sport Med. 2000;10:209–11.
3. Foundation ON. Guidelines for concussion management in adults 2018. Available from: https://onf.org/knowledge-mobilization/acquired-brain-injury/guidelines-for-concussion-and-moderate-to-severe-traumatic-brain-injury/.
4. Parachute Canada. Canadian guideline on concussion in sport. Toronto: Parachute Toronto; 2017. p. 2017.
5. McCrory P, Meeuwisse W, Dvorak J, Aubry M, Bailes J, Broglio S, et al. Consensus statement on concussion in sport-the 5th international conference on concussion in sport held in Berlin, October 2016. Br J Sports Med. 2017;51:838–47.
6. McCrory P, Meeuwisse WH, Dvorak J, Echemendia RJ, Engebretsen L, Feddermann-Demont N, et al. 5th international conference on concussion in sport (Berlin). Br J Sports Med. 2017;51(11):837.

7. Oliver JM, Jones MT, Kirk KM, Gable DA, Repshas JT, Johnson TA, et al. Serum neurofilament light in American football athletes over the course of a season. J Neurotrauma. 2016;33(19):1784–9.
8. Shahim P, Zetterberg H, Tegner Y, Blennow K. Serum neurofilament light as a biomarker for mild traumatic brain injury in contact sports. Neurology. 2017;88(19):1788–94.
9. Taghdiri F, Multani N, Ozzoude M, Tarazi A, Khodadadi M, Wennberg R, et al. Neurofilament-light in former athletes: a potential biomarker of neurodegeneration and progression. Eur J Neurol. 2020;27(7):1170–7.
10. Vasilevskaya A, Taghdiri F, Burke C, Tarazi A, Naeimi SA, Khodadadi M, et al. Interaction of APOE4 alleles and PET tau imaging in former contact sport athletes. Neuroimage Clin. 2020;26:102212.
11. Panwar J, Hsu CC, Tator CH, Mikulis D. Magnetic resonance imaging criteria for post-concussion syndrome: a study of 127 post-concussion syndrome patients. J Neurotrauma. 2020;37(10):1190–6.
12. Mansur A, Hauer TM, Hussain MW, Alatwi MK, Tarazi A, Khodadadi M, et al. A nonliquid crystal display screen computer for treatment of photosensitivity and computer screen intolerance in post-concussion syndrome. J Neurotrauma. 2018;35(16):1886–94.
13. Ellis MJ, Leddy J, Willer B. Multi-disciplinary management of athletes with post-concussion syndrome: an evolving pathophysiological approach. Front Neurol. 2016;7:136.
14. Manley G, Gardner AJ, Schneider KJ, Guskiewicz KM, Bailes J, Cantu RC, et al. A systematic review of potential long-term effects of sport-related concussion. Br J Sports Med. 2017;51(12):969–77.
15. Smith DH, Johnson VE, Trojanowski JQ, Stewart W. Chronic traumatic encephalopathy — confusion and controversies. Nat Rev Neurol. 2019;15(3):179–83.
16. Hazrati LN, Tartaglia MC, Diamandis P, Davis KD, Green RE, Wennberg R, et al. Absence of chronic traumatic encephalopathy in retired football players with multiple concussions and neurological symptomatology. Front Hum Neurosci. 2013;7:222.

Charles H. Tator, OC, MD, PhD, FRCSC
Krembil Brain Institute
Toronto Western Hospital, Toronto, ON, Canada

ThinkFirst Canada and Parachute Canada
Canadian Concussion Centre, Toronto Western Hospital
Toronto, ON, Canada

Professor of Neurosurgery
University of Toronto
ON, Canada

Preface

Concussion can have a devastating impact on people's lives. Even if not devastating, concussion can impair the experience of life for weeks, months, years, and decades. We have discovered that concussion affects more people than we ever realized. We have also discovered that concussion impacts different people in different ways. And finally, we have discovered that access to evidence-informed care is variable. Variability means that there are real opportunities to mitigate the impact of concussion on individuals and populations.

I am a critical care physician, and my co-editor is a cognitive/imaging neuroscientist. The idea for this book came after noting there were very few books that aimed to summarize the general topic of concussion, that could be accessible to healthcare professionals, and researchers eager to learn something about this field. This book aims to address this by bringing together the best evidence that may inform our care of the patient experiencing the effects of concussion. This book spans the basic mechanics and pathophysiology of concussion, through its assessment, management, and complications. It aims to provide a primer of the various domains of study and application, with the main purpose being to provide practitioners insights and evidence to bring to their care of the person.

Over the last 20 years or so, there has been an exponential growth in the science of concussion and concussion care. The science is revealing the surprising extent of the issue and thus gives the impression of an epidemic. The emerging awareness of the high incidence combined with the potential for enduring impact has brought the issue of concussion into a high priority in public health. While prevention is always better, this book is about when prevention fails or the unexpected happens.

Concussion may have suffered from anecdotal evidence, from inadequate science, and from biases and assumptions partly related to the impression of the absence of objective and/or imaging findings. These historical disadvantages combined with the increased awareness of the magnitude of the issue is compelling background for a commitment to the science of concussion care. The authors of this book have understood and absorbed this challenge and dedicated their expert careers to this science. We both learned a lot editing this book and hope that it will not only provide a primer in the various domains of study but will also inspire a commitment to this science.

Toronto, ON, Canada Andrew J. Baker
Toronto, ON, Canada Tom A. Schweizer

Contents

Contributors

Andrew J. Baker, MD, FRCPC Departments of Critical Care and Anesthesia, St. Michael's Hospital, Unity Health Toronto and University of Toronto, Toronto, ON, Canada

Mark T. Bayley, MD, FRCPC Division of Physical Medicine and Rehabilitation, Department of Medicine, Toronto Rehabilitation Institute-University Health Network and University of Toronto, Toronto, ON, Canada

Amanda K. Ceniti, HBSc Arthur Sommer Rotenberg Suicide & Depression Studies Program, St. Michael's Hospital, Toronto, ON, Canada

Nathan W. Churchill, PhD Keenan Research Centre, St. Michael's Hospital, Unity Health Toronto, Toronto, ON, Canada

Angela Colantonio, PhD, OT Reg. (Ont.) Department of Occupational Science and Occupational Therapy, University of Toronto, Toronto, ON, Canada

Edward D. Hall, PhD Spinal Cord & Brain Injury Research Center (SCoBIRC), University of Kentucky College of Medicine, Lexington, KY, USA

Michael G. Hutchison, PhD Faculty of Kinesiology and Physical Education, University of Toronto, Toronto, ON, Canada

Sidney H. Kennedy, MD, FRCPC, FCAHS Arthur Sommer Rotenberg Suicide & Depression Studies Program, St. Michael's Hospital, Toronto, ON, Canada

Mehmet Kurt, PhD Department of Mechanical Engineering, University of Washington, Seattle, WA, USA

Kaveh Laksari, PhD Department of Biomedical Engineering, University of Arizona, Tucson, AZ, USA

Ian R. Mackenzie, MD, LMCC, FRCPC Department of Pathology, Vancouver General Hospital, Vancouver, BC, Canada

Department of Pathology and Laboratory Medicine, University of British Columbia, Vancouver, BC, Canada

Shawn Marshall, MD, MSc, FRCPC Division of Physical Medicine and Rehabilitation, Department of Medicine, Ottawa Hospital Research Institute, Ottawa, ON, Canada

Bruyere Research Institute, Ottawa, ON, Canada

University of Ottawa, Ottawa, ON, Canada

McKyla McIntyre, MD, MSc, FRCPC Physical Medicine and Rehabilitation, Toronto Rehabilitation Institute, Toronto, ON, Canada

David G. Munoz, MD, FRCPC, MSc Keenan Research Centre for Biomedical Research, The Li Ka Shing Knowledge Institute, and Division of Pathology, St. Michael's Hospital, Toronto, ON, Canada

Department of Laboratory Medicine and Pathobiology, University of Toronto, Toronto, ON, Canada

Linda Papa, MD, MSc Department of Emergency Medicine, Orlando Health, Orlando Regional Medical Center, Orlando, FL, USA

Eugene Park, PhD Keenan Research Centre, Li Ka Shing Knowledge Institute at St. Michael's Hospital, Toronto, ON, Canada

Sakina J. Rizvi, PhD Arthur Sommer Rotenberg Suicide & Depression Studies Program, St. Michael's Hospital, Toronto, ON, Canada

Jacqueline van Ierssel, PT, PhD Research Institute, Children's Hospital of Eastern Ontario, Ottawa, ON, USA

Lyndia Chun Wu, PhD Department of Mechanical Engineering, The University of British Columbia, Vancouver, BC, Canada

Chapter 1
Concussion Mechanism: Biomechanical Perspectives

Kaveh Laksari, Mehmet Kurt, and Lyndia Chun Wu

Abbreviations

AFM	Atomic force microscopy
BAM	Brain angle metric
BrIC	Brain injury criterion
bTBI	Blast traumatic brain injury
CasPr	Contactin-associated protein
CDE	Common data elements
CSF	Cerebrospinal fluid
CT	Computed tomography
DTI	Diffusion tensor imaging
FE	Finite elements
FITBIR	Federal interagency traumatic brain injury research
HIC	Head injury criterion
HITS	Head impact telemetry system
IMU	Inertial measurement unit
MEMS	Microelectromechanical systems
MRE	Magnetic resonance elastography
MRI	Magnetic resonance imaging
mTBI	Mild traumatic brain injury
NHTSA	National Highway Traffic Safety Association

K. Laksari
Department of Biomedical Engineering, University of Arizona, Tucson, AZ, USA

M. Kurt
Department of Mechanical Engineering, University of Washington, Seattle, WA, USA

L. C. Wu (✉)
Department of Mechanical Engineering, The University of British Columbia, Vancouver, BC, Canada
e-mail: lwu@mech.ubc.ca

T. A. Schweizer, A. J. Baker (eds.), *Tackling the Concussion Epidemic*,
https://doi.org/10.1007/978-3-030-93813-0_1

NOCSAE	National Operating Committee on Standards for Athletic Equipment
OCT	Optical coherence tomography
PIV	Particle image velocimetry
SI	Severity index
TBI	Traumatic brain injury
UBrIC	Universal brain injury criterion

Concussion is an invisible injury, which does not present with wounds or fractures that are apparent in other types of traumatic injury. This also makes it difficult to pinpoint the mechanism of concussion, despite the knowledge that it results from a mechanical insult to the head. Understanding the injury mechanism is key to the further development of diagnostic, prevention, and treatment strategies. In this chapter, we will describe current knowledge regarding the characteristics of concussive head loading (section "Head Loading Biomechanics") and the tissue and microstructural injury biomechanics (section "Brain Tissue and Microstructural Biomechanics"), along with current efforts and emerging technologies in concussion mechanism research (section "Emerging Studies in Injury Mechanisms"). We will discuss loading scenarios such as blunt impact and blast exposure, as well as sensing and modeling tools that can further help understand the tissue and microstructural mechanics in concussion (Fig. 1.1). As shown in the figure, concussion mechanism research has mainly focused on relating the biomechanics of head and brain loading to the risk of concussion.

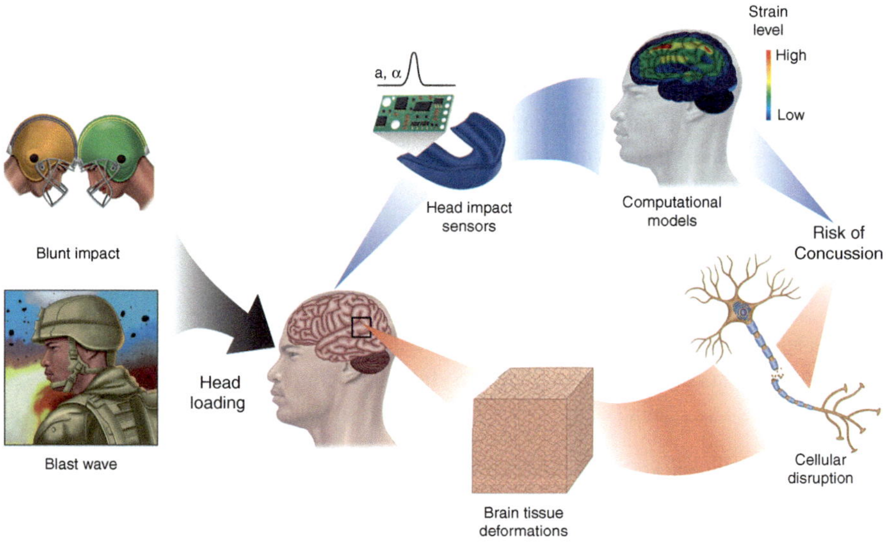

Fig. 1.1 Concussion mechanism overview. Concussions may be caused by blunt head impact, inertial head accelerations without impact, or blast overpressure. These external mechanical inputs to the head subsequently generate tissue deformations in the brain, followed and accompanied by cellular disruption. Currently, researchers are applying tools to further investigate human injury mechanisms, including wearable head impact sensors to measure head kinematics and computational brain models to estimate brain deformations

Head Loading Biomechanics

Direct Head Contact and Whiplash-Induced Head Loading

Concussions can occur from direct or indirect head loading in scenarios such as motor vehicle collisions, falls, sports, and recreational activities [1]. In motor vehicle accidents, the head may come in direct contact with surfaces in the vehicle or experience a jolting motion due to whiplash from sudden deceleration or acceleration of the vehicle. In falls, there is often direct head impact with the ground. In contact and collision sports, players may experience helmet–helmet impacts, head–object impacts, or whiplash-induced head motions due to body collisions. Although the brain is protected within the skull, head loading that results in damage, deformation, or accelerations of the skull could lead to subsequent brain loading and deformations. Some of these scenarios lead to severe traumatic brain injuries (TBIs) involving skull fracture or hematoma, which typically show clear lesions or structural damage in brain imaging using computed tomography (CT) or magnetic resonance imaging (MRI). Concussions, often referred to as mild traumatic brain injury (mTBI), usually present no observable structural changes using standard clinical imaging [2]. Past research has attempted to establish "injury risk functions" to quantify the probability of sustaining a concussion given head/brain loading inputs. Most studies have relied on a clinical diagnosis of concussion as the end point for the injury risk functions, due to the lack of agreed-upon objective and quantitative imaging, and physiological or functional biomarkers for concussion.

In an injury scenario involving direct head impact or whiplash, there is an impulsive force transfer to the head, which is thought to be the direct cause of concussion [2]. In closed head injury, the skull may experience transient local deformation, as well as sudden acceleration or deceleration [3]. Subsequently, the brain may experience focal or inertial loading. For most concussion cases where the head does not undergo substantial focal loading, it is thought that inertial skull loading is the main contributor to brain deformations [3]. Such skull loading is often characterized by linear (translational) and rotational (angular) skull accelerations lasting from a few milliseconds to tens of milliseconds [4, 5].

One main debate in concussion mechanism research is whether concussion risk is better correlated with the linear or rotational component of skull kinematics [3]. Head injury risk functions that dominate current safety regulation standards (e.g., helmet testing standards) are calculated from linear accelerations of the skull [6]. On the other hand, in 1943, Holbourn hypothesized that rotational skull kinematics could cause greater brain deformations in concussion scenarios [7]. He argued that since the brain is more or less incompressible with very low shear modulus (modulus of rigidity), it behaves like a low-rigidity gel and would deform substantially upon rotational skull accelerations but not linear skull accelerations.

A set of primate experiments led by Ommaya and Gennarelli tested Holbourn's theory in the 1960s to 1970s [8, 9]. It was shown that both linear and rotational accelerations of the skull contribute to focal lesions in the brain, while only head rotation could produce diffuse lesions in the brain [9]. The authors concluded that what was defined as "cerebral concussion" at the time could only result from rotational head accelerations and not linear acceleration alone, while if direct head contact were also involved, the injury tended to be more severe due to additional focal effects. From this work, Ommaya and Gennarelli proposed the "Centripetal Theory" of cerebral concussions, stating that concussion is caused by mechanically induced strains, which begin at the surfaces of the brain in the mild cases and extend inwards to affect deeper brain centers in the severe cases [9]. The depth of brain deformations and severity of trauma are thought to be influenced by the magnitude (i.e., energy level), direction, and duration of acceleration impulses experienced by the skull.

Investigating the direction dependence of injury, one study showed that primate head rotation in the coronal plane (which would be anatomically equivalent to human coronal head motion) produced more severe diffuse axonal injury (DAI) compared to other planes of head rotation [10]. In addition, Ommaya et al. investigated the effect of impulse duration, and similar to earlier research by Kornhauser [11] on structure failure analysis, they argued that short duration impacts have rotational velocity-dependent brain injury risk, while long duration impacts have peak rotational acceleration-dependent brain injury risk, where the critical duration may be around a 10–50 ms range [8]. In the context of real-world impacts, the short duration impacts may involve an unpadded head impacting with hard surfaces, while longer duration impacts may involve a padded head impacting soft surfaces or whiplash-type events.

Blast-Induced Head Loading

Another common concussion scenario more often found in military activities is blast-induced traumatic brain injuries (bTBI), which are mainly caused by high-pressure blast waves entering the brain through the skull and its cavities [12–14]. In his seminal 1950 publication, Benzinger described blast as a "shot without a bullet" [15]. In fact, acute bTBI could be a more complex injury that involves three mechanisms: primary injury, involving the overpressure wave propagating through the brain and its results on deforming the brain tissue; secondary injury, occurring due to penetrating or non-penetrating head impact with projectiles produced from the blast; and tertiary injury, occurring through head impact or acceleration as the body moves due to the blast wind [16–18]. The secondary and tertiary injury mechanisms involving head accelerations have similar mechanics as those discussed in section "Direct Head Contact and Whiplash-Induced Head Loading."

Much of the research in bTBI has focused on the primary mechanism, in which propagation of blast overpressure wave leads to the relative motion of brain tissue components. According to [19], there have been over 12,000 blast injury experiments involving 14 different in vivo mammalian models ranging in size from mice to cattle reported in the literature, with small rodents being the predominant experimental model. To briefly illustrate the primary injury mechanism, we can break down the sequence of events during an explosion. As a result of an explosion, a volume of highly compressed gas expands rapidly to occupy a volume that is orders of magnitude larger. This wave, i.e., the leading front of the expanding gases, travels with enormous speed (close to 1.5 times the speed of sound in laboratory shock tube experiments and as high as 20 times the speed of sound in open-field explosions) in the form of a sphere, a process known as the blast wave overpressure propagation [16]. Pioneering work in the 1950–70s showed that the severity of damage due to blast wave depends on the peak and duration of the overpressure, the density of transmitting medium (air or water), and the distance from the epicenter of the explosion [20–24]. While higher peak and longer duration of the overpressure wave could be more devastating, the strength of the wave diminishes inversely with distance. As the high-pressure shockwaves enter the brain, they disrupt its function either directly—through compression, tension, and shearing the brain tissue and its microstructure, or indirectly—through cavitation and introduction/ bursting of bubbles [25–27].

Measurement of Head Loading

While past theoretical and animal research has generated injury mechanism hypotheses, there is a substantial lack of human data to test these hypotheses. The recent advancements in microelectromechanical system (MEMS) inertial sensors have driven the development of wearable head impact sensors, which are typically deployed to gather data in scenarios such as contact and collision sports. Inertial measurement units (IMUs) including accelerometers and gyroscopes are typically used in wearable head impact sensors. Accelerometers measure linear accelerations, while gyroscopes measure rotational velocity, from which rotational accelerations can be derived by differentiation. If a large number of participants could be instrumented, human data from both injury and noninjury scenarios could be efficiently gathered given the relatively high incidence of concussions in sports.

Mounting these sensors on existing athletic headgear such as the helmet is a convenient option and was already used by early human studies [28]. In fact, the most widely used head impact sensor to date is the head impact telemetry system (HITS), which was developed in the early 2000s and deployed in youth and collegiate populations across multiple sports [5, 29, 30]. However, validation studies have shown that helmet sensors may dislocate from the head during impact, leading to large measurement errors [31, 32]. In recent years, other types of wearable sensors have been developed, including those mounted on skin and inside mouthguards [33–35]. Based on recent validation work, mouthguard-based sensing may provide tighter skull coupling due to direct mounting on the maxilla (upper jaw) and enable more accurate measurements of skull kinematics during head impact, compared to other mounting options prone to soft tissue artifacts and headgear dislocation [36, 37].

IMUs are most commonly used in commercial and research head impact sensors, enabling studies of the acceleration magnitude, direction, and duration factors in injury risk. Other sensing modalities have also been applied to estimate the force input to the head [38, 39] as opposed to the resultant accelerations. In addition, current inertial head impact sensors typically have limited sensing ranges of up to 200 g and 40 rad/s, with sensor bandwidths of 300–2000 Hz and 100–200 Hz for the accelerometer and gyroscope, respectively [40]. While such sensors are suited for measuring blunt head impacts with durations of tens of milliseconds, shorter duration impacts (e.g., bare-head impact with stiff surface) may require higher bandwidth sensors [40]. Such IMUs are also not suited for measuring blast injury scenarios, and blast overpressure sensors remain under development mainly in military research [41]. Considering blast overpressures that could result from explosions, as described in the last section, the blast sensors have design requirements of high-pressure sensing range up to 1000 kPa, fast response time within 1 ms, and high bandwidth up to a few thousand Hz [42, 43], which present significant engineering challenges.

Injury Criteria

One of the most important applications of concussion mechanism research is to reliably predict when injury has occurred for the screening and diagnosis of individuals with suspected traumatic brain injury. As such, a main research focus has been to identify biomechanical metrics (e.g., head kinematics) and develop injury risk functions based on these metrics. Such metrics are also called injury criteria, as their main purpose is to provide estimates of the likelihood of brain injury in a particular loading scenario. For example, one may develop an injury risk function based on peak linear acceleration of the head and associate the level of acceleration with the percentage of concussion risk based on empirical data. From such a risk function, an injury risk threshold can then be identified (e.g., above a certain head acceleration level, there is over 50% probability of sustaining a concussion). The injury risk threshold may be applied to set safety regulation standards or screen for injuries if head accelerations can be measured through wearable sensors and compared against this threshold.

In the 1960s, the rapid rise in motor vehicle accidents and deaths led to safety standards with the aim of reducing skull fractures. As a result, traffic deaths have been on a steady decline since the introduction of passive safety measures such as seat belts, interior padding, and airbags [44]. To address this, one of the first injury risk curves for intracranial pressure and skull fracture was developed at the Wayne State University based on results from cadaver head impact data and animal studies including hammer strike/air blast data on dog heads [45, 46]. Combined with these tests, additional experimental results from human acceleration/deceleration experiments led to the development of the Gadd severity index (SI) [47], which was the precursor to the currently used head impact criterion (HIC) [48]. Today, the National Highway Traffic Safety Administration (NHTSA) and the National Operating Committee on Standards for Athletic Equipment (NOCSAE) use HIC to evaluate injury risk based on translational acceleration in safety regulations and helmet design, with a HIC threshold of 1000 for 50% risk of skull fracture [6]. Although HIC and similar metrics have been effective in predicting severe traumatic brain injuries [49], their predictive value for milder injuries including concussion has been less certain. Studies have attempted to fit these injury risk functions using head impact data collected from clinically diagnosed concussion cases and found vastly different concussion risk thresholds. In one study, only 10% concussion risk was found for HIC = 700 [50], while another study found a HIC threshold of 250 for 50% risk of concussion [51].

As mentioned in section "Direct Head Contact and Whiplash-Induced Head Loading," theoretical and animal research has confirmed the importance of rotational head kinematics, which are not included in HIC or SI, in predicting concussion risk. Ommaya proposed rotational velocity and rotational acceleration injury criteria with human concussion thresholds of 50 rad/s for short duration inputs (<20 ms) and 1800 rad/s^2 for long duration inputs (>20 ms) scaled from primate data [8]. From the work of Margulies and colleagues in the late 1980s combining primate, analytical, and physical models, human injury tolerance for diffuse axonal injury (where concussion is considered a milder form of axonal injury) was found to range from 46.5 to 100 rad/s for short duration inputs and 7000–16,000 rad/s^2 for long duration inputs [52]. More recently, NHTSA developed the brain injury criterion (BrIC) to relate head rotational velocity to critical brain strains [53, 54]. BrIC is a composite measure of the ratios of head rotational velocity in each direction and corresponding critical values, which are 66.3, 53.8, and 41.5 rad/s for coronal, sagittal, and axial rotations, respectively [53]. Different from the prior rotational criteria, BrIC breaks down the resultant rotational velocity into directional components and builds in directional differences in injury tolerance.

While the early injury criteria are promising measures of head loading severity, they are mostly derived from correlations between head kinematics measures and surrogate measures that loosely approximate human injury outcomes, most notably cadaver skull fracture. Thus, such criteria, prone to underlying assumptions relating surrogates to the live human, are fundamentally an oversimplification of the complex mechanics of the human brain responding to inertial loads. Recognizing this limitation, recently developed metrics expand on the basic idea behind BrIC and focus more on brain mechanics, such as the brain angle metric (BAM) [55] and universal BrIC (UBrIC) [56], which approximate the brain's relative displacements with respect to the skull or more complex brain deformation metrics. These recent injury criteria, while still directly computed from skull kinematics, take into account the dynamical characteristics of the brain–skull system and are developed based on correlations with the most promising brain tissue response metrics. In the next section, we discuss tissue and microstructural mechanics during concussion as a more direct estimate of brain injury risk.

Brain Tissue and Microstructural Biomechanics

Rationale and Approach for Tissue-Based Biomechanics Metrics

As an alternative approach to skull kinematics-based injury criteria, researchers have developed brain biomechanics metrics, which are arguably more direct measures of the mechanical input transferred to the brain. Unlike head kinematics, it may not be possible to directly observe brain response in live human subjects during injury scenarios. As such, these metrics are estimated based on analytical and computational mechanical models of the skull–brain system, which enable estimation of brain response from skull inputs. Most of these approaches either use discrete mechanical elements, i.e., mass, spring, and dampers, to give a rigid body estimate of brain's relative motion with respect to the skull [55, 57–59], or more complex finite element (FE) models that include detailed representations of neuroanatomy and microstructure, and can simulate the local brain deformation and interaction with the stiff bony or membranous structures [60–62]. From these models, tissue metrics including displacements, strains, and pressures can be extracted and applied in similar manners as skull kinematics-based criteria (e.g., HIC) to estimate injury risk probability.

Analytical and Computational Models of Brain

Lumped Parameter Models

Lumped parameter models, which use a combination of zero-dimensional or discrete elements such as rigid body mass, spring, and damper systems, are widely applied in different engineering fields. This approach offers computationally efficient solutions for the dynamical behavior of otherwise highly complex systems. One of the first studies that applied this method to brain injury was performed by Kornhauser [11], who considered the brain–skull system as a linear second-order spring–mass system with base excitation from the skull and proposed that the relative brain displacement from the skull may be used as a measure for injury risk. Using a similar approach with different lump model representations, other simplified models of the brain have been developed to simulate the brain's response to external head loading [55, 57–59]. These models generally assume that the brain is a rigid mass attached to the skull via springs and dampers, and the amount of brain motion relative to the skull depends on the stiffness and damping parameters of the system. Both rotational and translational brain models have been developed, with varying levels of model complexity achieved by incorporating different combinations of inertias (discrete masses), and linear and torsional springs, and dampers [63, 64].

Such lumped parameter models require experimental measurements of brain–skull dynamics for parameter fitting and validation. One of the first attempts to combine in vivo experimental measurements with analytical modeling to study frequency-dependent behavior of the human brain was carried out in a recent study [58]. The authors developed a dynamic model of the skull–brain based on tagged MRI data with live human brain displacements from mild frontal head impacts [65]. It was found that for mild impacts, an underdamped second-order system could be fit to represent the brain–skull dynamics [58].

The simplicity and low computational cost of lumped models allow for the real-time analysis of measured kinematics from wearable sensors to estimate brain displacements and injury risks. Furthermore, these models easily lend themselves to parametric analyses to explore the relationship between input head kinematics and output brain response. However, such models cannot provide localized estimates of brain trauma, and run the risk of oversimplifying the nonlinear biomechanical effects of impacts on the brain. While researchers have found that rigid displacements of the brain in lumped parameter models could be correlated with injury, a more detailed analysis of injury mechanisms requires a model incorporating greater complexity.

Finite Element Models

Finite element (FE) models have been developed to incorporate a higher degree of complexity in model geometry, material behavior, and boundary conditions. Parallel to the advances in measurement techniques using wearable sensors as described above, modern FE head models have enabled researchers to use the measured human head kinematics as model inputs and investigate how they generate brain tissue deformation [60–62].

FE models of the brain allow for modeling the complex geometry of the skull and brain, including anatomical structures such as the scalp, meninges, bridging veins, and ventricles, and also differentiate between white and gray matter, which have distinct biomechanical properties. Furthermore, the material properties in FE brain models cover a range of linear elastic and nonlinear hyperelastic material models and rate-dependent viscoelastic characteristics [61, 66–68]. These material properties are based on mechanical testing of ex vivo tissue from cadaveric and animal specimens [66, 69, 70], or modern elastography techniques to extract in vivo tissue response from mild vibrations to the head [71–73]. Current FE models are generally based on representative adult brain/skull geometries [74], with a more recent move toward subject-specific models based on MRI scans [75–78].

While finite element models are useful for obtaining detailed models of the regional effects of concussion on the brain, finite element simulations are computationally expensive, with current models requiring hours of simulation time for a single impact. Recent research has sought to reduce computational costs through the development of precomputed atlases where a comprehensive test matrix of simulations is precomputed and interpolated to estimate the outputs of new impacts [79] and through deep learning architectures where peak strains are predicted using neural networks given input kinematics [80]. These models are also benefitting from more recent approaches to augment head kinematic data [81] to provide large training datasets required for neural network algorithms. Despite the computational cost, finite element models are still the best tools available at present to provide the most detailed picture of the mechanics of the brain during concussion.

Strain, Strain Rate, and Pressure

When applying analytical and computational models of the brain, it is of paramount importance to define output metrics that are correlated with brain injury. In the lumped parameter models, this has typically been the brain angle or displacement itself, but for finite element models, there is a greater diversity of biomechanical metrics that we correlate with brain injury. Historically, tissue strain has been studied as one of the main metrics to correlate with injury. In early in vitro experiments, it was found that excessive tissue stretch could lead to drastic physiological changes. Experiments on squid giant axons resulted in a graded depolarization response to impulsive stretching with incrementally increased strain and strain rate, where the neurons could not recover to normal membrane potential beyond a strain level of 20% [82]. This and other studies [83] clearly illustrated that physiological dysfunction could result from tissue strains and laid the foundation for using peak tissue strain as a measure of brain injury risk.

Aside from peak strain, other composite strain metrics have been proposed, such as the cumulative strain damage measure (CSDM), which describes the total volume fraction of brain tissue that undergoes strain values larger than a prescribed threshold (commonly 0.15 and 0.25 [62]) to provides an estimate for the overall brain deformation levels. Although complete consensus has not been reached on an exact strain value cutoff for concussion, various studies have reached comparable approximations for injury-prone conditions. Maximum principal strain values of 20–30% in the corpus callosum have been proposed as injury thresholds corresponding to 50 percent risk of injury [84]. In addition, motivated by structural and failure mechanics, a range of other FE-based tissue injury prediction metrics have been studied, including von Mises stress and the product of strain and strain rate [85].

Pressure is another promising parameter that is extensively cited as an injury metric in blast injury but has also been implicated in blunt impact/whiplash concussion. Injury-prone pressure and duration levels reported in the literature for humans under blast loading range from 30 to 200 kPa and 20 to 200 ms [13, 18, 25]. For blunt impacts, a recent study in collegiate American football found that intracranial pressure was the best predictor of concussions [86]. Other studies have similarly found correlations between the pressure level generated inside the brain and the resulting injury conditions in simulated blunt impacts [84, 87]. While pressure has been implicated in both blast and blunt impacts, it is worth mentioning that the pressure values derived from blunt impact simulations, unlike blast simulations, are in fact a function of the multiaxial stress response induced in the brain tissue rather than external pressure wave entering the skull.

Microstructural Metrics

The reliability of model-based brain injury studies critically depends on the fidelity of the head injury model and impact simulation, particularly in regions of interest. Based on early experimental studies such as the in vitro squid giant axon study [82], axonal strain was observed to directly cause physiological dysfunction. Since the brain's axons are mostly organized in an anisotropic manner through fiber tracts within white matter, it is thought that the inclusion of white matter tracts and their anisotropic material properties are necessary to accurately predict injury using finite element models. Indeed, the significance of including white matter tracts on brain deformation under head impacts has been confirmed recently [88–93].

Although scarce information exists on the effect of these fiber tracts on mechanical behavior of the tissue, indirect measurements of human brain fiber orientation and dispersion are becoming readily available through high angular resolution diffusion tensor imaging (DTI). To estimate loadings on these fibers, models may simply project the isotropic deformation metrics such as peak principal strain onto the fiber directions [90, 94], or in the case of more advanced models, the fiber anisotropy may be incorporated into models of brain tissue's biomechanical response [91] or account for full tractographic structure incorporated into biomechanical models [95]. Recently, various functions involving strain (S), strain rate (SR), and the multiplication of the two (S × SR) along the axon fibers have been proposed as predictors of injury [87]. In addition, empirical cumulative distribution functions involving S, SR, and S × SR have been proposed, such as Pop90 in which 90% of the brain exceeds a predefined value according to computations. Using statistical analysis, the best predictors of traumatic axonal injury were found in one animal study to be the tract-oriented strain (S > 6–7%), strain rate (SR > 38–40s^{-1}), and strain times strain rate (S × SR > 1.3–1.8 s^{-1}), when the thresholds were exceeded by 90% of the brain volume [87].

Region-Dependent Mechanical Vulnerability

As the human brain is a soft and complex biomaterial with membranes and fluid–structure interactions, complicated mechanical phenomena such as deformation localization, friction, and shear wave reflection occur during rapid head motion. This will inevitably make certain brain substructures and regions more vulnerable to deformation and damage than others. Both computational and experimental studies of TBI biomechanics have been utilized to address the question of regional mechanical vulnerability. The corpus callosum and brainstem have been the main regions of interest in this endeavor, since they frequently show pathological changes and lesions either through postmortem studies [96, 97] or through longitudinal imaging studies [98, 99]. Deep brain structures including the corpus callosum and the thalamus have indeed been shown to be mechanically vulnerable, with finite element simulation studies reporting high stress/strain concentrations in [84, 100, 101]. Some studies have also chosen to separately model these vulnerable substructures. For example, a nonlinear reduced order model of corpus callosum was developed under coronal excitation and was used to demonstrate that the corpus callosum exhibited strongly nonlinear behavior during high-amplitude impact [102].

The tentorium and falx are also hypothesized to be contributing factors in the localization of tissue deformations in deep brain regions. Mechanical neuroimaging studies have postulated that shear wave scattering occurs along these dural folds [103, 104], which might contribute to higher pressure and deformation seen in deep brain structures [104–107]. Other locations where these folds are believed to exacerbate mechanical deformations are the cerebellum–cerebrum interface and the brainstem [106, 108].

Blast waves can also cause regional mechanical damage in the brain, with a particular research emphasis on the effects of cavitation mechanisms [109]. Finite element simulations provide evidence that cavitation collapse causes localized high-pressure fields and therefore damage to regions adjacent to cerebrospinal fluid (CSF), including the periventricular regions [110]. In forward blast wave simulations, CSF cavitation has been associated with increased strain levels in the cerebral cortex, brainstem, and periventricular regions [110, 111], whereas simulation of backward blast was associated with largest strains in the occipital lobe [111].

Dynamical Considerations

As mentioned in section "Direct Head Contact and Whiplash-Induced Head Loading," impact duration is considered an important factor in estimating injury risk, which points toward a potential frequency dependence of brain injury risk. Despite the strongly damped nature of the brain–skull system dynamics [58], an impact to the brain can induce oscillatory shear wave behavior in the brain tissue due to the extremely soft nature of the human brain. If the brain–skull system is considered underdamped, these oscillations could be amplified at certain frequencies depending on how the brain's material properties are parametrized, creating the effect commonly referred to as resonance. Therefore, it is essential to understand the frequency response of the brain during head impacts and methods that have been used to characterize this phenomenon.

In initial attempts to estimate the frequency–response characteristics, researchers have used simplified models of the brain–skull system, including linear and nonlinear lumped parameter and continuum surrogate models, i.e., continuous reduced order models of the brain–skull system with simplified geometries and material models. Simple lumped models of the brain–skull system demonstrated that in different loading regimes, injury could be more sensitive to peak acceleration or maximum change in velocity or a combination of both [11]. In the time and frequency domain, the authors found that the brain exhibits resonance-like behavior near 20 Hz, and a troubling finding since American football impacts was found to have dominant input frequency near this frequency [58]. In a continuum brain surrogate model, the induced maximum strain was shown to peak around 25 Hz with respect to the frequency of the applied motion [112].

While lumped parameter or simplified models mentioned previously can be used to assess the frequency response of the whole brain, more complex frequency-based methods, such as advanced modal analysis, can begin to both identify and localize higher-order frequency dependencies in the finite element models of the human brain. Modal analysis has been utilized to extract the most dominant modal behavior (i.e., inherent dynamic characteristics of the brain in terms of vibration frequencies, mode shapes, and damping factors) of the brain's deformation by simulating a finite element model of the human brain with football head impacts [101]. Examination of temporal modes in the brain demonstrated that the brain's deformation is most sensitive in low-frequency regimes close to 30 Hz, and that for most subconcussive head impacts, the dynamics of brain deformation is dominated by a single global mode. This led to the postulation that the existence of localized modes and multimodal behavior in the brain as a hyperviscoelastic medium led to strain concentration patterns, particularly in deep brain regions, which is consistent with reported concussion pathology.

Since the frequency response of the brain is largely dependent on skull kinematics, there have also been efforts to study the frequency-dependent behavior of the human brain through idealized kinematics approaches. The effect of peak angular velocity amplitude and duration on the brain strain distribution was among the factors considered in one study, and it was shown that the profile of the angular velocity, which is directly correlated with the instantaneous frequency profile, can result in more than 20% change in the brain strain distribution [113]. In another study, idealized coronal rotations in a reduced order model of the corpus callosum have shown the existence of higher harmonics (around 45 Hz) near deep brain regions [102].

Blast scenarios induce a broadband mechanical response in the brain and therefore necessitate the consideration of brain tissue behavior over a large strain amplitude and a broad frequency range [114]. The incorporation of high-frequency material loading ranges of the brain (up to 10 MHz) has been proposed to increase the accuracy of numerical simulations of blast TBI [115]. The frequency spectrum of the blast wave has also been considered to be an important factor as it not only affects the frequency response of the brain, but also of the skull deformation. It has been speculated that transmission of low-frequency blast waves into the brain is more effective than high-frequency waves [116], although interactions between low and high-frequency modes could lead to complex deformation fields [117] and high-frequency skull flexures could lead to elevated levels of axonal deformation [118].

Emerging Studies in Injury Mechanisms

Sensor-Based Understanding

In recent years, there has been a substantial drive to collect more human data containing both impact biomechanics and concussion outcome information, such that concussion mechanism hypotheses can be tested, and human concussion thresholds can be confirmed. From past animal work, some injury criteria and thresholds have been extrapolated for humans, e.g., 50 rad/s rotational velocity and 1800 rad/s^2 rotational acceleration thresholds scaled from primate experiments to predict "cerebral concussion" involving loss of consciousness (LOC) in humans [8]. In the past two decades, a number of studies have used the HITS to measure impact biomechanics associated with concussions in helmeted sports including American football and ice hockey [5, 29, 30]. These studies have typically reported the head impact kinematics associated with concussions to range from tens of g to nearly 200 g in linear acceleration, and thousands to over 15,000 rad/s^2 in rotational acceleration [119–121].

While most of the concussions involve high accelerations as expected, these impact kinematic values, when compared with noninjury data, match poorly with previously identified thresholds of clinical concussion, and there is substantial overlap between noninjury events and concussion events in kinematic distributions [122]. From the limited human data available, many head impacts surpassing the proposed kinematic thresholds did not lead to clinical diagnoses of concussions [123], while some concussions were estimated to involve much lower skull kinematics (e.g., a 25th percentile concussive impact had 17.9 rad/s rotational velocity [122]). Some researchers have commented on the elusiveness of potential injury thresholds based on current data [124]. Other studies have suggested that directional kinematic parameters [123] and analytical model or FE-based brain deformation criteria are better predictors of injury outcomes [55, 123]. In more recent studies, it has been suggested that multiple impact accumulation, in addition to single severe impacts, may be an important mechanism of concussion [125–127]. Furthermore, the inconsistent findings in concussion risk thresholds may be due to the inconsistent definition of brain injury outcome in risk function research, which ranges from cadaver skull fracture to animal DAI/cerebral concussion (involving LOC) in earlier decades and have evolved to broader modern clinical concussion definitions in recent studies.

Given the need to standardize data collection, in 2018, the US National Institutes of Health (NIH) organized a working group to develop common data elements (CDE) for Biomechanical Devices in Head Impact and Blast Exposure Dynamics, with the aim of standardizing and aggregating biomechanical data collection for TBI research. The NIH has also established the "Federal Interagency Traumatic Brain Injury Research" (FITBIR) database for researchers to share TBI data. The current push to gather a large amount of human data should help generate a dataset that contains real-world head kinematics information associated with injury, which can then be used for further investigation of the biomechanical mechanism of injury.

Imaging-Based Understanding

As mentioned in section "Brain Tissue and Microstructural Biomechanics," computational models can estimate regional tissue-level brain mechanics to help understand and predict human concussion. One of the particular challenges with understanding TBI biomechanics is the lack of in vivo and/or in situ observations of human brain deformation and trauma during impacts to the head. Although various factors contribute to this poor state of knowledge, the inability to physically image brain motion in real-time during an impact has been a major limitation so far. Recent advances in neuroimaging have now started to provide promising solutions to this problem.

Magnetic resonance elastography (MRE) is one such advance that has begun to reveal the dynamic behavior of the brain in vivo. MRE is a noninvasive tool that allows researchers to probe the brain tissue, measure its dynamical response to oscillations of various frequencies in vivo, and process the shear wave datasets to obtain maps quantifying regional mechanical properties of the brain (i.e., elastograms*)* [128]. MRE has been used to understand how shear waves can propagate within the brain during an impact, by inducing "micro-impacts" to the brain [129] and studying shear wave reflections along dural folds [103] and between the cerebellum and cerebrum [130]. MRE has also been used to determine the regional material properties of brain tissue in healthy volunteers [72, 73, 131–133], and animal models [134, 135], which can be utilized to create more biofidelic materials in finite element models.

Other quantitative MR imaging methods such as tagged MRI have also been proposed and utilized to study the in vivo motion of the human brain under mild sagittal head bumps [65] and axial rotation [136]. Due to the impossibility of tracking brain motion under "high-amplitude" impacts, however, more recent studies have surgically embedded sonomicrometry crystals inside the brain tissue of a fresh cadaver to study three-dimensional dynamic deformation of the brain during pure rotational motion [137]. Here, it was determined the brain motion was around 20 Hz and lasted between 100 and 200 ms after the initial rotation, suggesting the presence of sustained deformations during this time period. Regional variances in the brain motion were also observed both in tagged MRI and in sonomicrometry experiments, which provides in vivo and postmortem evidence that mechanical deformation in the brain might be localized, such as in deep tissue structures including the corpus callosum. These datasets can also be further utilized to validate current finite element models under similar rotational loads.

For bTBI, imaging efforts for identification of the immediate brain biomechanics remain limited due to practical limitations (e.g., temporal resolution). Head surrogates and cadaveric samples have been primarily utilized to both images of the skull deformation and the surrogate brain model during blast exposure. Combinations of high-speed X-ray imaging with high-speed video tracking [117, 138] and time-resolved particle image velocimetry (PIV) [139] have been proposed to image the surrogate skull and brain motions, respectively. Recently, shear shock wavefronts in an ex vivo porcine brain were also observed, for the first time, through ultrasound imaging [140].

Micromechanical imaging has also recently emerged to study brain substructure mechanics, which will provide additional rich datasets for fundamental understanding and model validation [141]. For instance, the mechanical response of the brain–skull interface during indentation, with a particular focus on the pia–arachnoid complex, has been studied in animal models through optical coherence tomography (OCT) [142, 143] and atomic force microscopy (AFM) [144, 145]. Emergence of novel micromechanical and multiscale imaging techniques will likely advance our understanding of how the brain responds during an impact in the near future.

Multiscale Mechanical-Based Understanding

While the focus of brain injury mechanisms thus far has been at the tissue level, it is thought that trauma at the tissue level corresponds to axonal damage at the cellular level, which disrupts the ability of neurons to transmit information. Thus, novel experimental and computational approaches have been implemented to study the multiscale propagation of damage from the tissue level to the cellular and subcellular levels [146, 147]. Studies examining the transfer of tissue strain to cellular level have utilized contactin-associated protein (CasPr) expressed at the nodes of Ranvier in a chick embryo spinal cord model and identified the locations and numbers of broken axons [148, 149]. Uncoupled axons were found to have higher strain thresholds for breaking than coupled axons. In another experimental attempt to capture tissue-to-cell mechanical damage transmission, a novel 3D in vitro neuronal compression model interfaced with confocal fluorescence imaging was proposed to investigate the importance of tissue-level strain and strain rate on cellular viability, lifetime, and pathomorphology for neurons [150]. The 3D neuronal compression model demonstrated that cellular injury was driven by the strain on the entire cell rather than cytoskeleton only.

From a subcellular perspective, efforts have been made to model the mechanics of microtubules and binding tau proteins during axonal stretching [151, 152]. Rate-dependent breaking of microtubules has been associated with the viscoelastic properties of tau proteins [153]. Computational efforts are currently emerging to utilize the prior experimental efforts for multiscale characterizations of the brain and brain–skull interface to examine the whole scale response (i.e., all the way from skull down to the subcellular scale response) [154–156].

Multiscale modeling of bTBI has also been a recent focus in the community to study the effects and mechanism of primary micro-damage. Mechanical models of primary injury to axonal–oligodendrocyte structures, as well as damage to axonal networks and microtubules, were introduced and presented as a potential model for post-acute response of the brain microstructure [157–159]. In bTBI, due to the importance of transmission of loads from skull all the way to the axonal level, skull mechanics were also considered in multiscale mechanics, a direct contrast to blunt TBI. Skull flexures, for instance, were recently postulated to be a potential culprit for increased axonal deformation [160] and high strain rates obtained in finite element simulations were attributed to flexural displacements of the skull [118].

In summary, while concussion mechanism is still an active area of research, much understanding has been developed over the past decades to characterize injury-causing head loading and brain mechanics. With increasing efforts to gather human injury data and advances in novel imaging and modeling techniques, emerging studies will help gain a deeper understanding of the biomechanics and mechanobiology of concussion.

References

1. Cassidy JD, Carroll LJ, Peloso PM, Borg J, von Holst H, Holm LW, et al. Incidence, risk factors and prevention of mild traumatic brain injury: results of the WHO Collaborating Centre Task Force on Mild Traumatic Brain Injury. J Rehabil Med. 2004;36(June):28–60.
2. McCrory P, Meeuwisse W, Dvořák J, Aubry M, Bailes J, Broglio S, et al. Consensus statement on concussion in sport—The 5th international conference on concussion in sport held in Berlin, October 2016. Br J Sports Med. 2017;51(11):838–47.
3. Ommaya A, Thibault L, Bandak FA. Mechanisms of impact head injury. Int J Impact Eng. 1994;15(4):535–60.
4. Broglio SP, Sosnoff JJ, Shin S, He X, Alcaraz C, Zimmerman J. Head impacts during high school football: a biomechanical assessment. J Athl Train. 2009;44(4):342–9.
5. Rowson S, Brolinson G, Goforth M, Dietter D, Duma SM. Linear and angular head acceleration measurements in collegiate football. J Biomech Eng. 2009;131(6):061016.
6. Hertz E. A note on the head injury criterion (HIC) as a predictor of the risk of skull fracture. In: Proceedings: Association for the Advancement of automotive medicine annual conference. United States, TX: San Antonio; 1993. p. 303–12.
7. Holbourn. Mechanics of Head Injury. Lancet. 1943;438–41.
8. Ommaya AK, Hirsch AE. Tolerances for cerebral concussion from head impact and whiplash in primates. J Biomech. 1971;4:13–21.
9. Ommaya AK, Gennarelli TA. Cerebral concussion and traumatic unconsciousness. Brain. 1974;97(1):633–54.
10. Gennarelli TA, Thibault LE, Adams JH, Graham DI, Thompson CJ, Marcincin RP. Axonal injury and traumatic coma in the primate. Ann Neurol. 1982;12(6):564–74.
11. Kornhauser M. Prediction and evaluation of sensitivity to transient acceleration.pdf. J Appl Mech. 1954;21:371–80.
12. Chavko M, Watanabe T, Adeeb S, Lankasky J, Ahlers ST, McCarron RM. Relationship between orientation to a blast and pressure wave propagation inside the rat brain. J Neurosci Methods. 2011;195(1):61–6.
13. DePalma RG, Burris DG, Champion HR, Hodgson MJ. Blast injuries. N Engl J Med. 2005;352(13):1335–42.
14. Benzinger T, Brody D. Blast-related brain injury: imaging for clinical and research applications: report of the 2008 St. Louis workshop. J Neurotrauma. 2009;2144(December):2127–44.
15. Benzinger T. Physiological effects of blast in air and water. Ger Aviat Med World War II. 1950;2:1225–59.
16. Cernak I. Understanding blast-induced neurotrauma: how far have we come? Concussion. 2017;2(3):CNC42. https://doi.org/10.2217/cnc-2017-0006.
17. Goldstein LE, Fisher AM, Tagge CA, Zhang X-L, Velisek L, Sullivan JA, et al. Chronic traumatic encephalopathy in blast-exposed military veterans and a blast neurotrauma mouse model. Sci Transl Med. 2012;4(134):134ra60.
18. Chen Y, Huang W. Non-impact, blast-induced mild TBI and PTSD: concepts and caveats. Brain Inj. 2011;25(7–8):641–50.
19. Panzer MB, Wood GW, Bass CR. Scaling in neurotrauma: how do we apply animal experiments to people? Exp Neurol. 2014;261:120–6.
20. Friedlander FG. Propagation of a pulse in an inhomogeneous medium. New York University, Institute of Mathematical Sciences, Division of …; 1955.

21. Bowen I, Fletcher E, Richmond D. Estimate of man's tolerance to the direct effects of air blast. Washington, DC: Defense Atomic Support Agency; 1968.
22. Bowen IG, Fletcher ER, Richmond DR, Hirsch FG, White CS. Biophysical mechanisms and scaling procedures applicable in assessing responses of the thorax energized by air-blast overpressures or by nonpenetrating missiles. Ann N Y Acad Sci. 1968;152(1):122–46.
23. Richmond DR, Damon EG, Fletcher ER, Bowen IG, White CS. The relationship between selected blast-wave parameters and the response of mammals exposed to air blast. Lovelace Foundation for Medical Education and Research. Unied States, NM: Albuquerque; 1966.
24. Blair WC, Schulze FW. Psychological and physiological effects of muzzle and breech blast. Human Engineering Lab Aberdeen Proving Ground, United States, MD; 1956.
25. Cernak I, Noble-Haeusslein LJ. Traumatic brain injury: an overview of pathobiology with emphasis on military populations. J Cereb Blood Flow Metab. 2010;30(2):255–66.
26. Nakagawa A, Manley GT, Gean AD, Ohtani K, Armonda R, Tsukamoto A, et al. Mechanisms of primary blast-induced traumatic brain injury: insights from shock-wave research. J Neurotrauma. 2011;28(6):1101–19.
27. Crum LA, Mao Y. Acoustically enhanced bubble growth at low frequencies and its implications for human diver and marine mammal safety. J Acoust Soc Am. 1996;99(5):2898–907.
28. Pincemaille Y, Trosseille X, Mack P, Tarriere C, Breton F, Renault B, et al. Some new data related to human tolerance obtained from volunteer boxers. SAE Tech Pap. 1989;(892435):177–90.
29. Daniel RW, Rowson S, Duma SM. Head impact exposure in youth football: middle school ages 12 to 14 years. J Biomech Eng. 2014;1:3–8.
30. Wilcox BJ, Beckwith JG, Greenwald RM, Chu JJ, Mcallister TW, Flashman LA, et al. Head impact exposure in male and female collegiate ice hockey players. J Biomech. 2014;47(1):109–14.
31. Jadischke R, Viano DC, Dau N, King AI, McCarthy J. On the accuracy of the Head Impact Telemetry (HIT) System used in football helmets. J Biomech. 2013;46(13):2310–5.
32. Siegmund GP, Guskiewicz KM, Marshall SW, DeMarco AL, Bonin SJ. Laboratory validation of two wearable sensor systems for measuring head impact severity in football players. Ann Biomed Eng. 2015;44(4):1257–74.
33. Reynolds BB, Patrie J, Henry EJ, Goodkin HP, Broshek DK, Wintermark M, et al. Quantifying head impacts in collegiate lacrosse. Am J Sports Med. 2016;44(11):2947–56.
34. Camarillo DB, Shull PB, Mattson J, Shultz R, Garza D. An instrumented mouthguard for measuring linear and angular head impact kinematics in American football. Ann Biomed Eng. 2013;41(9):1939–49.
35. Wu LC, Kuo C, Loza J, Kurt M, Laksari K, Yanez LZ, et al. Detection of American football head impacts using biomechanical features and support vector machine classification. Sci Rep. 2018;8(1):1–14.
36. Wu LC, Nangia V, Bui K, Hammoor B, Kurt M, Hernandez F, et al. In vivo evaluation of wearable head impact sensors. Ann Biomed Eng. 2016;44(4):1234–45.
37. Rich AM, Filben TM, Miller LE, Tomblin BT, Van Gorkom AR, Hurst MA, et al. Development, validation and pilot field deployment of a custom mouthpiece for head impact measurement. Ann Biomed Eng. 2019;47(10):2109–21.
38. Rose SC, Yeates KO, Fuerst DR, Ercole PM, Nguyen JT, Pizzimenti NM. Head impact burden and change in neurocognitive function during a season of youth football. J Head Trauma Rehabil. 2019;34(2):87–95.
39. Wong RH, Wong AK, Bailes JE. Frequency, magnitude, and distribution of head impacts in Pop Warner football: the cumulative burden. Clin Neurol Neurosurg. 2014;118:1–4.
40. Wu LC, Laksari K, Kuo C, Luck JF, Kleiven S, 'Dale' Bass CR, et al. Bandwidth and sample rate requirements for wearable head impact sensors. J Biomech. 2016;49(13):2918–24.
41. Carr W, Stone JR, Walilko T, Young LA, Snook TL, Paggi ME, et al. Repeated low-level blast exposure: a descriptive human subjects study. Mil Med. 2016;181(5S):28–39.

42. Chu JJ, Beckwith JG, Leonard DS, Paye CM, Greenwald RM. Development of a multimodal blast sensor for measurement of head impact and over-pressurization exposure. Ann Biomed Eng. 2012;40(1):203–12.
43. Dionne J-P, Wong D, Halpin S, Levine J, Makris A. Helmet-mounted blast dosimeter for the military: electronics and signal processing challenges. In: 2010 3rd international symposium on applied sciences in biomedical and communication technologies (ISABEL 2010). IEEE, United States, MD; 2010. p. 1–5.
44. Nahum AM, Melvin JW. In: Nahum AM, Melvin JW, editors. Accidental injury. New York: Springer New York; 2002.
45. Gurdjian ES, Lissner HR, Latimer FR, Haddad BF, Webster JE. Quantitative determination of acceleration and intracranial pressure in experimental head injury; preliminary report. Neurology. 1953;3(6):417–23.
46. Gurdjian ES, Roberts VL, Thomas LM. Tolerance curves of acceleration and intracranial pressure and protective index in experimental head injury. J Trauma. 1966;6(5):600–4.
47. Gadd CW. Use of a weighted impulse criterion for estimating injury hazard. In: Proceedings of the10th stapp car crash conference. United States, NM: Hollomon Air Force Base; 1966. p. 164–74.
48. Newman JA. On the use of the Head Injury Criterion (HIC) in protective headgear evaluation. SAE Tech Pap. 1975. https://doi.org/10.4271/751162.
49. Bambach MR, Mitchell RJ, Grzebieta RH, Olivier J. The effectiveness of helmets in bicycle collisions with motor vehicles: a case-control study. Accid Anal Prev. 2013;53:78–88.
50. Funk JR, Rowson S, Daniel RW, Duma SM. Validation of concussion risk curves for collegiate football players derived from HITS data. Ann Biomed Eng. 2012;40(1):79–89.
51. Pellman EJ, Viano DC, Tucker AM, Casson IR. Concussion in professional football, part 1: reconstruction of game impacts and injuries. Neurosurgery. 2003;53(4):796.
52. Margulies SS, Thibault LE. A proposed tolerance criterion for diffuse axonal injury in man. J Biomech. 1992;25(8):917–23.
53. Takhounts EG, Craig MJ, Moorhouse K, McFadden J, Hasija V. Development of brain injury criteria (BrIC). Stapp Car Crash J. 2013;57(November):243–66.
54. Takhounts EG, Ridella SA, Rowson S, Duma SM. Kinematic rotational brain injury criterion (BRIC). In Proceedings of the 22nd Enhanced Safety of Vehicles Conference. United States, Washington, DC. 2010;11(0263):1–10.
55. Laksari K, Fanton M, Wu L, Nguyen T, Kurt M, Giordano C, et al. Multi-directional dynamic model for traumatic brain injury detection. J Neurotrauma. 2020;37(7). https://doi.org/10.1089/neu.2018.6340.
56. Gabler LF, Crandall JR, Panzer MB. Development of a metric for predicting brain strain responses using head kinematics. Ann Biomed Eng. 2018;46(7):972–85.
57. Low T, Stalnaker R. A lumped parameter approach to simulate the rotational head motion. Int Res Counc Biomechanics of Injury. 1987;15:203–15.
58. Laksari K, Wu LCLC, Kurt M, Kuo C, Camarillo DCDB. Resonance of human brain under head acceleration. J R Soc Interface. 2015;12(108):20150331.
59. Gabler LF, Joodaki H, Crandall JR, Panzer MB. Development of a single-degree-of-freedom mechanical model for predicting strain-based brain injury responses. J Biomech Eng. 2018;140(March):1–13.
60. Ji S, Ghadyani H, Bolander RP, Beckwith JG, Ford JC, McAllister TW, et al. Parametric comparisons of intracranial mechanical responses from three validated finite element models of the human head. Ann Biomed Eng. 2014;42(1):11–24.
61. Kleiven S. Evaluation of head injury criteria using a finite element model validated against experiments on localized brain motion, intracerebral acceleration, and intracranial pressure. Int J Crashworthiness. 2006;11(1):65–79.
62. Takhounts EG, Ridella SA, Hasija V, Tannous RE, Campbell JQ, Malone D, et al. Investigation of traumatic brain injuries using the next generation of simulated injury monitor (SIMon) finite element head model. Stapp Car Crash J. 2008;52(November):1–31.

63. Kornhauser M. Prediction and evaluation of sensitivity to transient accelerations. J Appl Mech. 1954;21:371–80.
64. Zou H, Kleiven S, Schmiedeler JP. The effect of brain mass and moment of inertia on relative brain-skull displacement during low-severity impacts. Int J Crashworthiness. 2007;12(4):341–53.
65. Sabet AA, Christoforou E, Zatlin B, Genin GM, Bayly PV. Deformation of the human brain induced by mild angular head acceleration. J Biomech. 2008;41(2):307–15.
66. Laksari K, Shafieian M, Darvish K. Constitutive model for brain tissue under finite compression. J Biomech. 2012;45(4):642–6.
67. Giudice JS, Zeng W, Wu T, Alshareef A, Shedd DF, Panzer MB. An analytical review of the numerical methods used for finite element modeling of traumatic brain injury. Ann Biomed Eng. 2019;47(9):1855–72.
68. Eskandari F, Shafieian M, Aghdam MM, Laksari K. Tension strain-softening and compression strain-stiffening behavior of brain white matter. Ann Biomed Eng. 2020:1–11.
69. Laksari K, Sadeghipour K, Darvish K. Mechanical response of brain tissue under blast loading. J Mech Behav Biomed Mater. 2014;32:132–44.
70. Budday S, Sommer G, Birkl C, Langkammer C, Haybaeck J, Kohnert J, et al. Mechanical characterization of human brain tissue. Acta Biomater. 2017;48:319–40.
71. Kurt M, Wu L, Laksari K, Ozkaya E, Suar ZM, Lv H, et al. Optimization of a multifrequency magnetic resonance elastography protocol for the human brain. J Neuroimaging. 2019;00:1–7.
72. Weickenmeier J, Kurt M, Wintermark M, Pauly KB, Weickenmeier J, Kurt M, et al. Magnetic resonance elastography of the brain: a comparison between pigs and humans. J Mech Behav Biomed Mater. 2018;77:702–10.
73. Johnson CL, McGarry MDJ, Gharibans AA, Weaver JB, Paulsen KD, Wang H, et al. Local mechanical properties of white matter structures in the human brain. NeuroImage. 2013;79:145–52.
74. Dixit P, Liu GR. A review on recent development of finite element models for head injury simulations. Arch Comput Methods Eng. 2017;24:979–1031.
75. Ho J, von Holst H, Kleiven S. Automatic generation and validation of patient-specific finite element head models suitable for crashworthiness analysis. Int J Crashworthiness. 2009;14(6):555–63.
76. Hu J, Jin X, Lee JB, Zhang L, Chaudhary V, Guthikonda M, et al. Intraoperative brain shift prediction using a 3D inhomogeneous patient-specific finite element model. J Neurosurg. 2007;106(1):164–9.
77. McAllister TW, Ford JC, Ji S, Beckwith JG, Flashman LA, Paulsen K, et al. Maximum principal strain and strain rate associated with concussion diagnosis correlates with changes in corpus callosum white matter indices. Ann Biomed Eng. 2012;40(1):127–40.
78. Miller LE, Urban JE, Stitzel JD. Development and validation of an atlas-based finite element brain model. Biomech Model Mechanobiol. 2016;15(5):1–14.
79. Ji S, Zhao W. A pre-computed brain response atlas for instantaneous strain estimation in contact sports. Ann Biomed Eng. 2015;43(8):1877–95.
80. Cai Y, Wu S, Zhao W, Li Z, Wu Z, Ji S. Concussion classification via deep learning using whole-brain white matter fiber strains. PLoS One. 2018;13(5):1–21.
81. Arrue P, Toosizadeh N, Babaee H, Laksari K. Low-rank representation of head impact kinematics: a data-driven emulator. Front Bioeng Biotechnol. 2020;8(September):1–11.
82. Galbraith JA, Thibault LE, Matteson DR. Mechanical and electrical responses of the squid giant axon to simple elongation. J Biomech Eng. 1993;115(1):13–22.
83. Smith DH, Wolf JA, Lusardi TA, Lee VM, Meaney DF. High tolerance and delayed elastic response of cultured axons to dynamic stretch injury. J Neurosci. 1999;19(11):4263–9.
84. Kleiven S. Predictors for traumatic brain injuries evaluated through accident reconstructions. Stapp Car Crash J. 2007;51(October):81–114.

85. Beckwith JG, Zhao W, Ji S, Ajamil AG, Bolander RP, Chu JJ, et al. Estimated brain tissue response following impacts associated with and without diagnosed concussion. Ann Biomed Eng. 2018;46(6):819–30.
86. Beckwith JG, Zhao W, Ji S, Ajamil AG, Bolander RP, Chu JJ, et al. Estimated brain tissue response following impacts associated with and without diagnosed concussion. Ann Biomed Eng. 2018;46(6):1–12.
87. Sullivan S, Eucker SA, Gabrieli D, Bradfield C, Coats B, Maltese MR, et al. White matter tract-oriented deformation predicts traumatic axonal brain injury and reveals rotational direction-specific vulnerabilities. Biomech Model Mechanobiol. 2015;14(4):877–96.
88. Giordano C, Cloots RJ, van Dommelen JA, Kleiven S. The influence of anisotropy on brain injury prediction. J Biomech. 2014;47(5):1052–9.
89. Zhao W, Ji S. White matter anisotropy for impact simulation and response sampling in traumatic brain injury. J Neurotrauma. 2019;36(2):250–63.
90. Ji S, Zhao W, Ford JC, Beckwith JG, Bolander RP, Greenwald RM, et al. Group-wise evaluation and comparison of white matter fiber strain and maximum principal strain in sports-related concussion. J Neurotrauma. 2014;14(603):1–43.
91. Giordano C, Zappalà S, Kleiven S. Anisotropic finite element models for brain injury prediction: the sensitivity of axonal strain to white matter tract inter-subject variability. Biomech Model Mechanobiol. 2017;16(4):1269–93.
92. Eskandari F, Shafieian M, Aghdam M, Laksari K. Structural anisotropy vs. mechanical anisotropy: the contribution of axonal fibers to the material properties of brain white matter. Ann Biomed Eng. 2021;49(3):991–9.
93. Eskandari F, Shafieian M, Aghdam MM, Laksari K. The importance of axonal directions in the brainstem injury during neurosurgical interventions. Injury. 2021;52(6):1271–6.
94. Sahoo D, Deck C, Willinger R. Brain injury tolerance limit based on computation of axonal strain. Accid Anal Prev. 2016;92:53–70.
95. Zhao W, Ford JC, Flashman LA, McAllister TW, Ji S. White matter injury susceptibility via fiber strain evaluation using whole-brain tractography. J Neurotrauma. 2015;14:(accepted).
96. Smith DH, Nonaka M, Miller R, Leoni M, Chen XH, Alsop D, et al. Immediate coma following inertial brain injury dependent on axonal damage in the brainstem. J Neurosurg. 2000;93(2):315–22.
97. Gurdjian E, Hodgson V, Thomas L, Patrick L. Significance of relative movements of scalp, skull, and intracranial contents during impact injury of the head*. J Neurosurg. 1968;29:70–2.
98. Wilde E, Chu Z, Bigler ED, Hunter JV, Fearing MA, Hanten G, et al. Diffusion tensor imaging in the corpus callosum in children after moderate to severe traumatic brain injury. J Neurotrauma. 2006;23(10):1412–26.
99. Dennis EL, Ellis MU, Marion SD, Jin Y, Moran L, Olsen A, et al. Callosal function in pediatric traumatic brain injury linked to disrupted white matter integrity. J Neurosci. 2015;35(28):10202–11.
100. Zhao W, Cai Y, Li Z, Ji S. Injury prediction and vulnerability assessment using strain and susceptibility measures of the deep white matter. Biomech Model Mechanobiol. 2017;16(5):1709–27.
101. Laksari K, Kurt M, Babaee H, Kleiven S, Camarillo D. Mechanistic insights into human brain impact dynamics through modal analysis. Phys Rev Lett. 2018;120(13):138101.
102. Mojahed A, Abderezaei J, Kurt M, Bergman LA, Vakakis AF. A nonlinear reduced-order model of the corpus callosum under planar coronal excitation. J Biomech Eng. 2020;142(9):0910090.
103. Clayton EH, Genin GM, Bayly PV. Transmission, attenuation and reflection of shear waves in the human brain. J R Soc Interface. 2012;9(76):2899–910.
104. Lu YC, Daphalapurkar NP, Knutsen AK, Glaister J, Pham DL, Butman JA, et al. A 3D computational head model under dynamic head rotation and head extension validated using live human brain data, including the falx and the tentorium. Ann Biomed Eng. 2019;47(9):1923–40.

105. Abderezaei J, Zhao W, Grijalva CL, Fabris G, Ji S, Laksari K, et al. Nonlinear dynamical behavior of the deep white matter during head impact. Phys Rev Appl. 2019;12(1):014058.
106. Ho J, Zhou Z, Li X, Kleiven S. The peculiar properties of the falx and tentorium in brain injury biomechanics. J Biomech. 2017;60:243–7.
107. Hernandez F, Giordano C, Goubran M, Parivash S, Grant G, Zeineh M, et al. Lateral impacts correlate with falx cerebri displacement and corpus callosum trauma in sports-related concussions. Biomech Model Mechanobiol. 2019;18(3):631–49.
108. Fijalkowski RJ, Yoganandan N, Zhang J. A finite element model of region-specific response for mild diffuse brain injury. Stapp Car Crash J. 2009;53:193–213.
109. Goeller J, Wardlaw A, Treichler D, O'Bruba J, Weiss G. Investigation of cavitation as a possible damage mechanism in blast-induced traumatic brain injury. J Neurotrauma. 2012;29(10):1970–81.
110. Panzer MB, Myers BS, Capehart BP, Bass CR. Development of a finite element model for blast brain injury and the effects of CSF cavitation. Ann Biomed Eng. 2012;40(7):1530–44.
111. Zhang L, Makwana R, Sharma S. Brain response to primary blast wave using validated finite element models of human head and advanced combat helmet. Front Neurol. 2013;4:88.
112. Margulies SS, Thibault LE. An analytical model of traumatic diffuse brain injury. J Biomech Eng. 1989;111(3):241–9.
113. Zhao W, Ji S. Brain strain uncertainty due to shape variation in and simplification of head angular velocity profiles. Biomech Model Mechanobiol. 2017;16(2):449–61.
114. Pervin F, Chen WW. Dynamic mechanical response of bovine gray matter and white matter brain tissues under compression. J Biomech. 2009;42(6):731–5.
115. Chafi MS, Ganpule S, Gu L, Chandra N. Dynamic response of brain subjected to blast loadings: influence of frequency ranges. Int J Appl Mech. 2011;3(4):803–23.
116. Säljö A, Mayorga M, Bolouri H, Svensson B, Hamberger A. Mechanisms and pathophysiology of the low-level blast brain injury in animal models. NeuroImage. 2011;54(Suppl. 1):S83–8.
117. Ouellet S, Philippens M. The multi-modal responses of a physical head model subjected to various blast exposure conditions. Shock Waves. 2018;28(1):19–36.
118. Garimella HT, Kraft RH, Przekwas AJ. Do blast induced skull flexures result in axonal deformation? PLoS One. 2018;13(3):e0190881.
119. Liao S, Lynall RC, Mihalik JP. The effect of head impact location on day of diagnosed concussion in college football. Med Sci Sports Exerc. 2016;48(7):1239–43.
120. Duhaime A-C, Beckwith JG, Maerlender AC, McAllister TW, Crisco JJ, Duma SM, Brolinson PG, Rowson S, Flashman LA, Chu JJ, Greenwald RM. Spectrum of acute clinical characteristics of diagnosed concussions in college athletes wearing instrumented helmets: clinical article. J Neurosurg. 2012;117(6):1092–9.
121. Guskiewicz KM, Ph D, Marshall SW. Measurement of head impacts in collegiate football players: impact biomechanics and acute clinical outcome after concussion. Neurosurgery. 2007;61(6):1244–53.
122. Rowson S, Duma SM, Beckwith JG, Chu JJ, Greenwald RM, Crisco JJ, et al. Rotational head kinematics in football impacts: an injury risk function for concussion. Ann Biomed Eng. 2012;40(1):1–13.
123. Hernandez F, Wu LC, Yip MC, Laksari K, Hoffman AR, Lopez JR, et al. Six degree-of-freedom measurements of human mild traumatic brain injury. Ann Biomed Eng. 2015;43(8):1918–34.
124. Guskiewicz KM, Mihalik JP. Biomechanics of sport concussion: quest for the elusive injury threshold. Exerc Sport Sci Rev. 2011;39(1):4–11.
125. Beckwith JG, Greenwald RM, Chu JJ, Crisco JJ, Rowson S, Duma SM, et al. Head impact exposure sustained by football players on days of diagnosed concussion. Med Sci Sports Exerc. 2013;45(4):737–46.
126. Broglio SP, Lapointe AP, O'Connor KL, McCrea M. Head impact density: a model to explain the elusive concussion threshold. J Neurotrauma. 2017;9:neu.2016.4767.

127. Stemper BD, Shah AS, Harezlak J, Rowson S, Mihalik JP, Duma SM, et al. Comparison of head impact exposure between concussed football athletes and matched controls: evidence for a possible second mechanism of sport-related concussion. Ann Biomed Eng. 2019;47(10):2057–72.
128. Kruse SA, Rose GH, Glaser KJ, Manduca A, Felmlee JP, Jack CR, et al. Magnetic resonance elastography of the brain. NeuroImage. 2008;39(1):231–7.
129. Mccracken P, Manduca A, Ehman R. MR Elastography for studying the biomechanics of traumatic brain injury. Int Soc Magn Reson Med. 10–16 July 2003. Toronto.
130. Okamoto RJ, Romano AJ, Johnson CL, Bayly PV. Insights into traumatic brain injury from MRI of harmonic brain motion. J Exp Neurosci. 2019;13:117906951984044.
131. Sack I, Beierbach B, Wuerfel J, Klatt D, Hamhaber U, Papazoglou S, et al. The impact of aging and gender on brain viscoelasticity. NeuroImage. 2009;46(3):652–7.
132. Kurt M, Wu L, Laksari K, Ozkaya E, Suar ZM, Lv H, et al. Optimization of a multifrequency magnetic resonance elastography protocol for the human brain. J Neuroimaging. 2019;29(4):440–6.
133. Lv H, Kurt M, Zeng N, Ozkaya E, Marcuz F, Wu L, et al. MR elastography frequency–dependent and independent parameters demonstrate accelerated decrease of brain stiffness in elder subjects. Eur Radiol. 2020;30(12):6614–23.
134. Clayton EH, Garbow JR, Bayly PV. Frequency-dependent viscoelastic parameters of mouse brain tissue estimated by MR elastography. Phys Med Biol. 2011;56(8):2391–406.
135. Weickenmeier J, Kurt M, Ozkaya E, de Rooij R, Ovaert TC, Ehman RL, et al. Brain stiffens post mortem. J Mech Behav Biomed Mater. 2018;84(April):88–98.
136. Knutsen AK, Magrath E, McEntee JE, Xing F, Prince JL, Bayly PV, et al. Improved measurement of brain deformation during mild head acceleration using a novel tagged MRI sequence. J Biomech. 2014;47(14):3475–81.
137. Alshareef A, Giudice JS, Forman J, Salzar RS, Panzer MB. A novel method for quantifying human in situ whole brain deformation under rotational loading using sonomicrometry. liebertpub.com. 2018;35(5):780–9.
138. Armiger RS, Otake Y, Iwaskiw AS, Wickwire AC, Ott KA, Voo LM, et al. Biomechanical response of blast loading to the head using 2D-3D cineradiographic registration. In: Conference Proceedings of the Society for Experimental Mechanics Series. Cham: Springer; 2014. p. 127–34.
139. Mejia-Alvarez R, Willis AM, Tartis M, Morgan RV. Understanding the role of fluid-structure interactions in traumatic brain injury [Internet]. tsfp-conference.org. 2017;48823.
140. Espíndola D, Lee S, Pinton G. Shear shock waves observed in the brain. Phys Rev Appl. 2017;8(4):044024.
141. Babu PKV, Radmacher M. Mechanics of brain tissues studied by atomic force microscopy: a perspective. Front Neurosci. 2019;13:600. https://doi.org/10.3389/fnins.2019.00600.
142. Scott GG, Coats B. Microstructural characterization of the pia-arachnoid complex using optical coherence tomography. IEEE Trans Med Imaging. 2015;34(7):1452–9.
143. Choi WJ, Wang RK. Optical coherence tomography imaging of cranial meninges post brain injury in vivo. Chin Opt Lett. 2017;15(9):090005.
144. Fabris G, Suar ZM, Kurt M. Micromechanical heterogeneity of the rat pia-arachnoid complex. Acta Biomater. 2019;100:29–37.
145. Suar ZM, Fabris G, Kurt M. Isolation and immunofluorescent staining of fresh rat pia–Arachnoid complex tissue for micromechanical characterization. Curr Protoc Neurosci. 2019;89(1):e83.
146. Meaney DF, Morrison B, Dale BC. The mechanics of traumatic brain injury: a review of what we know and what we need to know for reducing its societal burden. J Biomech Eng. 2014;136(2):021008.
147. Eskandari F, Shafieian M, Aghdam MM, Laksari K. A knowledge map analysis of brain biomechanics: current evidence and future directions. Clin Biomech. 2020;75:105000.

148. Singh S, Pelegri AA, Shreiber DI. Estimating axonal strain and failure following white matter stretch using contactin-associated protein as a fiduciary marker. J Biomech. 2017;51:32–41.
149. Singh S, Pelegri AA, Shreiber DI. Characterization of the three-dimensional kinematic behavior of axons in central nervous system white matter. Biomech Model Mechanobiol. 2015;14(6):1303–15.
150. Bar-Kochba E, Scimone MT, Estrada JB, Franck C. Strain and rate-dependent neuronal injury in a 3D in vitro compression model of traumatic brain injury. Sci Rep. 2016;6:30550.
151. Ahmadzadeh H, Smith DH, Shenoy VB. Mechanical effects of dynamic binding between tau proteins on microtubules during axonal injury. Biophys J. 2015;109(11):2328–37.
152. Tang-Schomer MD, Patel AR, Baas PW, Smith DH. Mechanical breaking of microtubules in axons during dynamic stretch injury underlies delayed elasticity, microtubule disassembly, and axon degeneration. FASEB J. 2010;24(5):1401–10.
153. Ahmadzadeh H, Smith DH, Shenoy VB. Viscoelasticity of tau proteins leads to strain rate-dependent breaking of microtubules during axonal stretch injury: predictions from a mathematical model. Biophys J. 2014;106(5):1123–33.
154. Montanino A, Kleiven S. Utilizing a structural mechanics approach to assess the primary effects of injury loads onto the axon and its components. Front Neurol. 2018;9:643. https://doi.org/10.3389/fneur.2018.00643.
155. Kraft RH, Mckee PJ, Dagro AM, Grafton ST. Combining the finite element method with structural connectome-based analysis for modeling neurotrauma: connectome neurotrauma mechanics. Sporns O, editor. PLoS Comput Biol. 2012;8(8):e1002619.
156. Panzer MB, Matthews KA, Yu AW, Morrison B, Meaney DF, Bass CR. A multiscale approach to blast neurotrauma modeling: part I – development of novel test devices for in vivo and in vitro blast injury models. Front Neurol. 2012;3(46):1–11.
157. Panzer MB, Matthews KA, Yu AW, Morrison B, Meaney DF, Bass CR. A multiscale approach to blast neurotrauma modeling: part I – development of novel test devices for in vivo and in vitro blast injury models. Front Neurol. 2012;3(March):46.
158. Effgen GB, Hue CD, Vogel E, Panzer MB, Meaney DF, Bass CR, et al. A multiscale approach to blast neurotrauma modeling: part II: methodology for inducing blast injury to in vitro models. Front Neurol. 2012;3:1–10.
159. Gupta RK. Multiscale modelling of blast-induced TBI mechanobiology – from body to neuron to molecule. Def Life Sci J. 2017;2(1):3–13.
160. Moss W, King M, Blackman E. Skull flexure from blast waves: a mechanism for brain injury with implications for helmet design. Phys Rev Lett. 2009;103(10):108702.

Chapter 2
The Pathophysiology of Concussion

Eugene Park and Andrew J. Baker

The clinical diagnosis of concussion largely relies on external indicators of neurological injury for diagnosis. The ever-evolving definition of concussion has been described as "a clinical syndrome of biomechanically induced alteration of brain function typically affecting memory and orientation, which may involve loss of consciousness" [1, 2]. Despite the vague clinical description, advances in imaging technologies, the development of clinical biomarkers, and preclinical animal studies have been used to identify subtle but significant microstructural abnormalities and identified molecular cascades occurring in the absence of macroscopic injury to the brain, which collectively contribute to changes in cognitive status. These pathophysiological events highlight the contribution of concussions as an important contributor to the spectrum of traumatic brain injury (TBI) despite being classified as a mild traumatic event. As technologies evolve, the ability to reliably and accurately detect subtle cellular changes in cerebral structure and function gives hope for the more accurate diagnosis of injuries, with potential prognostic and therapeutic applications. Understanding the link between biophysical transduction of force from tissue to cellular and even ultrastructural scales for an injury type that is notoriously heterogeneous is challenging from a preclinical modeling perspective. In turn, this challenge is amplified when it comes to developing therapeutic strategies [3].

This chapter focuses on our current understanding of the pathophysiology associated with these processes and their contribution to the evolving pathophysiology of concussion. Although many of the mechanisms described below are also applicable to moderate and severe TBIs, it can be assumed that these cellular processes are also

E. Park (✉)
Keenan Research Centre, Li Ka Shing Knowledge Institute at St. Michael's Hospital, Toronto, ON, Canada
e-mail: eugene.park@unityhealth.to

A. J. Baker
Departments of Critical Care and Anesthesia, St. Michael's Hospital, Unity Health Toronto and University of Toronto, Toronto, ON, Canada

T. A. Schweizer, A. J. Baker (eds.), *Tackling the Concussion Epidemic*,
https://doi.org/10.1007/978-3-030-93813-0_2

shared with mild TBI albeit at a lesser degree. There are, however, notable differences between injury severities which further suggest that not all mechanisms of secondary injury are applicable or relevant across the injury spectrum of TBI. Preclinical models have helped to bridge the gaps in knowledge from the limitations of noninvasive methods in the clinic and provide a better understanding of the discrete molecular and cellular events unfolding within the brain. Although developing clinically relevant, accurate, and reliable models of clinical concussion remains a challenge, animal models, computer simulations, and in vitro systems remain an important component of understanding the pathophysiology of concussion [4, 5].

Normal and Abnormal Physiological States

It is interesting to consider that even under normal conditions there exists considerable pliability and tolerance within an otherwise static central nervous system (CNS). For example, the very act of bending the spine or rapid rotation of the head involves some substantial biomechanical force transduction to axons within the CNS. Body movement lends itself to some degree of nerve fiber stretching and bending or inertial motion of otherwise static tissues but ultimately returning to a normal state [6]. In vitro studies have demonstrated that axons have a high tolerance to stretch and exhibit a delayed elasticity phenomenon, whereby axons subject to up to 65% tensile strain generally recover their original shape afterward [7].

Similarly, the brain is largely protected from inertial acceleration by the fluidity of the brain encased in a rigid skull surrounded by cerebral spinal fluid (CSF). Dramatic and instantaneous acceleration or deceleration (<50 ms) beyond the cushioning capacity of the CSF generates substantial linear and rotational shear forces that result in axonal injury [8]. In particular, the anisotropic properties of white matter tissue in the brain lend itself to higher levels of strain and shear depending on the direction and magnitude of the force being applied [9–11]. Microcompartment in vitro studies further support the argument that axons or neurites are more susceptible to injury than the cell soma [12]. In the same vein, computational models of axons that incorporate cytoskeletal dynamics between actin and spectrin, as well as the anchoring protein, ankyrin, point to greater intrinsic stiffness in axons relative to the cell bodies which further imply increased susceptibility to injury relative to other brain structures [13]. While imaging studies of mTBI patients support the vulnerability of white matter in concussion, the underlying pathophysiology remains a complex issue to resolve (Fig. 2.1).

Imaging White Matter Injury in Concussion

Diffuse axonal injury (DAI) is a common component across all TBI severities [9, 14–16]. However, the dispersed and subtle structural alterations associated with concussion are inherently difficult to verify clinically, despite being the most likely

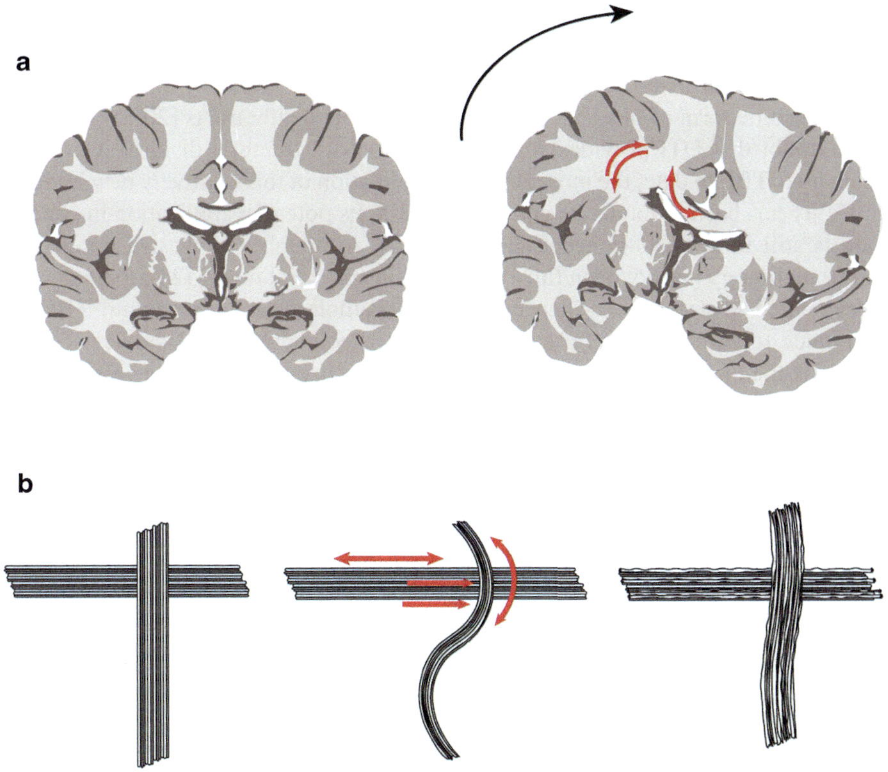

Fig. 2.1 (**a**) A schematic representation of the coronal section of a human brain. At the macroscopic level, rotational forces resulting from head trauma can translate into inertial movement of the brain resulting in deformation of tissues and brain structures. These forces include shear strain on white matter tracts that are particularly vulnerable to excessive inertial loading. (**b**) At a microscopic scale, the biomechanical transduction of forces on white matter tracts can result in rapid stretching of axons regardless of the orientation of fibers. Studies suggest that shear strain rarely results in primary axotomy. The majority of axons return to their normal state. However, a subset of axons develops morphological characteristics associated with diffuse axonal injury. These include abnormal morphology, such as axonal swelling and varicosities associated with activation of molecular pathways and disruption of cytoskeletal networks

contributor to morbidity after injury. Computed tomography (CT) and magnetic resonance imaging (MRI) are largely limited to the detection of macroscopic abnormalities and lack the sensitivity to detect the subtle white matter changes that may occur with concussions [17, 18]. Biomarker studies, improved imaging technologies, and animal models of TBI have demonstrated that the absence of overt changes in brain following mTBI and concussion does not exclude the occurrence of pathophysiological changes which contribute to morbidity.

Diffuse tensor imaging (DTI) is an advanced MRI sequence that provides a more sensitive technique for identifying injuries to white matter tracts [18]. The underlying principle is that DTI estimates rates of water diffusion in different spatial directions, with white matter tracts exhibiting highly directional, anisotropic diffusion parallel to fiber bundles [18, 19]. Its application in TBI involves the detection of changes in diffusion properties within axonal fiber bundles, which presumably indicate some form of physical anomaly in the white matter tracts [20]. While there is no doubt that DTI provides some information on structural disturbances to the brain after TBI, to date, the most effective application of this relatively new imaging modality has been in moderate and severe TBIs. Its potential application in concussion is still in proof-of-concept stage [21, 22]. Although noninvasive imaging of mTBI pathology has not been fully translated into clinical utility [23], there is well-documented postmortem evidence of axonal pathology associated with TBI that continues to be bolstered by complementary preclinical animal studies [5, 14]. Bridging the uncertainty of pathophysiological mechanisms in human cases of concussion with what is known from more severe forms of TBI and animal models will help further validate the use of imaging modalities in characterizing concussion pathophysiology.

Axonal Injury at the Microscopic Level

The collective contribution of subcellular structural alterations, metabolic changes, and ionic shifts in the brain results in impairment of neurotransmission that contributes to the morbidity associated with concussion [24]. Computational modeling conducted at the tissue, cellular [25, 26], and ultrastructural levels [27] has demonstrated the physical transfer of strain, shear, and stresses from macroscopic to microscopic scales in greater detail [8]. These studies are useful in conjunction with histological studies which highlight the complex physical relationship between individual axons within white matter bundles [28] and other constituents of the brain including glia, myelin sheathing, and vascular cells [6]. Differing viscoelastic properties among varying cell types and structures create a complex environment when transduction of biomechanical forces is taken into consideration.

The axonal membrane demonstrates a spectrum of injury responses dependent on the force of injury. For example, in fluid percussion injured cats, mTBI was shown not to result in disrupted axonal membranes but was still associated with neurofilament disruption and compaction. However, with increasing injury severity, axonal membrane permeability becomes a pathological feature of injury [29–31]. Similarly, in vitro studies have demonstrated that mammalian axonal membranes display a remarkable tolerance to stretch injury, with uptake of low molecular weight fluorescent dyes only in instances of higher strain rates resulting in primary axotomy [7]. A direct translation of critical injury thresholds between in vitro and in vivo injury models is difficult to implement, due to the simplicity of in vitro systems compared to the heterogeneity of cells and tissue composition of in vivo systems. However, both injury models agree that primary axotomy resulting from initial strain or impact is in fact a rare occurrence. Experimental evidence suggests that axonal injury or disconnection is an evolving and progressive condition [10,

28]. Human postmortem studies have confirmed that the development of neurofilament misalignment and the formation of retraction bulbs is a delayed process even in severe cases of TBI which suggests that primary axotomy is a rare occurrence at the time of impact [32]. Detailed in vitro investigation has demonstrated that stretch injury conditions that mimic in vivo injury forces result in a heterogeneous response among cytoskeletal elements. For example, early periodic breaks in the microtubule structure were reported [33], while fast transported beta-amyloid precursor protein (β-APP) and slow transported heavy neurofilament (NF200) demonstrated different morphological deposition patterns along injured axons [34]. The described process of events involves the disruption of microtubule networks which are generally regarded as rigid structure within axons and pathways for axonal transport [10, 14]. The misalignment and disruption of the microtubule network set into motion a molecular sequence of failure in axonal transport, resulting in the accumulation of proteins within axons and the formation of axonal varicosities. Collectively, these studies suggest that concussions are likely not a cause of primary axotomy, but rather a sequence of molecular events that unfolds after trauma leads to axonal dysfunction or disconnection. Understanding these processes and their course of evolution is paramount to the development of therapeutic strategies.

Despite a relative resilience of the axonal membrane to shear forces, at the subcellular level the complex organization of the cytoskeleton involves dynamic reorganization in response to structural disruption [6]. The response to cytoskeletal disruption can manifest in the form of axonal swellings. Similar to a motor accident on a highway creating congestion as incoming traffic continues to backlog, transported proteins accumulate within disrupted networks of microtubules and neurofilaments that no longer support continuous paths to traverse. This characteristic histopathology finding is a hallmark for diffuse axonal injury resulting in axonal swellings and eventually disconnection bulbs [35]. β-APP has been used as a general histopathology marker of axonal injury in this regard [36]. Mild TBI has been shown to result in deposition of β-APP in human postmortem studies [37]. These findings have also been replicated in a swine model of rotational acceleration injury [38]. Furthermore, these animal studies provide evidence that axonal injury can occur in the absence of loss of consciousness [38], leading to a greater appreciation of the sensitivity of the brain to mild traumatic forces.

Mechanisms of Ionic Dysregulation and Calcium Pathways

Based on the disruption of cytoskeletal networks and the relative impermeability of cell membranes to primary axotomy, the question arises as to how cytoskeletal disruption is initiated. Although primary physical disruption likely plays some role, it is the delayed evolution of axonal pathology that suggests active cellular mechanisms are involved in the development of axonal pathology. One major event in concussion at the cellular level is shifted in ionic equilibrium within neurons and glia which lead to dysregulation of calcium signaling pathways with detrimental consequences. Large early increases in extracellular glutamate and potassium post-trauma have been demonstrated in rat models of mTBI [39, 40]. The altered neurotransmitter levels and ion flux coincide with spreading waves of depolarization observed experimentally [41] as well as clinically [42, 43]. The increased concentration of extracellular glutamate is presumably responsible for large influxes of extracellular calcium mediated by ionotropic N-methyl D-aspartate (NMDA) receptors [44], which further contribute to cellular depolarization through voltage-gated sodium channels. Interestingly, sublethal stretch injury in vitro in primary neuronal cultures has been shown to increase susceptibility to subsequent NMDA challenge [45]. Although performed in vitro, these findings suggest a state of increased vulnerability of the brain to repeat concussion.

Alterations in ion concentrations within intact axons have also been shown in vitro to be due to the activation of mechano-sensitive voltage-gated sodium channels [46]. The proposed sequence of events involves influx of sodium resulting in subsequent activation of voltage-gated calcium channels as well as reversal of sodium–calcium (Na+-Ca2+) exchangers [47]. The net effect is an increase in intra-axonal calcium, which results in the activation of calcium-sensitive proteases, such as calpain, or further release of intracellular calcium stores from the mitochondria or endoplasmic reticulum leading to a feedforward process of calcium overload [48]. While these mechanisms hold true for more severe forms of axonal injury, other data indicate that intracellular calcium stores are likely involved in the initial calcium spike observed after mild trauma [49]. Moreover, there are biphasic responses to calcium dynamics in the cell, suggesting that the initial phase is a result of intracellular calcium, while subsequent calcium waves are propagated through primarily extracellular sources [49, 50]. These data highlight the complexity and ever-changing physiology of cell receptors and signaling in response to injury and further highlight the difficulties in treating these conditions.

Calcium plays a critical role in numerous cell functions and acts as a linchpin in the activation of pathophysiological processes linked to axonal and neuronal degeneration [51]. The numerous processes initiated by calcium include excessive protease activation, phosphatase activation, initiation of apoptosis, mitochondrial failure, and the reversal of sodium–calcium exchangers leading to membrane depolarization [48, 52]. Importantly, the route of calcium entry, also referred to as the "source specificity hypothesis," [44] determines subsequent activation of discrete downstream signaling pathways adding another layer of complexity to the calcium signaling process [48, 53].

One important consequence of calcium influx is the activation of calpains, a family of cysteine proteases with numerous cell functions including cytoskeletal remodeling, cell signaling, differentiation, and vesicular trafficking under physiological conditions [53, 54]. The two calpains relevant to diseases in the CNS are activated at micro- and millimolar concentrations of calcium [55], denoted as either μ-calpain, m-calpain, or calpain-1 and calpain-2, respectively [56]. While studies suggest that TBI results in influxes of calcium in the millimolar range [51], it has been difficult to parse out the individual contributions of each isoform due to overlapping cleavage targets and a lack of isoform-specific inhibitors [56]. However, several studies point toward μ-calpain as the mediator of neurodegeneration after insult [57]. Regardless, calpain activation is associated with the cleavage of alpha II spectrin, an important structural component of the axonal cytoskeleton [58]. The breakdown of intact 280 kDa alpha II spectrin to a lower molecular weight product of 145/150 kDa is a hallmark indicator of calpain-mediated axonal injury [59]. The specificity of calpain's cleavage site makes it an attractive target as a clinical biomarker of axonal injury as it has been shown to be detectable in CSF after injury [59]. Moreover, alpha II spectrin is further cleaved into a 120 kDa product associated with cleaved caspase-3 activity [60]. Although concussion typically does not result in significant neuronal loss, there is evidence that apoptosis can occur in some instances [61]. Thus, alpha II spectrin may be useful to delineate the underlying injury pathways of axonal injury or delayed cell death occurring after mTBI (Fig. 2.2).

Military Concussion

Military operations over the last two decades have highlighted a shift in warfare injuries, with a larger percentage attributable to survivable blast exposure and head trauma [62]. Blast exposures and mTBI contribute to a large segment of the brain-injured cohort and have become increasingly recognized as a potentially unique form of mTBI [63]. Evidence suggests that non-fatal blast exposure in the absence of contusion also results in axonal injury but with patterns distinct from conventional head trauma [64]. Data from animal models of subclinical blast trauma point toward significant and delayed alterations in heavy neurofilament expression, which is in turn associated with impaired electrophysiological function in white matter and behavioral impairments [65, 66]. In these studies of mild non-blunt force trauma, calpain-mediated breakdown of alpha II spectrin has been observed in the absence of neuronal cell loss [66, 67]. Mild blast studies have also demonstrated neurovascular changes including disruption of the blood–brain barrier (BBB) involving degeneration of astrocytic endfeet following blast exposure [68, 69]. There is also evidence to suggest that soldiers exposed to blast injuries develop deposits of phosphorylated tau, a hallmark indicator of chronic traumatic encephalopathy [70]. Although tau deposition is generally attributed to repetitive head trauma, there is well-documented postmortem and in vivo imaging studies describing tau deposition years following a single TBI event highlighting the complexity of the response to various modalities of TBI [71, 72]. It should also be noted that blast modeling in animals, particularly the defining criteria for mild or low-level blast and its correlation to human low-level blast exposure, is not well-established or standardized [73]. Moreover, blast injuries in human cases are often confounded with comorbidities including post-traumatic stress disorder (PTSD) and limited to imaging studies which provide little insight into the underlying molecular mechanisms of injury.

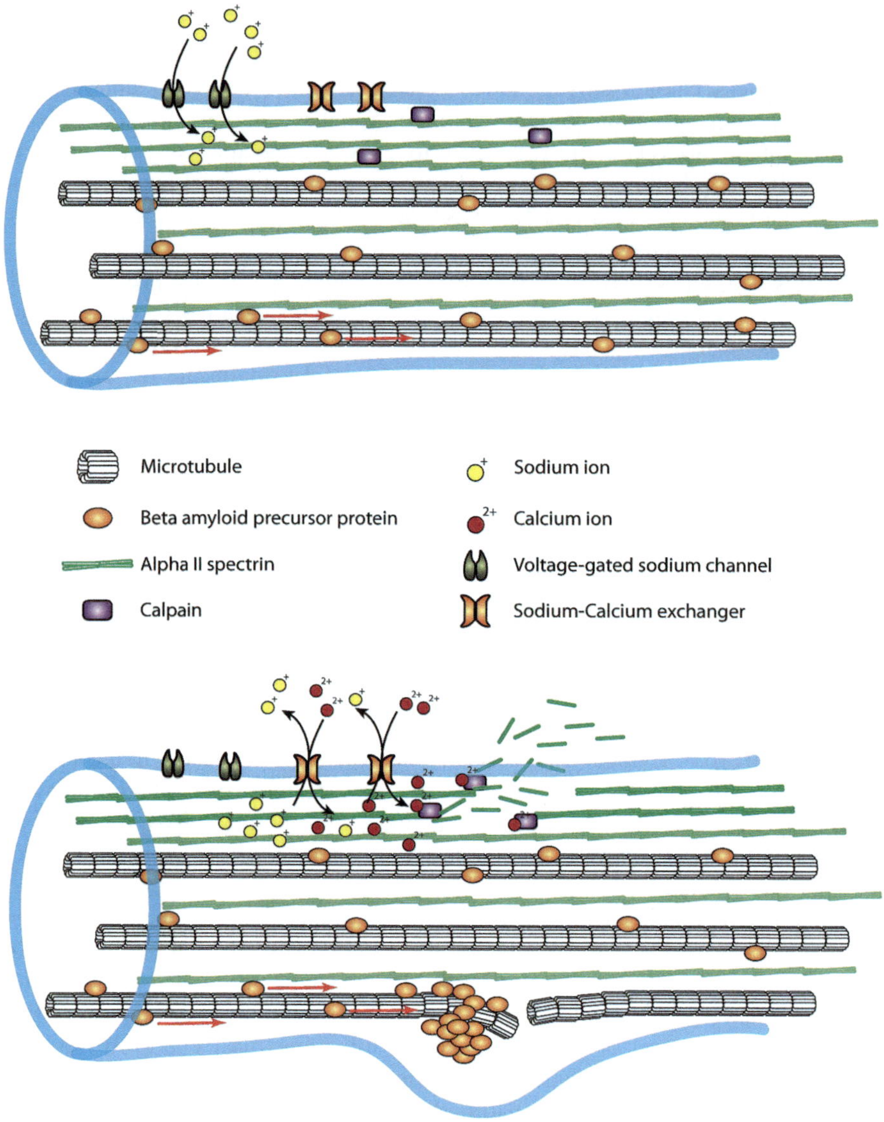

Fig. 2.2 Cartoon depicts some of the molecular mechanisms described in in vitro studies believed to contribute to secondary axonal injury after trauma. Mechano-sensitive voltage-gated sodium channels are activated by physical insult resulting in sodium ion influx into axons. The increased sodium ion concentration in axons causes reversal of sodium–calcium exchangers. Increased calcium ion influx can mobilize calpains resulting in proteolysis of cytoskeletal proteins such as alpha II spectrin. Shedding of spectrin into the interstitial spaces finds its way into peripheral circulation and can be detected in some instances as a serum biomarker. Physical damage to microtubule networks results in the disruption of axonal transport. Beta-amyloid precursor protein accumulates within axons due to transport failure resulting in axonal swelling. This has been shown to result in axonal disconnection

The Biology of Biomarkers for Concussion

Biomarkers have the potential to noninvasively diagnose the presence of concussion. However, their clinical application has been met with limited success. The shortcomings of biomarker utility in concussion are in part due to the broad clinical definition based on external symptoms, while the underlying physiological criteria have not been specifically defined [74]. This disconnect is also complicated by evidence pointing to the pathophysiology of concussion extending beyond the window of clinical symptoms [75–77]. These imply a disconnect between the current gold standard of clinical diagnosis for concussion and the molecular tools meant to supplant this standard. Despite these challenges, the principles of biomarker development are based on our understanding of cellular and subcellular changes occurring in neural tissues in response to trauma which are presumably important contributors to clinical outcome.

Cleavage substrates of calpain are presumably released from injured neurons or axotomized axons and find their way into the CSF and blood. These biomarkers provide insight into proteolytic cleavage targets and an opportunity for the development of a noninvasive diagnostic tool. Neurofilaments have also been evaluated clinically as biomarkers of axonal injury and provide some insight into the pathophysiology of axonal injury after concussion. Neurofilaments subtypes consisting of light, medium, and heavy chains are the largest contributors to the intermediate filament family that make up the neuronal cytoskeleton along with actin microfilaments and the larger caliber microtubules [78]. In animal models, there is a demonstrated correlation between injury severity and the amount of detectable serum and CSF phosphorylated heavy neurofilament [79]. Serum presence of neurofilament light (NFL) chain after repetitive head injury has been shown to correlate with injury severity in repetitively concussed athletes [80]. Serum presence of NFL chain also correlates with CSF levels. Interestingly, elevated NFL was detected up to 3 months post-injury in boxers [81], suggesting a prolonged shedding or injury to axons, which further supports that notion of concussion as an ongoing neurodegenerative process.

A common theme in studies of white matter injury is that heterogeneity exists among axons, not only in composition (e.g., myelinated vs unmyelinated), but also in their response to injury. In vitro studies have demonstrated that the neurofilament response to stretch varies depending on the degree of mechanical insult applied [82]. The varied response to different degrees of mechanical injury in vitro is consistent with in vivo observations which indicate differing immunoreactivity subtypes following TBI [83]. Similarly, neurofilament compaction, believed to be a result of phosphatase activity on the sidearm structures [84], does not occur in the same axons as those exhibiting microtubule destabilization and impaired axonal transport [85–87]. The varying axonal response to injury across these studies were reported under conditions of moderate to severe modeled TBI and whether these mechanisms are involved in concussion remains to be further elucidated. However, these studies are valuable for demonstrating the heterogenic axonal response to TBI.

Mild stretch injury induces increased neurofilament immunoreactivity in axons [82] which also correlates with increases in heavy neurofilament expression observed in fluid percussion models and mild blast in vivo models [66, 88]. Similarly, increases in phosphorylated heavy neurofilament have been detected in serum samples from boxers [89]. Phosphorylated heavy neurofilaments are predominantly localized in long axons [90], which make them a potential marker for axonal injury. Numerous clinical and animal studies have examined the prognostic and diagnostic value of serum and CSF biomarkers. However, their meaningful application in human concussion and mTBI remains to be further clarified due to technical hurdles in terms of thresholds of detection and whether the select measures are truly indicative of underlying pathophysiology [91, 92] (Fig. 2.3).

Biomarker studies have been useful in identifying some generalized cellular changes in addition to neurofilament shedding and disruption. One such example is the detection of glial fibrillary acidic protein (GFAP). GFAP is an astrocytic scaffolding protein whose presence in serum has recently been shown to correlate with MRI abnormalities [93] and has been detected in serum samples of mild and moderate TBI patients [94]. As with all biomarkers of TBI, the reliability of GFAP as an indicator of mTBI is not firmly established. For example, no reported changes in GFAP were found in serum samples from Olympic boxers, while phosphorylated tau, a suspected indicator of chronic traumatic encephalopathy, was detected [95]. There are numerous reasons for the discrepancies between these findings including timing of sampling, type of injuries sustained, and methods used for detection. Similar to the issues surrounding the clinical usefulness of DTI in the detection of concussion, further studies are required to better understand the physiology and temporal sequence of protein shedding into the blood from the CNS for the application of biomarkers to be both reliable and accurate [91, 96].

Repetitive Concussion and Implications for Neurodegenerative Diseases

Animal studies examining GFAP immunoreactivity after mTBI have demonstrated an increase in GFAP expression after injury and increased the presence of microglial activation [97]. Not surprisingly, repetitive head injury in rodents also demonstrates increased the presence of GFAP expression and immune reactivity in brain tissue [98]. In addition to increase in GFAP expression in reactive astrocytes, increased presence of phosphorylated tau (p-tau) protein in rodent models [98] is consistent with the neuropathology of chronic traumatic encephalopathy (CTE). CTE is the term used to describe the neurodegenerative condition characterized by perivascular neurofibrillary tangles of p-tau which frequently occurs in tandem with tau expressing astrocytic tangles seen in a significant number of repetitively head-injured cases [99, 100]. Although there is no clear consensus on whether CTE constitutes a distinct clinical condition [101] or is part of a larger spectrum of tauopathies, the diagnosis for CTE can only be confirmed through postmortem histological

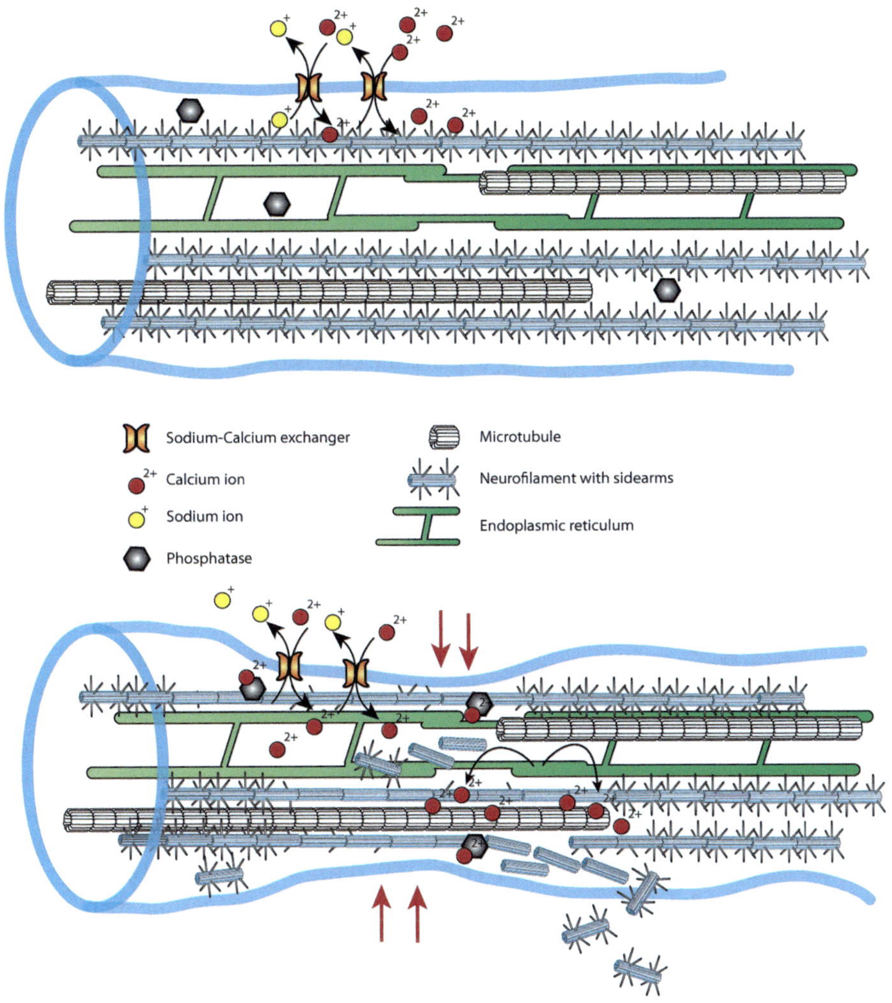

Fig. 2.3 In vitro studies suggest that axon membrane permeability is not a common occurrence in mTBI. Thus, calcium entry either occurs from reversal of sodium–calcium exchangers or secondary calcium release from the axonal endoplasmic reticulum. Other observed mechanisms of calcium-induced injury to axons include activation of phosphatases resulting in loss of sidearms projections in neurofilaments. Sidearm projections are believed to be responsible for maintaining axon caliber and their loss results in neurofilament compaction contributing to changes to axon diameter. Phosphorylated neurofilaments have been detected in serum samples from concussed patients. While it is assumed that membrane disruption is responsible for the shedding of proteins such as neurofilaments and GFAP, there remains the issue as to how these proteins end up in peripheral circulation given that concussed patients and mild TBI animal models demonstrated little evidence of axotomy

analysis and is graded in four stages based on the distribution and density of p-tau [100, 102]. Given that stages of CTE pathology can be categorized, the molecular course of events suggests an evolution of injury but also the potential for intervention at limiting the progression of CTE.

In addition to the association with neurodegenerative tauopathies, concussion and mTBI are also associated with an increased risk of developing dementias [103]. The link between concussion and dementia demonstrates an overlap in pathology and molecular pathways with some evidence pointing toward concussion as an accelerator for those at risk or increasing risk of development of neurodegenerative disorders such as Alzheimer's disease [104]. At a basic research level, understanding the progression of tauopathies has been difficult despite the recent development of numerous models of mild and repetitive brain trauma [105–107]. Although histological examination in these models demonstrates tau deposition [105], the time course of detection is relatively acute compared to the life-long development in human cases of concussion. The overlapping pathology of other neurodegenerative disorders poses a challenge in trying to parse the effects of tau deposition in the sole context of mTBI, considering the natural progression of dementias and AD. This necessitates longitudinal studies when attempting to establish risk of disease development. This further highlights the challenges in modeling human concussion in animals [4, 108] and extrapolating information between species and reconciling what are shared mechanisms of injury progression, regardless of temporal discrepancies.

Emerging Research

An area of important clinical significance in the acute management of brain trauma is the integrity of the BBB and its effects on cerebrovascular dynamics. While the majority of the work to date has focused on the effects on concussion on white matter injury, there is evidence from animal models that BBB and neurovascular disruption are also components of mTBI [106, 109, 110]. A recent mTBI case study indicates potential detection of neurovascular compromise suggesting that the neurovascular unit may also be at risk in addition to neurons and glia [111]. A critical limitation in the clinical evaluation of BBB is the lack of direct assessment on its integrity. Serum biomarkers are a surrogate measure of a presumed leakiness of the BBB, allowing passage of otherwise impermeable proteins into peripheral circulation and vice versa. However, as indicated, the reliability and the interpretation of serum biomarkers, particularly in light of the discovery of the perivascular glymphatic system [112] and a sinus-associated lymphatic system [113] in the brain poses new questions about the mechanism of protein leakage into circulation following TBI [112, 114].

These recent anatomical findings bring to light the complexity of the brain and our limited understanding, even in its native state. There is an obvious need for further study in understanding the interactions between anatomical structures at the cellular and molecular levels in the context of concussion. Advances in imaging technologies in parallel with advanced molecular techniques are on the verge of providing an integrated understanding of preclinical and clinical pathophysiological mechanisms underlying concussion.

References

1. Giza CC, Kutcher JS, Ashwal S, et al. Summary of evidence-based guideline update: evaluation and management of concussion in sports: report of the Guideline Development Subcommittee of the American Academy of Neurology. Neurology. 2013;80:2250–7.
2. Mullally WJ. Concussion. Am J Med. 2017;130:885–92.
3. Meaney DF, Morrison B, Dale BC. The mechanics of traumatic brain injury: a review of what we know and what we need to know for reducing its societal burden. J Biomech Eng. 2014;136:021008.
4. Park E, Baker AJ. Translational mild traumatic brain injury research: bridging the gap between models and clinical uncertainty. In: Wang KKW, editor. Neurotrauma. Oxford University Press; Madison Ave., New York, NY, USA. 2018.
5. Petraglia AL, Dashnaw ML, Turner RC, Bailes JE. Models of mild traumatic brain injury: translation of physiological and anatomic injury. Neurosurgery. 2014;75(Suppl 4):S34–49.
6. Kirkcaldie MT, Collins JM. The axon as a physical structure in health and acute trauma. J Chem Neuroanat. 2016;76:9–18.
7. Smith DH, Wolf JA, Lusardi TA, Lee VM, Meaney DF. High tolerance and delayed elastic response of cultured axons to dynamic stretch injury. J Neurosci. 1999;19:4263–9.
8. Meaney DF, Smith DH. Biomechanics of concussion. Clin Sports Med. 2011;30:19–31, vii.
9. Smith DH, Meaney DF, Shull WH. Diffuse axonal injury in head trauma. J Head Trauma Rehabil. 2003;18:307–16.
10. Smith DH. Neuromechanics and pathophysiology of diffuse axonal injury in concussion. Bridge (Wash D C). 2016;46:79–84.
11. Giordano C, Cloots RJ, van Dommelen JA, Kleiven S. The influence of anisotropy on brain injury prediction. J Biomech. 2014;47:1052–9.
12. Grevesse T, Dabiri BE, Parker KK, Gabriele S. Opposite rheological properties of neuronal microcompartments predict axonal vulnerability in brain injury. Sci Rep. 2015;5:9475.
13. Zhang Y, Abiraman K, Li H, Pierce DM, Tzingounis AV, Lykotrafitis G. Modeling of the axon membrane skeleton structure and implications for its mechanical properties. PLoS Comput Biol. 2017;13:e1005407.
14. Johnson VE, Stewart W, Smith DH. Axonal pathology in traumatic brain injury. Exp Neurol. 2013;246:35–43.
15. Adams JH, Doyle D, Ford I, Gennarelli TA, Graham DI, McLellan DR. Diffuse axonal injury in head injury: definition, diagnosis and grading. Histopathology. 1989;15:49–59.
16. Povlishock JT. Traumatically induced axonal injury: pathogenesis and pathobiological implications. Brain Pathol. 1992;2:1–12.
17. Bazarian JJ, Blyth B, Cimpello L. Bench to bedside: evidence for brain injury after concussion--looking beyond the computed tomography scan. Acad Emerg Med. 2006;13:199–214.
18. Shenton ME, Hamoda HM, Schneiderman JS, et al. A review of magnetic resonance imaging and diffusion tensor imaging findings in mild traumatic brain injury. Brain Imaging Behav. 2012;6:137–92.
19. Basser PJ, Mattiello J, LeBihan D. MR diffusion tensor spectroscopy and imaging. Biophys J. 1994;66:259–67.
20. Le TH, Mukherjee P, Henry RG, Berman JI, Ware M, Manley GT. Diffusion tensor imaging with three-dimensional fiber tractography of traumatic axonal shearing injury: an imaging correlate for the posterior callosal "disconnection" syndrome: case report. Neurosurgery. 2005;56:189.
21. Yuh EL, Hawryluk GW, Manley GT. Imaging concussion: a review. Neurosurgery. 2014;75(Suppl 4):S50–63.
22. Asken BM, DeKosky ST, Clugston JR, Jaffee MS, Bauer RM. Diffusion tensor imaging (DTI) findings in adult civilian, military, and sport-related mild traumatic brain injury (mTBI): a systematic critical review. Brain Imaging Behav. 2018;12:585–612.

23. Churchill NW, Hutchison MG, Richards D, Leung G, Graham SJ, Schweizer TA. The first week after concussion: blood flow, brain function and white matter microstructure. Neuroimage Clin. 2017;14:480–9.
24. Giza CC, Hovda DA. The new neurometabolic cascade of concussion. Neurosurgery. 2014;75(Suppl 4):S24–33.
25. Karami G, Grundman N, Abolfathi N, Naik A, Ziejewski M. A micromechanical hyperelastic modeling of brain white matter under large deformation. J Mech Behav Biomed Mater. 2009;2:243–54.
26. Abolfathi N, Naik A, Sotudeh Chafi M, Karami G, Ziejewski M. A micromechanical procedure for modelling the anisotropic mechanical properties of brain white matter. Comput Methods Biomech Biomed Engin. 2009;12:249–62.
27. Cloots RJ, van Dommelen JA, Nyberg T, Kleiven S, Geers MG. Micromechanics of diffuse axonal injury: influence of axonal orientation and anisotropy. Biomech Model Mechanobiol. 2011;10:413–22.
28. Povlishock JT. Pathobiology of traumatically induced axonal injury in animals and man. Ann Emerg Med. 1993;22:980–6.
29. Pettus EH, Christman CW, Giebel ML, Povlishock JT. Traumatically induced altered membrane permeability: its relationship to traumatically induced reactive axonal change. J Neurotrauma. 1994;11:507–22.
30. Kilinc D, Gallo G, Barbee KA. Mechanically-induced membrane poration causes axonal beading and localized cytoskeletal damage. Exp Neurol. 2008;212:422–30.
31. Kilinc D, Gallo G, Barbee KA. Mechanical membrane injury induces axonal beading through localized activation of calpain. Exp Neurol. 2009;219:553–61.
32. Christman CW, Grady MS, Walker SA, Holloway KL, Povlishock JT. Ultrastructural studies of diffuse axonal injury in humans. J Neurotrauma. 1994;11:173–86.
33. Tang-Schomer MD, Patel AR, Baas PW, Smith DH. Mechanical breaking of microtubules in axons during dynamic stretch injury underlies delayed elasticity, microtubule disassembly, and axon degeneration. FASEB J. 2010;24:1401–10.
34. Tang-Schomer MD, Johnson VE, Baas PW, Stewart W, Smith DH. Partial interruption of axonal transport due to microtubule breakage accounts for the formation of periodic varicosities after traumatic axonal injury. Exp Neurol. 2012;233:364–72.
35. Povlishock JT, Christman CW. The pathobiology of traumatically induced axonal injury in animals and humans: a review of current thoughts. J Neurotrauma. 1995;12:555–64.
36. Gentleman SM, Nash MJ, Sweeting CJ, Graham DI, Roberts GW. Beta-amyloid precursor protein (beta APP) as a marker for axonal injury after head injury. Neurosci Lett. 1993;160:139–44.
37. Blumbergs PC, Scott G, Manavis J, Wainwright H, Simpson DA, McLean AJ. Staining of amyloid precursor protein to study axonal damage in mild head injury. Lancet. 1994;344:1055–6.
38. Browne KD, Chen XH, Meaney DF, Smith DH. Mild traumatic brain injury and diffuse axonal injury in swine. J Neurotrauma. 2011;28:1747–55.
39. Katayama Y, Becker DP, Tamura T, Hovda DA. Massive increases in extracellular potassium and the indiscriminate release of glutamate following concussive brain injury. J Neurosurg. 1990;73:889–900.
40. Takahashi H, Manaka S, Sano K. Changes in extracellular potassium concentration in cortex and brain stem during the acute phase of experimental closed head injury. J Neurosurg. 1981;55:708–17.
41. Bouley J, Chung DY, Ayata C, Brown RH Jr, Henninger N. Cortical spreading depression denotes concussion injury. J Neurotrauma. 2019;36:1008–17.
42. Oka H, Kako M, Matsushima M, Ando K. Traumatic spreading depression syndrome. Review of a particular type of head injury in 37 patients. Brain. 1977;100:287–98.
43. Shaw NA. The neurophysiology of concussion. Prog Neurobiol. 2002;67:281–344.
44. Arundine M, Tymianski M. Molecular mechanisms of calcium-dependent neurodegeneration in excitotoxicity. Cell Calcium. 2003;34:325–37.
45. Arundine M, Chopra GK, Wrong A, et al. Enhanced vulnerability to NMDA toxicity in sublethal traumatic neuronal injury in vitro. J Neurotrauma. 2003;20:1377–95.

46. Iwata A, Stys PK, Wolf JA, et al. Traumatic axonal injury induces proteolytic cleavage of the voltage-gated sodium channels modulated by tetrodotoxin and protease inhibitors. J Neurosci. 2004;24:4605–13.
47. Wolf JA, Stys PK, Lusardi T, Meaney D, Smith DH. Traumatic axonal injury induces calcium influx modulated by tetrodotoxin-sensitive sodium channels. J Neurosci. 2001;21:1923–30.
48. Stirling DP, Stys PK. Mechanisms of axonal injury: internodal nanocomplexes and calcium deregulation. Trends Mol Med. 2010;16:160–70.
49. Staal JA, Dickson TC, Gasperini R, Liu Y, Foa L, Vickers JC. Initial calcium release from intracellular stores followed by calcium dysregulation is linked to secondary axotomy following transient axonal stretch injury. J Neurochem. 2010;112:1147–55.
50. Weber JT, Rzigalinski BA, Willoughby KA, Moore SF, Ellis EF. Alterations in calcium-mediated signal transduction after traumatic injury of cortical neurons. Cell Calcium. 1999;26:289–99.
51. Weber JT. Altered calcium signaling following traumatic brain injury. Front Pharmacol. 2012;3:60.
52. Buki A, Povlishock JT. All roads lead to disconnection?--Traumatic axonal injury revisited. Acta Neurochir. 2006;148:181–93; discussion 93–4.
53. Ma M. Role of calpains in the injury-induced dysfunction and degeneration of the mammalian axon. Neurobiol Dis. 2013;60:61–79.
54. Zatz M, Starling A. Calpains and disease. N Engl J Med. 2005;352:2413–23.
55. Dayton WR. Comparison of low- and high-calcium-requiring forms of the calcium-activated protease with their autocatalytic breakdown products. Biochim Biophys Acta. 1982;709:166–72.
56. Goll DE, Thompson VF, Li H, Wei W, Cong J. The calpain system. Physiol Rev. 2003;83:731–801.
57. Geddes JW, Saatman KE. Targeting individual calpain isoforms for neuroprotection. Exp Neurol. 2010;226:6–7.
58. Goodman SR, Zimmer WE, Clark MB, Zagon IS, Barker JE, Bloom ML. Brain spectrin: of mice and men. Brain Res Bull. 1995;36:593–606.
59. Pineda JA, Wang KK, Hayes RL. Biomarkers of proteolytic damage following traumatic brain injury. Brain Pathol. 2004;14:202–9.
60. Wang KK, Posmantur R, Nath R, et al. Simultaneous degradation of alphaII- and betaII-spectrin by caspase 3 (CPP32) in apoptotic cells. J Biol Chem. 1998;273:22490–7.
61. Riggs JE. Delayed diffuse neurodegeneration after cerebral concussion. Mil Med. 2001;166:1029–30.
62. Warden D. Military TBI during the Iraq and Afghanistan wars. J Head Trauma Rehabil. 2006;21:398–402.
63. Hoge CW, McGurk D, Thomas JL, Cox AL, Engel CC, Castro CA. Mild traumatic brain injury in U.S. Soldiers returning from Iraq. N Engl J Med. 2008;358:453–63.
64. Ryu J, Horkayne-Szakaly I, Xu L, et al. The problem of axonal injury in the brains of veterans with histories of blast exposure. Acta Neuropathol Commun. 2014;2:153.
65. Park E, Eisen R, Kinio A, Baker AJ. Electrophysiological white matter dysfunction and association with neurobehavioral deficits following low-level primary blast trauma. Neurobiol Dis. 2013;52:150–9.
66. Park E, Gottlieb JJ, Cheung B, Shek PN, Baker AJ. A model of low-level primary blast brain trauma results in cytoskeletal proteolysis and chronic functional impairment in the absence of lung barotrauma. J Neurotrauma. 2011;28:343–57.
67. Hernandez A, Tan C, Plattner F, et al. Exposure to mild blast forces induces neuropathological effects, neurophysiological deficits and biochemical changes. Mol Brain. 2018;11:64.
68. Shetty AK, Mishra V, Kodali M, Hattiangady B. Blood brain barrier dysfunction and delayed neurological deficits in mild traumatic brain injury induced by blast shock waves. Front Cell Neurosci. 2014;8:232.
69. Gama Sosa MA, De Gasperi R, Perez Garcia GS, et al. Low-level blast exposure disrupts gliovascular and neurovascular connections and induces a chronic vascular pathology in rat brain. Acta Neuropathol Commun. 2019;7:6.

70. McKee AC, Robinson ME. Military-related traumatic brain injury and neurodegeneration. Alzheimers Dement. 2014;10:S242–53.
71. Johnson VE, Stewart W, Smith DH. Widespread tau and amyloid-beta pathology many years after a single traumatic brain injury in humans. Brain Pathol. 2012;22:142–9.
72. Gorgoraptis N, Li LM, Whittington A, et al. In vivo detection of cerebral tau pathology in long-term survivors of traumatic brain injury. Sci Transl Med. 2019;11:eaaw1993.
73. Elder GA, Stone JR, Ahlers ST. Effects of low-level blast exposure on the nervous system: is there really a controversy? Front Neurol. 2014;5:269.
74. Asken BM. Concussion biomarkers: deviating from the garden path. JAMA Neurol. 2019;76:515–6.
75. Meier TB, Bellgowan PS, Singh R, Kuplicki R, Polanski DW, Mayer AR. Recovery of cerebral blood flow following sports-related concussion. JAMA Neurol. 2015;72:530–8.
76. Prichep LS, McCrea M, Barr W, Powell M, Chabot RJ. Time course of clinical and electrophysiological recovery after sport-related concussion. J Head Trauma Rehabil. 2013;28:266–73.
77. Vagnozzi R, Signoretti S, Floris R, et al. Decrease in N-acetylaspartate following concussion may be coupled to decrease in creatine. J Head Trauma Rehabil. 2013;28:284–92.
78. Perrot R, Berges R, Bocquet A, Eyer J. Review of the multiple aspects of neurofilament functions, and their possible contribution to neurodegeneration. Mol Neurobiol. 2008;38:27–65.
79. Anderson KJ, Scheff SW, Miller KM, et al. The phosphorylated axonal form of the neurofilament subunit NF-H (pNF-H) as a blood biomarker of traumatic brain injury. J Neurotrauma. 2008;25:1079–85.
80. Shahim P, Tegner Y, Marklund N, et al. Astroglial activation and altered amyloid metabolism in human repetitive concussion. Neurology. 2017;88:1400–7.
81. Zetterberg H, Hietala MA, Jonsson M, et al. Neurochemical aftermath of amateur boxing. Arch Neurol. 2006;63:1277–80.
82. Chung RS, Staal JA, McCormack GH, et al. Mild axonal stretch injury in vitro induces a progressive series of neurofilament alterations ultimately leading to delayed axotomy. J Neurotrauma. 2005;22:1081–91.
83. Johnson VE, Stewart W, Weber MT, Cullen DK, Siman R, Smith DH. SNTF immunostaining reveals previously undetected axonal pathology in traumatic brain injury. Acta Neuropathol. 2016;131:115–35.
84. Okonkwo DO, Pettus EH, Moroi J, Povlishock JT. Alteration of the neurofilament sidearm and its relation to neurofilament compaction occurring with traumatic axonal injury. Brain Res. 1998;784:1–6.
85. DiLeonardi AM, Huh JW, Raghupathi R. Impaired axonal transport and neurofilament compaction occur in separate populations of injured axons following diffuse brain injury in the immature rat. Brain Res. 2009;1263:174–82.
86. Stone JR, Singleton RH, Povlishock JT. Intra-axonal neurofilament compaction does not evoke local axonal swelling in all traumatically injured axons. Exp Neurol. 2001;172:320–31.
87. Marmarou CR, Povlishock JT. Administration of the immunophilin ligand FK506 differentially attenuates neurofilament compaction and impaired axonal transport in injured axons following diffuse traumatic brain injury. Exp Neurol. 2006;197:353–62.
88. Park E, Liu E, Shek M, Park A, Baker AJ. Heavy neurofilament accumulation and alpha-spectrin degradation accompany cerebellar white matter functional deficits following forebrain fluid percussion injury. Exp Neurol. 2007;204:49–57.
89. Neselius S, Zetterberg H, Blennow K, Marcusson J, Brisby H. Increased CSF levels of phosphorylated neurofilament heavy protein following bout in amateur boxers. PLoS One. 2013;8:e81249.
90. Sternberger LA, Sternberger NH. Monoclonal antibodies distinguish phosphorylated and nonphosphorylated forms of neurofilaments in situ. Proc Natl Acad Sci U S A. 1983;80:6126–30.
91. Agoston DV, Shutes-David A, Peskind ER. Biofluid biomarkers of traumatic brain injury. Brain Inj. 2017;31:1195–203.
92. Zetterberg H, Smith DH, Blennow K. Biomarkers of mild traumatic brain injury in cerebrospinal fluid and blood. Nat Rev Neurol. 2013;9:201–10.

93. Gill J, Latour L, Diaz-Arrastia R, et al. Glial fibrillary acidic protein elevations relate to neuroimaging abnormalities after mild TBI. Neurology. 2018;91:e1385–e9.
94. Papa L, Lewis LM, Falk JL, et al. Elevated levels of serum glial fibrillary acidic protein breakdown products in mild and moderate traumatic brain injury are associated with intracranial lesions and neurosurgical intervention. Ann Emerg Med. 2012;59:471–83.
95. Neselius S, Zetterberg H, Blennow K, et al. Olympic boxing is associated with elevated levels of the neuronal protein tau in plasma. Brain Inj. 2013;27:425–33.
96. Gan ZS, Stein SC, Swanson R, et al. Blood biomarkers for traumatic brain injury: a quantitative assessment of diagnostic and prognostic accuracy. Front Neurol. 2019;10:446.
97. Hylin MJ, Orsi SA, Zhao J, et al. Behavioral and histopathological alterations resulting from mild fluid percussion injury. J Neurotrauma. 2013;30:702–15.
98. Petraglia AL, Plog BA, Dayawansa S, et al. The pathophysiology underlying repetitive mild traumatic brain injury in a novel mouse model of chronic traumatic encephalopathy. Surg Neurol Int. 2014;5:184.
99. McKee AC, Cantu RC, Nowinski CJ, et al. Chronic traumatic encephalopathy in athletes: progressive tauopathy after repetitive head injury. J Neuropathol Exp Neurol. 2009;68:709–35.
100. McKee AC, Stein TD, Kiernan PT, Alvarez VE. The neuropathology of chronic traumatic encephalopathy. Brain Pathol. 2015;25:350–64.
101. LoBue C, Schaffert J, Cullum CM. Chronic traumatic encephalopathy: understanding the facts and debate. Curr Opin Psychiatry. 2020;33:130–5.
102. McKee AC, Stern RA, Nowinski CJ, et al. The spectrum of disease in chronic traumatic encephalopathy. Brain. 2013;136:43–64.
103. Pattinson CL, Gill JM. Risk of dementia after TBI – a cause of growing concern. Nat Rev Neurol. 2018;14:511–2.
104. Fakhran S, Yaeger K, Alhilali L. Symptomatic white matter changes in mild traumatic brain injury resemble pathologic features of early Alzheimer dementia. Radiology. 2013;269:249–57.
105. Kane MJ, Angoa-Perez M, Briggs DI, Viano DC, Kreipke CW, Kuhn DM. A mouse model of human repetitive mild traumatic brain injury. J Neurosci Methods. 2012;203:41–9.
106. Tagge CA, Fisher AM, Minaeva OV, et al. Concussion, microvascular injury, and early tauopathy in young athletes after impact head injury and an impact concussion mouse model. Brain. 2018;141:422–58.
107. Goldstein LE, Fisher AM, Tagge CA, et al. Chronic traumatic encephalopathy in blast-exposed military veterans and a blast neurotrauma mouse model. Sci Transl Med. 2012;4:134ra60.
108. Dewitt DS, Perez-Polo R, Hulsebosch CE, Dash PK, Robertson CS. Challenges in the development of rodent models of mild traumatic brain injury. J Neurotrauma. 2013;30:688–701.
109. Johnson VE, Weber MT, Xiao R, et al. Mechanical disruption of the blood-brain barrier following experimental concussion. Acta Neuropathol. 2018;135:711–26.
110. Abdul-Muneer PM, Schuetz H, Wang F, et al. Induction of oxidative and nitrosative damage leads to cerebrovascular inflammation in an animal model of mild traumatic brain injury induced by primary blast. Free Radic Biol Med. 2013;60:282–91.
111. Chan ST, Evans KC, Rosen BR, Song TY, Kwong KK. A case study of magnetic resonance imaging of cerebrovascular reactivity: a powerful imaging marker for mild traumatic brain injury. Brain Inj. 2015;29:403–7.
112. Iliff JJ, Wang M, Liao Y, et al. A paravascular pathway facilitates CSF flow through the brain parenchyma and the clearance of interstitial solutes, including amyloid beta. Sci Transl Med. 2012;4:147ra11.
113. Louveau A, Smirnov I, Keyes TJ, et al. Structural and functional features of central nervous system lymphatic vessels. Nature. 2015;523:337–41.
114. Plog BA, Dashnaw ML, Hitomi E, et al. Biomarkers of traumatic injury are transported from brain to blood via the glymphatic system. J Neurosci. 2015;35:518–26.

Chapter 3
Neuroimaging of Concussion

Nathan W. Churchill

Introduction

Concussion is a form of mild traumatic brain injury (TBI) that is associated with disturbances in physical function, cognition, mood, and quality of sleep, despite an absence of overt neuroanatomical lesions. At present, the diagnosis and clinical management of concussion are based on symptom assessments and brief evaluations of cognition and balance. It remains an ongoing challenge, however, to identify the alterations in brain physiology that underlie neuropsychological and behavioral symptoms. There is also growing concern that biological recovery lags behind clinical resolution and that the long-term effects of concussion may be compounded if individuals are re-injured after returning too soon to normal activities.

Neuroimaging can be used to investigate the changes in brain physiology that occur after a concussion, in a safe, noninvasive way. Historically, radiological imaging, including computed tomography (CT) and anatomical magnetic resonance imaging (MRI), has been used to rule out more severe injury but has limited use in assessing the "invisible" effects of concussion, which include microstructural injury, altered neurometabolism, and disrupted cerebral blood flow (CBF) [1]. Over the past two decades, however, advanced neuroimaging techniques have been developed and refined, leading to new insights into these subtle physiological changes. Arguably, most of the work in this area has been done with MRI, which is a technique that immerses the brain in a strong static magnetic field and uses a combination of magnetic gradients and radiofrequency pulses to obtain brain images. Its power comes from its versatility, as a diverse array of pulse sequences have been developed, producing images that are sensitive to different aspects of physiology. This is particularly important when studying the complex pathophysiology of

N. W. Churchill (✉)
Keenan Research Centre, St. Michael's Hospital, Unity Health Toronto,
Toronto, ON, Canada

T. A. Schweizer, A. J. Baker (eds.), *Tackling the Concussion Epidemic*,
https://doi.org/10.1007/978-3-030-93813-0_3

concussion, as different MRI sequences provide complementary information about the effects of concussive injury and subsequent recovery.

This chapter provides a high-level survey of the current state of knowledge in concussion neuroimaging for each of the key MRI sequences that are used in concussion research. This includes a review of (1) the aspects of brain physiology that it measures, (2) what it tells us about the initial physiological response to injury and recovery process, and (3) its relationship with clinical presentation, focusing mainly on severity of self-reported symptoms, which is a cornerstone of concussion assessment. At the end of each section, there is also (4) a discussion of new and underutilized imaging techniques that can provide new insights into concussion. As shown in Fig. 3.1, the key sequences include structural imaging, diffusion tensor imaging (DTI), task-based functional MRI (tb-fMRI), resting-state fMRI (rs-fMRI), arterial spin labeling (ASL), and magnetic resonance spectroscopy (MRS). The final section of this chapter will discuss opportunities and limitations in the application of MRI to the clinical management of concussion. A note on terminology: throughout this chapter, "concussion" and "mild TBI" will be used interchangeably. In addition, no distinction will be made

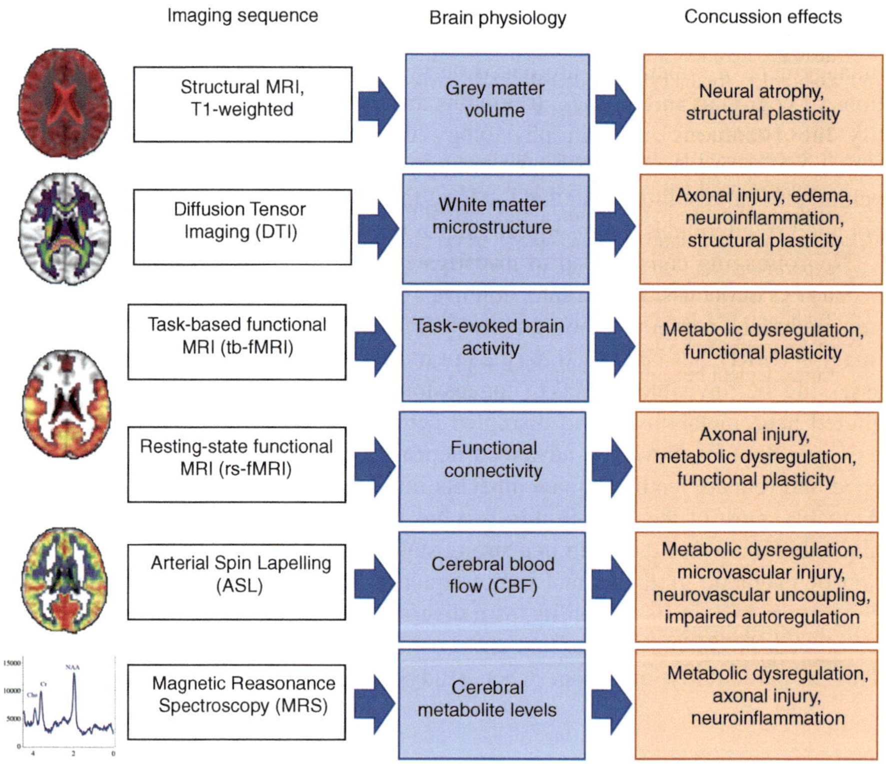

Fig. 3.1 Diagram summarizing the key magnetic resonance imaging (MRI) sequences used to investigate concussion, the measures of brain physiology they are used to interrogate, and the concussion effects they are most sensitive to

between the different criteria used by studies to determine concussion status, nor between concussion in sport, military, and civilian settings. Although they are likely to be important considerations, these issues have been largely unexplored in the neuroimaging literature, and as such, they are beyond the scope of this chapter.

Structural Imaging

Although concussion is primarily considered a functional injury [2], structural MRI sequences have been used in hospital settings to identify abnormalities that signify complex or more severe TBI. In this section, clinical imaging techniques are briefly summarized, but for a more comprehensive overview see, for example, Shenton et al. [3]. One of the oldest TBI assessment tools is CT, and it remains the dominant approach for acute assessment, given its accessibility, ease of use, and sensitivity to injuries requiring immediate intervention, such as hemorrhage, intra-cerebral swelling, and skull fracture [4]. After patient stabilization, structural MRI can be used to further assess for injuries, due to improved soft tissue contrast and sequences that detect specific tissue types and pathologies. In concussion, this includes injuries that do not require neurosurgical intervention, such as minor petechial hemorrhage, contusions, and edema [5, 6]. Among the standard sequences, T1-weighted imaging is often used to visualize neuroanatomy, with high signal in white matter, intermediate signal in gray matter, and low signal in cerebrospinal fluid (CSF). T2-weighted imaging has high signal in CSF and is used to detect lesions, which appear hyperintense due to edema in injured tissues. Fluid attenuation inversion recovery (FLAIR) is an alternative sequence where the CSF signal is nullified, creating hyperintensities that are specific to intraparenchymal lesion and hemorrhage. Susceptibility-weighted imaging (SWI) has more recently attained widespread use [7], as this sequence is sensitive to deoxyhemoglobin and cerebral microhemorrhage, which appear as hypointense "spots." The prevalence of findings on clinical MRI varies substantially between studies, ranging from infrequent (15–20%) to highly frequent (57%) [8–10], although these numbers are often based on limited sample sizes. In general, clinical imaging has limited utility in concussion research, with studies consistently reporting that structural MRI findings are nonpredictive of cognitive deficits and recovery time after injury [8, 9, 11, 12]. As a representative example, Fig. 3.2a shows anatomical scans for a patient with symptomatic concussion. Despite the patient exhibiting typical signs and symptoms of injury, no gross structural abnormalities are seen.

Beyond the acute identification of cerebral lesions, T1-weighted imaging has been combined with sophisticated algorithms to detect subtle long-term changes in brain morphology and tissue composition associated with concussion. This includes neuronal cell loss due to axonal injury and secondary biochemical cascades [13], which may be exacerbated by impaired neurovascular function and protracted neuroinflammatory response [14, 15], with potential long-term effects on cognition [16]. However, post-concussion changes in gray matter may also include

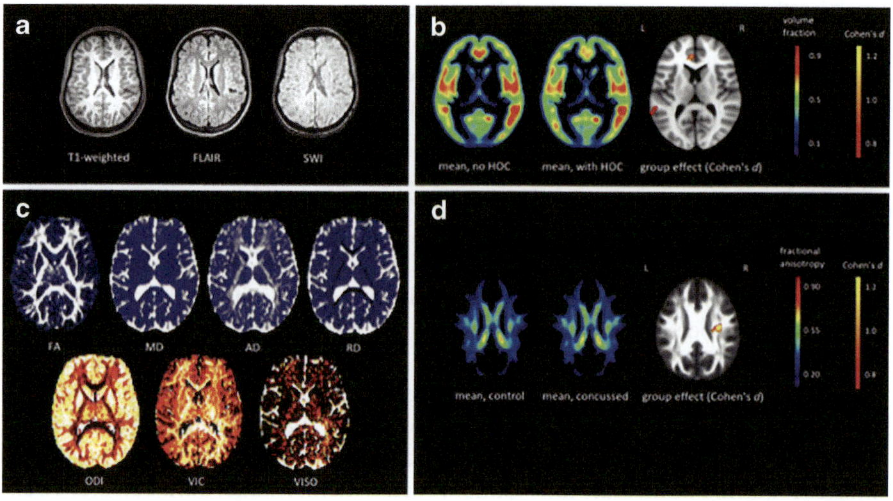

Fig. 3.2 Imaging sequences used to assess neuroanatomy and tissue microstructure after concussion. (**a**) common anatomical sequences include T1-weighted imaging, fluid attenuation inversion recovery (FLAIR) and susceptibility weighted imaging (SWI), shown for a participant with symptomatic concussion. No gross structural abnormalities are observed. (**b**) comparison of mean grey matter volume fraction, for groups without (N=24) and with (N=13) a history of concussion (HOC), i.e., occurring more than 1 year prior to imaging. Areas of substantial group difference (Cohen's d > 0.8) are also depicted. The HOC group shows reduced volume in the anterior cingulate and left temporal lobe. (**c**) diffusion-weighted parameter maps, shown for an individual with symptomatic concussion. Standard DTI measures include fractional anisotropy (FA), along with mean, axial and radial diffusivities (MD, AD, RD). Advanced NODDI measures include the orientation dispersion index (ODI), along with intracellular and isotropic water volume fractions (VIC, VISO). (**d**) comparison of mean FA in white matter tracts of control (N=37) and symptomatic concussed (N=22) groups. The concussed group shows increased FA in the right external capsule

neuroplastic reorganization to compensate for the effects of injury [17], and effects may be confounded by the normal processes of maturation and aging. Owing to transient physiological changes at early injury (e.g., tissue edema) and the delayed nature of structural changes, most MRI studies of gray matter morphology have been conducted long after clinical recovery, averaging between 9 months and 2 years post-injury [18–21]. Whereas more severe forms of TBI are associated with long-term functional deficits and correspondingly widespread declines in gray matter volume [22, 23], the effects of concussion tend to be far subtler. See, for example, Fig. 3.2b, in which a comparison of groups with and without a history of concussion shows spatially limited differences in grey matter volume.

Despite an absence of neurocognitive impairments or elevated symptom complaints at the time of imaging, concussed individuals often show reduced gray matter in the prefrontal cortex and hippocampus compared to individuals without prior concussion [18–21]. There is also more limited evidence that individuals with greater self-reported symptoms at the time of imaging have reduced prefrontal gray matter [18], and those with longer recovery times after their last concussion have reduced frontotemporal volumes [19]. Intriguingly, one study examined patients longitudinally, scanning them an average of 23 days post-injury and at 1-year follow-up [24]. They reported longitudinal declines in frontal and precuneal gray matter and increased symptoms of depression and anxiety. These findings indicate subtle concussion-related atrophy, with frontotemporal gray matter being particularly susceptible, likely caused by impacting on bony protuberances of the skull during injury [25]. Interestingly, these effects were seen despite an absence of impairments on standardized neurocognitive testing. This highlights the brain's ability to adapt to injury, likely through changes in white matter pathways and functional organization, as discussed in the DTI and fMRI sections below, respectively.

Nevertheless, these studies raise concerns about the long-term consequences of concussion, particularly after repeated injury. Retrospective studies of retired professional athletes have identified negative long-term outcomes of repeated head impacts, ranging from subtle cognitive impairment and mood dysregulation to an increased risk of neurodegenerative mortality [26–28]. This represents a uniquely high-risk population though, and it is unclear to what extent subtle post-concussion changes in gray matter seen in the general population, typically imaged 1–2 years after injury, affect their physical and mental health later in life. To address these issues, longitudinal studies are needed that extend beyond current post-injury time intervals, with an emphasis on tracking the effects of serial concussions. As an additional consideration, structural MRI studies of concussion have largely conducted univariate analyses in brain regions of interest. This neglects the fact that gray matter change in one region may affect gray matter in structurally and functionally connected regions. Structural covariance may be used to address this issue [29], as it identifies distributed patterns of gray matter that co-vary between individuals. The approach has been used in studies of acquired brain injury and pathological aging [30, 31] and represents a promising tool for concussion research.

Diffusion Tensor Imaging (DTI)

Moving beyond conventional structural imaging, DTI is used to investigate concussion-related changes in the microstructure of cerebral tissue. This technique produces images that are sensitive to diffusion, i.e., the random movement of water molecules in tissues. Its use comes from the fact that the speed and directionality of water diffusion depend on local tissue composition and geometry. During imaging, diffusion is measured in many spatial directions and used to fit a tensor that describes local diffusion behavior. This tensor can be summarized by its mean diffusivity (MD), characterizing the overall diffusion rate, and its fractional anisotropy (FA), characterizing the tendency for diffusion to move in a preferred direction. Some studies also examine axial diffusivity (AD), which measures diffusion along the preferred direction, and radial diffusivity (RD) which measures diffusion perpendicular to the preferred direction. Representative DTI parameter maps are shown for a single patient with symptomatic concussion in Fig. 3.2c. DTI studies typically focus on white matter tracts, where myelinated fiber bundles allow unhindered diffusion parallel to the tracts but limit diffusion perpendicular to the tracts. This produces high values of FA and low values of MD for healthy white matter tissue. In

TBI, impact-related shear forces cause diffuse injury to neural cells [32], including axotomy, cytoskeleton breakdown, impaired vesicular transport, and mitochondrial swelling, along with the demyelination of intact axons [13, 33]. Collectively, these injuries are associated with reduced FA and increased MD in white matter; there is also evidence that axonal pathology is mainly associated with reduced AD, while myelin loss is mainly associated with increased RD [34, 35]. The DTI parameters are often treated as proxy measures of white matter integrity for more severe TBI [36], where significant effects are seen in patients imaged months to years after an injury [23].

In concussion, the effects of axonal injury are more limited and other, potentially reversible, physiological changes are thought to predominate [37]. Among symptomatic individuals, imaged within the first week post-injury, concussion has been associated with decreases in FA and increases in MD and RD within white matter tracts [38–41]. These effects have been interpreted as being due to diffuse axonal injury, as the affected white matter regions are consistent with patterns of injury seen in more severe TBI, including infratentorial and subcortical white matter, internal capsule, corpus callosum, and fornix [41]. However, other likely contributors include vasogenic edema, caused by breakdown of the blood–brain barrier [42], and intracellular edema, stemming from transient ionic imbalances, with glial cells being most affected [42, 43]. Both may increase extra-axonal water, although there is evidence that cytotoxic effects predominate at the later stages of injury [44]. Neuroinflammatory response may also contribute to diffusivity changes, with glial hypertrophy increasing extra-axonal water volume contributions [45]. It is, however, worth noting that not all studies have reported the same direction of change, as early increases in FA and reductions in MD have also been identified within 1–2 weeks post-injury [46, 47]. This may reflect spatial heterogeneity in the effects of tissue edema; for example, cytotoxic effects may induce a greater relative amount of intra-axonal swelling within some cohorts and brain regions. As an example of enhanced post-concussion FA, Fig. 3.2d compares control and symptomatic concussed groups, with the latter showing elevated FA values in the right external capsule.

Longitudinal studies have often shown DTI effects lasting well beyond the early symptomatic phase of injury. The disturbances in FA and MD seen in the first week after concussion were typically unresolved at follow-up imaging, which ranged from 3 weeks to 6 months post-injury [39, 40, 48, 49], despite participants being generally asymptomatic at the later time points. More recently, we conducted one of the longest follow-up studies to date, where we imaged concussed varsity athletes in the first week of injury, at medical clearance (median: 27 days) and 1 year afterward. We found that white matter FA was reduced at early injury and at medical clearance but had resolved at 1-year follow-up. In contrast, MD was elevated at early injury, with ongoing effects at medical clearance and 1 year follow-up, despite participants having returned to normal daily activities without persistent clinical deficits. These findings suggest a subtle long-term neurobiological response that evolves over months to years post-injury, which is supported by cross-sectional studies showing DTI abnormalities in patients with a history of concussion, imaged months to years after injury [19, 41]. This may be a consequence of diffuse axonal injury, although the extent and magnitude of findings are surprising, given the normal course of recovery and absence of functional deficits. An alternate candidate is neuroinflammation, which has been identified months after insult [50], and can exacerbate blood–brain barrier permeability in self-reinforcing fashion [51]. This is supported by prior research, in which peripheral markers of neuroinflammation were seen in individuals years after their last concussion [52]. The long-term DTI findings also do not necessarily represent pathophysiology, but may instead derive from neuroplastic changes in response to injury [17] which may be adaptive or maladaptive in nature [53]. These processes also need not be mutually exclusive. For example, axonal injury may in turn mobilize local neuroinflammatory response and subsequent neuroplastic reorganization, all of which may contribute to a pattern of DTI abnormalities that appears relatively static over time.

Besides an indicator of concussion pathophysiology, changes in white matter microstructure have significant consequences for information transmission in the brain, which may contribute to both subjective symptoms and neurocognitive impairments. Elevated FA and reduced diffusivity at early injury have been correlated with more severe acute symptoms and longer recovery times [46, 54], potentially due to greater edema-related compression of intracellular spaces. Interestingly, among patients experiencing prolonged symptoms months after injury, these effects may be reversed, as more severe symptoms are associated with reduced FA [18, 55] and higher diffusivity [56, 57], although in some cases null effects have also been reported [58]. White matter injury may further contribute to neurocognitive impairments after a concussion. Although less well-studied, there is intriguing evidence that regional DTI abnormalities may correlate with deficits in specific cognitive domains; for example, lower frontal FA has been correlated with deficits in executive function [59], whereas higher FA of the corpus callosum has been correlated with deficits of attention [47].

A major limitation of DTI is its nonspecificity, as changes in FA and diffusivity can arise from many mechanisms, including changes in cell morphology and packing density, altered myelination status, and variations in fiber orientation. This is partly due to DTI's simplistic model of diffusion, which consists of a single water compartment with anisotropic Gaussian water diffusion. To improve imaging specificity, more sophisticated models of water diffusion are needed. Promising approaches include diffusion kurtosis imaging (DKI) [60], which quantifies deviations from the Gaussian model, and neurite orientation dispersion and density imaging (NODDI) [61], which obtains intra-neurite and extra-neurite water contributions, along with a more detailed model of fiber orientation; examples of NODDI parameter maps are shown for a patient with symptomatic concussion in Fig. 3.2c. We recently conducted a combined study of DTI and NODDI, in which we found that post-concussion decreases in FA and increases in diffusivity are driven mainly by reduced intra-neurite water [62]. In addition, individuals with prolonged recovery showed FA and diffusivity effects that were driven by changes in neurite orientation. These results emphasize that post-concussion changes in DTI parameters may arise from multiple mechanisms, which may be disentangled using advanced diffusion-based techniques.

Task-Based Functional MRI (tb-fMRI)

Blood-oxygenation-level-dependent (BOLD) fMRI provides a window into the effects of concussion on brain function. This technique exploits the fact that hemoglobin in red blood cells has differing magnetic properties depending on whether it is bound to oxygen. Oxy-hemoglobin is weakly paramagnetic, whereas deoxyhemoglobin is diamagnetic and distorts local magnetic fields, with decreased signal during imaging. This is used to infer brain activity, as increased neurometabolic activity triggers an influx of oxygenated blood, washing out deoxygenated blood, and causing the BOLD signal to increase. The BOLD response is sluggish, peaking roughly 4–6 seconds after neural activation, but can localize activity to millimeter-level spatial precision. Conventional task-based fMRI (tb-fMRI) uses BOLD fMRI to measure changes in brain activity between experimentally induced cognitive states. Images are acquired while participants perform a task that engages sensory, motor, or cognitive domains of interest, along with a "control" condition that does not engage these domains. By measuring the difference in BOLD signal between task and control states, we may obtain maps of task-specific brain activity. In TBI, mechanical injury causes impaired neural function, with effects that are well-studied using animal models. During an impact, forces exerted on neural cells cause mechanoporation and the unchecked release of ions and neurotransmitters [63]. The ionic imbalances, coupled with impaired CBF, cause neurometabolic dysregulation, while neurotransmitter release further impairs neural communication and induces excitotoxic effects [64, 65]. Collectively, these effects alter synaptic excitability, inhibition, and learning-related neuroplasticity [66, 67]. While the effects of TBI on neural function are well-studied at the cellular level, tb-fMRI can be used to measure changes in task-related activity at the whole-brain level. Studies of more severe TBI have reported spatially distributed changes in brain activity lasting months to years after injury [68, 69], with effects that are correlated with the severity of white matter injury [70].

In concussion, tb-fMRI studies have focused on aspects of higher-level cognition that are often impaired after injury. The most well-studied domain is working memory, which refers to the ability of the brain to temporarily retain and manipulate information in order to guide decision-making. Historically, tb-fMRI studies have been conducted beyond the initial symptomatic window of injury, with the goal of identifying subtle deficits in otherwise clinically recovered patients. A series of landmark studies compared working memory of controls and patients with mild TBI, imaged within 1 month of injury [71–73]. They found that both groups activated an extensive set of prefrontal, temporal, parietal, and cerebellar regions, but patients had greater amplitudes of activation and recruited more extensively, mainly from prefrontal and anterior cingulate areas. This was despite a general lack of impaired performance on working memory tasks. The mechanisms of hyperactivation are disputed, although a few explanations have been proposed [74–76]. This includes maladaptive response, with concussion-related disruptions in brain networks leading to the inefficient use of resources; and compensatory response, with recruitment of areas that are not normally involved in the task, in order to sustain performance; and the engagement of latent resources, with recruitment of areas that are available for the task, but not activated under normal cognitive loads. As the concussed individuals in these studies are not significantly impaired on task performance, and hyper-activation occurs in frontal regions engaged in healthy individuals under higher cognitive loads [72], the evidence favors a latent resource interpretation, but this question is not yet definitively resolved. Outside of working memory tasks, concussed patients have also shown enhanced frontoparietal activity during virtual reality navigation tasks, in the absence of performance deficits [77, 78]. However, concussed patients had reduced frontoparietal activity while performing auditory tasks of executive function and attentional control [79–81], indicating that hyper-activation is not a universal response and may depend on the cognitive task under investigation.

In longitudinal studies, disturbances in task-related brain function appear to emerge at early injury and persist beyond clinical recovery. Studies identifying altered brain function during working memory tasks in the first week of injury reported persistent frontal hyper-activation at 1–2 months post-injury [82–84], despite an absence of behavioral or symptom impairments at the later imaging time points. This is not restricted to working memory, as persistent functional disturbances were found in patients performing an oculomotor task, imaged at 1 week and 1 month post-injury [85]. To date, there have been no longitudinal studies extending beyond 2 months post-injury. However, cross-sectional tb-fMRI studies of asymptomatic individuals with a history of multiple concussions, imaged an average of 6 and 9 months post-injury, showed no functional abnormalities associated with a working memory task [86, 87], suggesting that resolution of functional disturbances occurs prior to this time interval. Collectively, these findings indicate that disturbances in task-related activity appear soon after the concussion event, likely in response to neural injury and neurometabolic dysregulation, and persist beyond clinical recovery with eventual resolution in the months afterward, supporting the adaptive, plastic nature of brain function.

The tb-fMRI studies provide direct evidence that task-related brain activity is affected by concussion, despite an absence of task-related behavioral impairments. It is less clear how these disturbances relate to acute symptom presentation and the course of clinical recovery. One study reported that, in the first week of injury, higher parietal activity correlated with lower cognitive and somatic symptoms, whereas greater premotor activity predicted a longer recovery time [82]. This suggests that the initial hyper-activation is not solely an adaptive response to injury but depends on the specific brain regions that are recruited. Similarly, among individuals with persistent post-concussion symptoms, a series of working memory studies conducted from 3 to 6 months post-injury found reduced activation within core prefrontal networks and greater recruitment of nontask regions, despite generally unimpaired task performance [88–91]. Based on these findings, hyper-activation is associated with both better and worse outcomes, depending on the affected brain regions, whereas hypo-activation is more broadly associated with symptom impairments. At present, these findings have been mainly evaluated in the context of working memory. Given the diversity of other functional domains affected by concussion, there is a need to extend research to other paradigms and establish the generalizability of these findings.

Although there is tight coupling between synaptic activity and neurovascular response in the uninjured brain, it should be emphasized that BOLD fMRI only indirectly measures neuronal activity. The BOLD response is driven by changes in neural metabolism, CBF, and cerebral blood volume, and does not have an interpretation in absolute physiological units. Studies typically interpret concussion-related changes in BOLD response as altered neural activity, which is supported by the spatial specificity and task load-dependent nature of the effects [71, 92]. Nevertheless, the BOLD response may also be influenced by post-concussion changes in basal perfusion and neurovascular uncoupling [93]. One way to address this ambiguity is through the use of quantitative BOLD, where the controlled delivery of blended gases is used to measure BOLD changes under different hemodynamic states, thereby estimating changes in O_2 metabolism [94]. Another solution is to acquire fMRI concurrently with electroencephalography (EEG) or positron emission tomography (PET), which measure electrophysiological activity and glucose metabolism, respectively; both have been used to identify neural dysfunction after mild TBI [95, 96]. Future work may benefit from combining imaging modalities, although this is nontrivial to implement and will require creative solutions to avoid placing undue burden on patients with concussion, particularly in the early symptomatic phase of injury.

Resting-State fMRI (rs-fMRI)

More recently in the field of fMRI, there has been a growing body of research focusing on spontaneous BOLD fluctuations in the resting brain. This approach is based on the discovery that, even at rest, functionally related brain regions tend to have synchronized BOLD signal fluctuations [97]. By mapping the temporal correlations (or "functional connectivity") between regions, we may thereby characterize the functional organization of the brain. This may be described in terms of functional networks, which are ensembles of brain regions that have high inter-regional connectivity, are stable over time and between individuals, and are implicated in specific sensory, motor, or cognitive domains [98, 99]. Termed resting-state fMRI (rs-fMRI), this method is appealing for studies of TBI and concussion, as it only requires that patients rest quietly in the scanner. Due to the minimal setup and compliance demands, rs-fMRI has largely supplanted tb-fMRI as the functional paradigm of choice in concussion research. The ease of implementation comes at a cost though, as "rest" represents an unknown, unconstrained brain state [100], and we do not have concurrent measures of cognition and behavior. Nevertheless, given the evidence for TBI-related disruptions in the white matter connections between brain regions and in local brain function, rs-fMRI provides an appealing method for interrogating the effects of concussion on inter-regional communication. Studies of more severe TBI often identify functional hyper-connectivity as a signature of injury, but are generally conducted months post-injury and focus on the neural correlates of persistent cognitive deficits [101, 102].

In concussion, studies in the early symptomatic phase of injury remain somewhat limited and have reported variable findings. Most of the research to date has focused on the default mode network (DMN), which is a highly conserved network of regions that includes the posterior cingulate, precuneus, medial prefrontal cortex, and angular gyrus. The network is implicated in inward-directed aspects of cognition (e.g., introspection, emotion processing, and memory consolidation) and shows reduced activity during externally directed tasks [103]. Examples of the DMN are shown for a patient with symptomatic concussion in Fig. 3.3a, using both seed-based and component-based techniques. Of the few DMN studies, a mixture of effects have been reported, with reduced posterior connectivity but enhanced frontal connections, within the first 24 hours post-injury and at 10 days post-injury [104, 105]; conversely, another study reported uniformly reduced connectivity at 7 days post-injury [106]. However, focusing on a single functional brain network limits our ability to discern changes occurring elsewhere in the brain. As an alternative, our group has studied global connectivity, in which we calculate the average connectivity of each brain region with all others, providing an index of total integration; Fig. 3.3a shows a representative global connectivity map. We have found enhanced global connectivity after concussion, primarily in dorsal and frontal brain regions [38, 40, 54]. However, there is also evidence that both DMN and global connectivity values decline over the first week of injury [38, 106]. These findings further highlight the complex, volatile nature of functional connectivity changes in the early phase of injury, which may be driven by a combination of neurometabolic dysregulation, axonal injury, and cerebrovascular impairment, along with compensatory changes in functional network organization [107].

Longitudinal rs-fMRI studies have reported persistent effects beyond symptom resolution, which most consistently manifest as hyper-connectivity. A study found evidence of emerging hyper-connectivity between the DMN and areas implicated in attention and emotion processing from 1 to 30 days post-injury, corresponding to the partial resolution of depressive symptoms [108]. This change is corroborated by cross-sectional studies of concussed patients, which show DMN hyper-connectivity at 1–2 months post-injury [109, 110], despite symptom resolution at the later time points. In terms of global connectivity, we recently conducted a longitudinal study with imaging in the first week post-injury, at medical clearance (median: 27 days), and 1-year follow-up. In this study, we found acute dorsal hyper-connectivity, which remained elevated at medical clearance, although it had dissipated at 1-year follow-up [54]. This is also supported by a study of history of concussion, showing an absence of significant functional connectivity effects over 9 months post-injury [111]. Collectively, these findings indicate that a key component of the concussion response is functional hyper-connectivity, which has been observed in many forms of acquired brain injury and is thought to sustain brain function in the presence of neural injury, at the expense of reduced overall network efficiency [74, 107]. Consistent with tb-fMRI findings, this enhanced

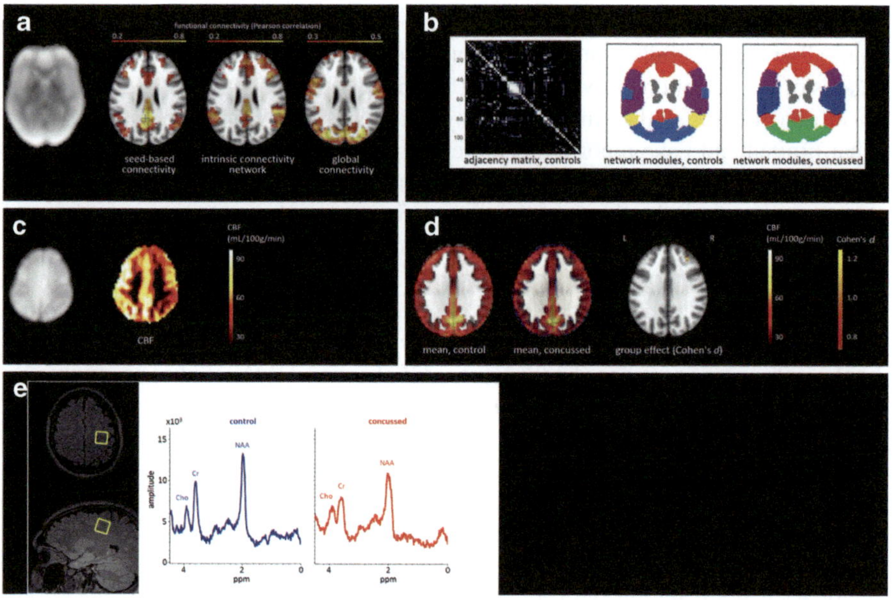

Fig. 3.3 Imaging sequences used to assess cerebral blood flow, neural activity and metabolism after concussion. (**a**) functional connectivity maps, shown for an individual with symptomatic concussion. Techniques include seed-based connectivity (i.e., correlation with a posterior cingulate region of interest), intrinsic connectivity networks (i.e., data-driven component estimation) and global connectivity (i.e., mean connectivity of each voxel to all others in the brain). (**b**) comparison of graph theoretic network measures for control (N=37) and symptomatic concussed (N=22) groups. The brain was parcelled into 116 regions (using the Automated Anatomical Labeling atlas) and thresholded to include the top 10% strongest connections. The mean adjacency graph is shown for controls and network modules shown for control and concussed groups (obtained via Louvain algorithm). For each module, brain regions are represented by a unique colour; the concussed group shows a distinct community organization compared to controls. (**c**) cerebral blood flow (CBF) map obtained from a participant with symptomatic concussion. (**d**) comparison of mean CBF in grey matter of control (N=37) and symptomatic concussed (N=22) groups. Areas of substantial group difference (Cohen's d > 0.8) are also depicted. The concussed group shows reduced CBF in the right frontal lobe. (**e**) single-voxel spectroscopy measures obtained within the right motor cortex. Metabolite spectra are shown for control and concussed individuals, with peaks corresponding to choline (Cho), creatine (Cr) and N-acetyl aspartate (NAA). The concussed participant shows reduced spectral peak amplitudes, particularly for the NAA peak

response persists beyond medical clearance, but appears to have recovered by the following year.

There is a substantial body of research examining the clinical correlates of functional connectivity. In the early symptomatic phase of injury, greater symptom severity and more prolonged subsequent recovery have been linked to lower functional connectivity within the DMN [112, 113] and lower global integration, particularly in frontotemporal networks [54, 114], although symptom severity has also been linked to *greater* connectivity between the DMN and other networks [112–114]. These findings suggest that, while greater overall functional integration is a sign of better outcome at early injury, a failure of networks to segregate appropriately may lead to functional impairments. Among patients with prolonged symptoms, greater symptom severity has also been linked to reduced frontal connectivity, at 1 and 6 months post-injury [115, 116]. Investigations of neurocognitive impairments are rarer, although one study found that reduced segregation between the DMN and other networks leads to both greater self-reported memory problems and worse performance on tests of memory [117].

One of the major challenges of rs-fMRI involves selecting brain regions of interest for connectivity analysis. Moreover, focusing on pairwise connectivity between regions may provide an overly simplistic description of concussion effects, as integration occurs at multiple spatial scales, and the network changes due to injury and compensatory response may be distributed throughout the brain. One solution to this dilemma comes from graph theory. This framework represents functional connectivity in the brain as a "graph," consisting of nodes (brain regions) that are connected by edges (functional connectivity values). In this framework, topological properties have been identified that underlie healthy brain function, including small-world organization (i.e., nodes form highly connected clusters, with a short average inter-nodal path length), network modularity (i.e., nodes are grouped into "modules," where within-module connections are dense but between-module connections are sparse) and network hubs (i.e., a few nodes have a disproportionately high number of inter-nodal connections) [118]. Studies of moderate-to-severe TBI have found evidence of disruptions in all these network properties, reflecting changes in information flow within the brain [102, 119, 120]. To date, there has been limited examination of concussion, although a study showed that at 2 weeks post-injury, local network efficiency and hub behavior are correlated with symptom severity [121], and a study of patients with persistent symptoms found no abnormalities at 1–3 weeks post-injury but reduced network modularity at 6 months [116]. Fig. 3.3b compares the functional brain networks of control and symptomatic concussed groups, showing group differences in modular brain organization. These findings provide encouraging evidence that graph theoretic measures may lead to a richer description of the changes in functional connectivity that occur after a concussion.

Arterial Spin Labeling (ASL)

ASL is used to quantify CBF, which is a critical component of brain function, given its role in delivering oxygen and glucose to brain tissues. For this technique, a radiofrequency pulse is used to magnetically label arterial blood water that is inflowing through the basal arteries. The labeled water molecules diffuse through the brain and eventually transit into the arterioles and capillaries. There, they alter local tissue magnetization, producing a detectable change in MRI signal. The "labeled" brain map is then contrasted against a "control" map where no labeling pulse was performed. The difference in signal intensity due to labeled water is proportional to CBF and can be converted into absolute units using a simplified kinetic model of blood flow [122], with an example map shown for a patient with symptomatic concussion in Fig. 3.3c. CBF must be tightly regulated to meet time-varying neurometabolic demands, and it is highly sensitive to neuropathology. In animal models, CBF is often impaired after a brain injury [123, 124], which causes an uncoupling with metabolism, as it fails to meet the increased demand of restoring ionic homeostasis in injured neural cells [125]. This leads to an energy crisis and the subsequent oxidative stress weakens the blood–brain barrier, with further negative effects on cerebrovascular functioning [126, 127]. Similar effects of blood–brain barrier weakening are also induced by the post-concussion neuroinflammatory response [128]. In humans, more severe TBI is associated with significant reductions in CBF at early injury, which may cause ischemic injury of brain tissues [129].

In studies of concussion, the observed changes in CBF are far milder, and the direction of effect is less consistent between studies. Different studies have reported elevated CBF relative to controls [54], reduced values [130], and nonsignificant differences [38], all within the first week of injury. The variable nature of early CBF may be partly driven by the effects of time post-injury, as a longitudinal study reported significant within-subject declines in CBF from 1 to 8 days post-injury [131], and a cross-sectional study similarly found elevated CBF for patients imaged 1–3 days post-injury, but reduced CBF for those imaged 5–7 days post-injury [38]. These findings are congruent with animal models, where acute hypermetabolism and subsequent energy crisis lead to declines in CBF over time [1]. However, other factors may contribute to the variable CBF response, including inter-individual differences in the extent of microvascular injury [132]. Damage to the neurovascular unit also impairs autoregulatory capacity, i.e., the ability of the brain to maintain consistent CBF in the presence of fluctuations in systemic blood pressure [133, 134]. These mechanisms may influence the CBF response to varying degrees, with effects that likely depend on pre-injury vascular health, concussion biomechanics, and the time interval from injury to imaging, leading to the observed variability between studies. As an example of heterogeneous post-concussion effects, see Fig. 3.3d, in which a comparison of control and symptomatic concussed groups shows spatially limited declines in CBF of the frontal cortex.

In terms of longitudinal recovery, the limited ASL literature indicates a similarly complex time course of recovery for CBF. A landmark study imaged concussed athletes 1 day, 1 week, and 1 month post-injury [130] and reported focal reductions in CBF of the insular and superior temporal cortex at 1 day, with a return to normal between 1 week and 1 month post-injury. This timeline followed the resolution of cognitive and neuropsychiatric symptoms. More recently, we conducted a longitudinal study with imaging in the first week post-injury, at medical clearance (median: 27 days) and at 1-year follow-up. The acute disturbances in CBF had dissipated at medical clearance; however, at 1-year follow-up, new frontal reductions in CBF had appeared. These findings are corroborated by cross-sectional studies conducted months to years after injury [19, 135, 136], which also found reduced frontal and subcortical CBF. This suggests that, following the resolution of acute CBF disturbances, there are distinct mechanisms causing chronic disturbances in CBF. This may reflect the evolution of processes initiated at acute injury, including the emergence of chronic neuroinflammatory response [50] and axonal injury of networks connections involved in blood flow regulation [137]. The chronic declines in CBF may also be accompanied by subtle reductions in frontal gray matter [19], which further reduce regional blood flow demand for individuals with a history of concussion.

In terms of associations between CBF and clinical presentation, this is also a relatively understudied aspect of concussion. Previous studies have indicated that higher posterior CBF is associated with greater symptom severity in the first week of injury [138] and at 1 month post-injury [139]. In contrast, lower frontal CBF is associated with greater symptom severity and longer time to recovery the first week of injury [54], and lower insular perfusion at 1 month post-injury is associated with longer recovery times and acute symptom severity [130]. These findings suggest a possible anterior-to-posterior gradient of perfusion effects, where more frontal elevations in CBF are associated with better outcome. Intriguingly, one study found evidence that different symptom subtypes correlate with different patterns of CBF response [138], as disturbances in posterior CBF were correlated with predominantly somatic complaints, whereas disturbances in frontal CBF were related to predominantly cognitive complaints.

The ASL literature has focused mainly on the effects of concussion on resting CBF. However, cerebrovascular reactivity (CVR) is also of substantial interest; this refers to the ability of blood vessels to alter CBF via changes in vascular tone. These regulatory mechanisms are highly sensitive to TBI and often show dysfunction after injury [140, 141]. MRI-based assessments of CVR typically involve the acquisition of BOLD fMRI during a task that modulates intravascular CO_2, which is a potent vasodilator [142, 143]. The modulation of CO_2 may be achieved using externally guided changes in respiration [144] or blended gas delivery systems [145, 146], with the latter requiring more specialized equipment but providing absolute quantification of CVR. The manipulation of CO_2 alters CBF without affecting cerebral metabolism, and the corresponding changes in BOLD signal can be used to index CVR. In a recent investigation of acute injury, we examined a cohort scanned in the first week post-injury, using a respiratory task paradigm [147]. In this study, concussed individuals had an enhanced negative BOLD response to hypocapnic challenge, which was interpreted as a dysfunctional vasoconstrictive response. More recently, we showed that significant CVR impairments were only detected in the first week of injury, but subtler longitudinal changes could be measured up to 1 year after medical clearance [148]. These findings highlight the presence of subtle but long-lasting alterations in CVR that are distinct from those seen in resting CBF.

Magnetic Resonance Spectroscopy (MRS)

Magnetic resonance spectroscopy (MRS), which typically refers to proton MRS (^{1}H MRS), is used to study neurochemistry *in vivo*, by quantifying the concentrations of different cerebral metabolites. This is achieved by using a radiofrequency pulse to stimulate protons at their resonance frequency and measuring the resulting signal.

Protons resonate at different frequencies depending on the molecular structure which they are part of, which produces a spectrum of "shifted" frequencies corresponding to the signal contributions of different molecules. As water is the most abundant source of protons in the brain, it dominates the spectrum; after subtracting its signal contribution, we may then localize the spectral peaks of other molecules. Due to the relatively minor contributions of non-water molecules to the MRS signal, the spectrum is typically acquired within a single, large voxel region, with averaging over multiple acquisitions to improve signal detection. Afterward, spectral peaks are identified, and relative metabolite contributions are derived from the areas under the curve, which are calculated using specialized curve-fitting techniques.

Only a few cerebral metabolites have spectral peaks that are within detectable ranges and sufficiently resolved from other metabolites, when using conventional MRS sequences. This includes macromolecules of N-acetyl aspartate (NAA), creatine (Cr), choline (Cho), and myo-inositol. NAA is found almost exclusively within neurons, and it is considered a marker of cell integrity, mitochondrial function, and myelin synthesis [149]. Cr levels are indicative of cellular energy metabolism [150], while Cho levels are associated with cell membrane synthesis and breakdown [151]. Myo-inositol is located mainly within glial cells and plays a role in the regulation of osmotic balance and in membrane synthesis and turnover [152]. More infrequently, glutamate is investigated, as it is a neurotransmitter that is released during concussive injury [1], but it is more difficult to reliably detect using standard MRS protocols. It should also be noted that MRS techniques often fail to distinguish between chemically similar metabolites that have highly overlapping spectral peaks; for example, the NAA spectral peak also includes contributions from N-acetylaspartylglutamate (NAAG). Fig. 3.3e shows representative spectral curves of control and symptomatic concussed individuals, obtained within the right motor cortex, with a lower NAA peak seen for the concussed patient.

To date, most of the MRS literature has focused on NAA, as it is the largest peak of the water-suppressed spectrum and it is highly sensitive to neuropathology. Reduced NAA levels in more severe TBI are interpreted as a sign of neural loss [153], whereas concussion-related changes in NAA are thought to reflect reversible changes in brain physiology. A landmark study with imaging at 3, 15, and 30 days post-injury found impaired NAA at 3 days post-injury despite symptom resolution at this time [154], which was related to mitochondrial impairment in a concurrent animal model study [155]. Individuals without second injury fully recovered at 30 days, whereas those who were re-injured before NAA recovery had more prolonged recovery of both symptom and NAA after the second injury. These findings highlight that NAA is a sensitive marker of post-concussion metabolic derangement, and moreover, incomplete recovery may signify a period of metabolic vulnerability, where re-injury leads to protracted recovery. However, subsequent studies have obtained variable NAA findings, with one reporting elevated NAA in motor and prefrontal gray matter from 6 days to 6 months post-injury [156], while others found no significant NAA effects in the corpus callosum from 3 days to 2 months after injury [157] and in the primary motor cortex from 1 week post-injury to 1 year after medical clearance [158]. These studies indicate that the NAA response is highly variable and depends on cohort demographics and the brain regions being imaged.

In most studies, the relative metabolite concentrations are used, and interpretation depends on the reference values of choice; however, there is also some evidence that standard references such as Cr and Cho may also be affected by brain injury [159]. While techniques are available to use absolute values, it is unclear which offers superior sensitivity to concussion outcomes. The second major limitation is the dependence on localization. For a standard MR protocol, it is prohibitive to test more than a few regions, and the biomechanics of injury and the spatial distribution of effects remain largely unknown. Techniques are available that expand the field of view, such as chemical shift imaging (CSI), but they come at the cost of a reduced signal-to-noise ratio [160]. In terms of future methodology, MRS techniques that target other non-hydrogen nuclei have also been under-utilized. In particular, ^{31}P-MRS is a promising idea, as it is relatively abundant in the brain and can target membrane phospholipids and high-energy metabolic products, but research has been largely limited to animal studies [161, 162].

Clinical Utility of Concussion Neuroimaging: Promise and Challenges

Neuroimaging promises novel insights into the clinical determination of concussive injury and the management of patient recovery. At present, diagnosis is primarily based on self-reported symptoms, where the willingness and ability to self-disclose may be confounded by a variety of sociocultural factors [163–165]. Neuroimaging findings may help to refine existing diagnostic criteria and to develop low-cost objective biomarkers supplementing standard clinical assessments. Among individuals with prolonged symptoms, there is also growing recognition that different clinical profiles (i.e., predominantly somatic, cognitive, mood, or sleep-related) may correspond to injury "subtypes," requiring different approaches to patient management [166]. In this context, neuroimaging may identify biological subtypes, paving the way toward new evidence-based patient interventions. Relatedly, the determination of safe return to activities is largely dependent on symptom presentation [2], with evidence that biological recovery likely extends beyond clinical recovery [167]. Objective imaging data may serve to refine minimal "stand-down" times and to inform clinicians about individual variability in the course of biological recovery. Finally, there are concerns about the cumulative effects of repeated concussion on long-term brain health and cognition [26–28]. In this respect, imaging may provide objective measures of the biological "dose-response" of concussion on brain tissue, thereby contributing to guidelines about the relative risk of multiple concussions and the identification of factors putting individuals at greater risk of long-term health issues.

At present, however, these clinical applications remain largely unfulfilled. While MRI has the potential to clarify biological mechanisms of injury and recovery after a concussion, its interpretation has been hampered by inconsistent findings between studies. This is partly because many landmark studies are based on sample sizes of less than 20, limiting our ability to detect all but the largest effects and, paradoxically, increasing the risk of overestimating effect sizes for statistically significant relationships; for a discussion of these issues, see Button et al. [168]. This may be mitigated by emerging large-scale multisite studies, conducted in civilian, military, and athlete cohorts [169–171]. Larger samples are only part of the solution, though, as there is evidence of substantial patient variability in demographic characteristics and in the features of concussive injury, both clinical and pathophysiological [172]. Studies that do not account for these issues may, at minimum, underestimate effects of concussion on brain physiology due to unmodeled variance. At worst, combining groups with differing concussion responses may "cancel out" effects and produce null findings, irrespective of the sample size.

Given the limited current understanding of how concussion effects vary across cohorts, more broadly inclusive studies are needed, with comprehensive modeling of demographic factors in a single cohort (e.g., age, sex, concussion history, pre-injury physical, and mental health). Variables such as age, sex, and history of concussion have been previously studied but mainly in piecemeal fashion [105, 173, 174], making it difficult to judge their relative importance in concussion outcome. An important complement to demographic modeling lies in the use of prospective studies, where pre- and post-injury neuroimaging data are collected, allowing researchers to directly account for variations in pre-injury brain physiology. In these studies, large cohorts of high-risk individuals are imaged, e.g., from athlete and military populations, with follow-up imaging of those who are later concussed. This has been a growing area of investigation [175–177], but given the challenges of obtaining sufficient sample sizes, more research is needed. The importance is highlighted by a recent study, where we showed that cross-sectional and prospective MRI analyses of concussion have comparable sensitivity but tend to identify non-overlapping patterns of brain abnormalities [178]. More broadly, there is a need to quantify the reproducibility of neuroimaging findings [179]. This includes a need for replication studies, particularly for earlier findings obtained from small samples, given the subtle, heterogeneous effects of concussion. Replications should also draw from different cohorts to determine how widely concussion findings generalize, between athlete, military, and civilian injury, different socioeconomic strata, and differences in pre-injury health status.

Conclusions

As illustrated in this chapter, MRI-based neuroimaging of concussion provides a powerful technique for investigating the complex changes in brain physiology that occur after concussion. Neuroimaging findings have validated and expanded upon previous work using animal models and in vitro studies. They highlight that concussion is a complex, evolving process of injury and recovery, with both acute and long-term changes in tissue microstructure, metabolism, and cerebrovascular function. Moreover, the changes have been linked to neuropsychological and behavioral symptoms of injury, providing evidence of identifiable biological underpinnings for these deficits. Interestingly, there is also evidence that physiological changes extend months to years beyond clinical recovery, although the long-term health implications are not yet entirely understood. Although we have made a great deal of progress, there remain major gaps in our understanding, but the emergence of new imaging techniques and large-scale, multisite studies may accelerate the discovery process. With these ongoing developments, we hope that neuroimaging will soon provide us with new advances in evidence-based diagnosis and management of patients.

References

1. Giza CC, Hovda DA. The new neurometabolic cascade of concussion. Neurosurgery. 2014;75(Suppl_4):S24–33.
2. McCrory P, Meeuwisse W, Dvorak J, Aubry M, Bailes J, Broglio S, et al. Consensus statement on concussion in sport—The 5th international conference on concussion in sport held in Berlin, October 2016. Br J Sports Med. 2017;51(11):838–47.
3. Shenton ME, Hamoda H, Schneiderman J, Bouix S, Pasternak O, Rathi Y, et al. A review of magnetic resonance imaging and diffusion tensor imaging findings in mild traumatic brain injury. Brain Imaging Behav. 2012;6(2):137–92.
4. Davis PC. Head trauma. Am J Neuroradiol. 2007;28(8):1619–21.
5. Gallagher CN, Hutchinson PJ, Pickard JD. Neuroimaging in trauma. Curr Opin Neurol. 2007;20(4):403–9.
6. Smits M, Hunink M, Van Rijssel D, Dekker H, Vos P, Kool D, et al. Outcome after complicated minor head injury. Am J Neuroradiol. 2008;29(3):506–13.
7. Haacke EM, Xu Y, Cheng YCN, Reichenbach JR. Susceptibility weighted imaging (SWI). Magn Reson Med. 2004;52(3):612–8.
8. Hughes DG, Jackson A, Mason DL, Berry E, Hollis S, Yates DW. Abnormalities on magnetic resonance imaging seen acutely following mild traumatic brain injury: correlation with neuropsychological tests and delayed recovery. Neuroradiology. 2004;46(7):550–8.
9. Hofman PA, Stapert SZ, van Kroonenburgh MJ, Jolles J, de Kruijk J, Wilmink JT. MR imaging, single-photon emission CT, and neurocognitive performance after mild traumatic brain injury. Am J Neuroradiol. 2001;22(3):441–9.
10. Mittl R, Grossman R, Hiehle J, Hurst RW, Kauder DR, Gennarelli TA, et al. Prevalence of MR evidence of diffuse axonal injury in patients with mild head injury and normal head CT findings. Am J Neuroradiol. 1994;15(8):1583–9.
11. Lee H, Wintermark M, Gean AD, Ghajar J, Manley GT, Mukherjee P. Focal lesions in acute mild traumatic brain injury and neurocognitive outcome: CT versus 3T MRI. J Neurotrauma. 2008;25(9):1049–56.
12. Iverson GL. Complicated vs uncomplicated mild traumatic brain injury: acute neuropsychological outcome. Brain Inj. 2006;20(13–14):1335–44.
13. Povlishock JT, CHRISTMAN CW. The pathobiology of traumatically induced axonal injury in animals and humans: a review of current thoughts. J Neurotrauma. 1995;12(4):555–64.
14. Kou Z, VandeVord PJ. Traumatic white matter injury and glial activation: from basic science to clinics. Glia. 2014;62(11):1831–55.
15. Sullivan DR. A cerebrovascular hypothesis of neurodegeneration in mTBI. J Head Trauma Rehabil. 2019;34(3):E18.

16. Milman A, Rosenberg A, Weizman R, Pick C. Mild traumatic brain injury induces persistent cognitive deficits and behavioral disturbances in mice. J Neurotrauma. 2005;22(9):1003–10.
17. Povlishock J, Erb D, Astruc J. Axonal response to traumatic brain injury: reactive axonal change, deafferentation, and neuroplasticity. J Neurotrauma. 1992;9:S189–200.
18. Dean PJ, Sato JR, Vieira G, McNamara A, Sterr A. Long-term structural changes after mTBI and their relation to post-concussion symptoms. Brain Inj. 2015;29(10):1211–8.
19. Churchill N, Hutchison M, Richards D, Leung G, Graham S, Schweizer TA. Brain structure and function associated with a history of sport concussion: a multi-modal magnetic resonance imaging study. J Neurotrauma. 2017;34(4):765–71.
20. Singh R, Meier TB, Kuplicki R, Savitz J, Mukai I, Cavanagh L, et al. Relationship of collegiate football experience and concussion with hippocampal volume and cognitive outcomes. JAMA. 2014;311(18):1883–8.
21. Meier TB, Bellgowan PS, Bergamino M, Ling JM, Mayer AR. Thinner cortex in collegiate football players with, but not without, a self-reported history of concussion. J Neurotrauma. 2016;33(4):330–8.
22. Sidaros A, Skimminge A, Liptrot MG, Sidaros K, Engberg AW, Herning M, et al. Long-term global and regional brain volume changes following severe traumatic brain injury: a longitudinal study with clinical correlates. NeuroImage. 2009;44(1):1–8.
23. Bendlin BB, Ries ML, Lazar M, Alexander AL, Dempsey RJ, Rowley HA, et al. Longitudinal changes in patients with traumatic brain injury assessed with diffusion-tensor and volumetric imaging. NeuroImage. 2008;42(2):503–14.
24. Zhou Y, Kierans A, Kenul D, Ge Y, Rath J, Reaume J, et al. Mild traumatic brain injury: longitudinal regional brain volume changes. Radiology. 2013;267(3):880–90.
25. Graham D, Gennarelli T, McIntosh T. Trauma. Greenfield's neuropathology. New York: Oxford University Press; 2002.
26. Control CD. Prevention. National Football League players mortality study. Cincinnati: NIOSH; 1994.
27. Gavett BE, Stern RA, Cantu RC, Nowinski CJ, McKee AC. Mild traumatic brain injury: a risk factor for neurodegeneration. Alzheimers Res Ther. 2010;2(3):18.
28. Lehman EJ, Hein MJ, Baron SL, Gersic CM. Neurodegenerative causes of death among retired National Football League players. Neurology. 2012;79(19):1970–4.
29. Alexander-Bloch A, Giedd JN, Bullmore E. Imaging structural co-variance between human brain regions. Nat Rev Neurosci. 2013;14(5):322–36.
30. Wang C, Zhao L, Luo Y, Liu J, Miao P, Wei S, et al. Structural covariance in subcortical stroke patients measured by automated MRI-based volumetry. Neuroimage Clin. 2019;22:101682.
31. Novellino F, López ME, Vaccaro MG, Miguel Y, Delgado ML, Maestu F. Association between hippocampus, thalamus, and caudate in mild cognitive impairment APOEε4 carriers: a structural covariance MRI study. Front Neurol. 2019;10:1303.
32. Ommaya AK, Gennarelli T. Cerebral concussion and traumatic unconsciousness: correlation of experimental and clinical observations on blunt head injuries. Brain. 1974;97(4):633–54.
33. Armstrong RC, Mierzwa AJ, Marion CM, Sullivan GM. White matter involvement after TBI: clues to axon and myelin repair capacity. Exp Neurol. 2016;275:328–33.
34. Song S-K, Sun S-W, Ju W-K, Lin S-J, Cross AH, Neufeld AH. Diffusion tensor imaging detects and differentiates axon and myelin degeneration in mouse optic nerve after retinal ischemia. NeuroImage. 2003;20(3):1714–22.
35. Song S-K, Yoshino J, Le TQ, Lin S-J, Sun S-W, Cross AH, et al. Demyelination increases radial diffusivity in corpus callosum of mouse brain. NeuroImage. 2005;26(1):132–40.
36. Alexander AL, Lee JE, Lazar M, Field AS. Diffusion tensor imaging of the brain. Neurotherapeutics. 2007;4(3):316–29.
37. Kraus MF, Susmaras T, Caughlin BP, Walker CJ, Sweeney JA, Little DM. White matter integrity and cognition in chronic traumatic brain injury: a diffusion tensor imaging study. Brain. 2007;130(10):2508–19.

38. Churchill NW, Hutchison MG, Richards D, Leung G, Graham SJ, Schweizer TA. The first week after concussion: blood flow, brain function and white matter microstructure. Neuroimage Clin. 2017;14:480–9.
39. Murugavel M, Cubon V, Putukian M, Echemendia R, Cabrera J, Osherson D, et al. A longitudinal diffusion tensor imaging study assessing white matter fiber tracts after sports-related concussion. J Neurotrauma. 2014;31(22):1860–71.
40. Churchill NW, Hutchison MG, Richards D, Leung G, Graham SJ, Schweizer TA. Neuroimaging of sport concussion: persistent alterations in brain structure and function at medical clearance. Sci Rep. 2017;7(1):8297.
41. Inglese M, Makani S, Johnson G, Cohen BA, Silver JA, Gonen O, et al. Diffuse axonal injury in mild traumatic brain injury: a diffusion tensor imaging study. J Neurosurg. 2005;103(2):298–303.
42. Unterberg A, Stover J, Kress B, Kiening K. Edema and brain trauma. Neuroscience. 2004;129(4):1019–27.
43. Marmarou A. A review of progress in understanding the pathophysiology and treatment of brain edema. Neurosurg Focus. 2007;22(5):1–10.
44. Barzó P, Marmarou A, Fatouros P, Hayasaki K, Corwin F. Contribution of vasogenic and cellular edema to traumatic brain swelling measured by diffusion-weighted imaging. J Neurosurg. 1997;87(6):900–7.
45. Streit WJ, Mrak RE, Griffin WST. Microglia and neuroinflammation: a pathological perspective. J Neuroinflammation. 2004;1(1):14.
46. Wilde E, McCauley S, Hunter J, Bigler E, Chu Z, Wang Z, et al. Diffusion tensor imaging of acute mild traumatic brain injury in adolescents. Neurology. 2008;70(12):948–55.
47. Mayer A, Ling J, Mannell M, Gasparovic C, Phillips J, Doezema D, et al. A prospective diffusion tensor imaging study in mild traumatic brain injury. Neurology. 2010;74(8):643–50.
48. Meier TB, Bergamino M, Bellgowan PS, Teague T, Ling JM, Jeromin A, et al. Longitudinal assessment of white matter abnormalities following sports-related concussion. Hum Brain Mapp. 2016;37(2):833–45.
49. Henry LC, Tremblay J, Tremblay S, Lee A, Brun C, Lepore N, et al. Acute and chronic changes in diffusivity measures after sports concussion. J Neurotrauma. 2011;28(10):2049–59.
50. Patterson ZR, Holahan MR. Understanding the neuroinflammatory response following concussion to develop treatment strategies. Front Cell Neurosci. 2012;6:58.
51. Marchi N, Bazarian JJ, Puvenna V, Janigro M, Ghosh C, Zhong J, et al. Consequences of repeated blood-brain barrier disruption in football players. PLoS One. 2013;8(3):e56805.
52. Di Battista AP, Rhind SG, Richards D, Churchill N, Baker AJ, Hutchison MG. Altered blood biomarker profiles in athletes with a history of repetitive head impacts. PLoS One. 2016;11(7):e0159929.
53. Tomaszczyk JC, Green NL, Frasca D, Colella B, Turner GR, Christensen BK, et al. Negative neuroplasticity in chronic traumatic brain injury and implications for neurorehabilitation. Neuropsychol Rev. 2014;24(4):409–27.
54. Churchill NW, Hutchison MG, Graham SJ, Schweizer TA. Mapping recovery of brain physiology after concussion: from acute injury to one year after medical clearance. Neurology. 2019;93(21):e1980–92.
55. Smits M, Houston GC, Dippel DW, Wielopolski PA, Vernooij MW, Koudstaal PJ, et al. Microstructural brain injury in post-concussion syndrome after minor head injury. Neuroradiology. 2011;53(8):553–63.
56. Messé A, Caplain S, Paradot G, Garrigue D, Mineo JF, Soto Ares G, et al. Diffusion tensor imaging and white matter lesions at the subacute stage in mild traumatic brain injury with persistent neurobehavioral impairment. Hum Brain Mapp. 2011;32(6):999–1011.
57. Cubon V, Putukian M, Boyer C, Dettwiler A. A diffusion tensor imaging study on the white matter skeleton in individuals with sports-related concussion. J Neurotrauma. 2011;28(2):189–201.

58. Lange RT, Iverson GL, Brubacher JR, Mädler B, Heran MK. Diffusion tensor imaging findings are not strongly associated with postconcussional disorder 2 months following mild traumatic brain injury. J Head Trauma Rehabil. 2012;27(3):188–98.
59. Lipton ML, Gulko E, Zimmerman ME, Friedman BW, Kim M, Gellella E, et al. Diffusion-tensor imaging implicates prefrontal axonal injury in executive function impairment following very mild traumatic brain injury. Radiology. 2009;252(3):816–24.
60. Jensen JH, Helpern JA, Ramani A, Lu H, Kaczynski K. Diffusional kurtosis imaging: the quantification of non-gaussian water diffusion by means of magnetic resonance imaging. Magn Reson Med. 2005;53(6):1432–40.
61. Zhang H, Schneider T, Wheeler-Kingshott C, Alexander D. NODDI: practical in vivo neurite orientation dispersion and density imaging of the human brain. NeuroImage. 2012;61(4):1000–16.
62. Churchill NW, Caverzasi E, Graham SJ, Hutchison MG, Schweizer TA. White matter during concussion recovery: comparing diffusion tensor imaging (DTI) and neurite orientation dispersion and density imaging (NODDI). Hum Brain Mapp. 2019;40(6):1908–18.
63. Katayama Y, Becker DP, Tamura T, Hovda DA. Massive increases in extracellular potassium and the indiscriminate release of glutamate following concussive brain injury. J Neurosurg. 1990;73(6):889–900.
64. Nilsson P, Hillered L, Ponten U, Ungerstedt U. Changes in cortical extracellular levels of energy-related metabolites and amino acids following concussive brain injury in rats. J Cereb Blood Flow Metab. 1990;10(5):631–7.
65. Faden AI, Loane DJ. Chronic neurodegeneration after traumatic brain injury: Alzheimer disease, chronic traumatic encephalopathy, or persistent neuroinflammation? Neurotherapeutics. 2015;12(1):143–50.
66. Witgen B, Lifshitz J, Smith M, Schwarzbach E, Liang S-L, Grady M, et al. Regional hippocampal alteration associated with cognitive deficit following experimental brain injury: a systems, network and cellular evaluation. Neuroscience. 2005;133(1):1–15.
67. Sick TJ, Pérez-Pinzón MA, Feng Z-Z. Impaired expression of long-term potentiation in hippocampal slices 4 and 48 h following mild fluid-percussion brain injury in vivo. Brain Res. 1998;785(2):287–92.
68. Scheibel RS, Newsome MR, Steinberg JL, Pearson DA, Rauch RA, Mao H, et al. Altered brain activation during cognitive control in patients with moderate to severe traumatic brain injury. Neurorehabil Neural Repair. 2007;21(1):36–45.
69. Perlstein WM, Cole MA, Demery JA, Seignourel PJ, Dixit NK, Larson MJ, et al. Parametric manipulation of working memory load in traumatic brain injury: behavioral and neural correlates. J Int Neuropsychol Soc. 2004;10(5):724–41.
70. Palacios EM, Sala-Llonch R, Junque C, Roig T, Tormos JM, Bargallo N, et al. White matter integrity related to functional working memory networks in traumatic brain injury. Neurology. 2012;78(12):852–60.
71. McAllister TW, Saykin A, Flashman L, Sparling M, Johnson S, Guerin S, et al. Brain activation during working memory 1 month after mild traumatic brain injury: a functional MRI study. Neurology. 1999;53(6):1300. https://doi.org/10.1212/WNL.53.6.1300.
72. McAllister TW, Sparling MB, Flashman LA, Guerin SJ, Mamourian AC, Saykin AJ. Differential working memory load effects after mild traumatic brain injury. NeuroImage. 2001;14(5):1004–12.
73. McAllister TW, Flashman LA, McDonald BC, Saykin AJ. Mechanisms of working memory dysfunction after mild and moderate TBI: evidence from functional MRI and neurogenetics. J Neurotrauma. 2006;23(10):1450–67.
74. Medaglia JD. Functional neuroimaging in traumatic brain injury: from nodes to networks. Front Neurol. 2017;8:407.
75. Hillary FG. Neuroimaging of working memory dysfunction and the dilemma with brain reorganization hypotheses. J Int Neuropsychol Soc. 2008;14(4):526–34.
76. Bryer E, Medaglia J, Rostami S, Hillary FG. Neural recruitment after mild traumatic brain injury is task dependent: a meta-analysis. J Int Neuropsychol Soc. 2013;19(7):751.

77. Slobounov SM, Zhang K, Pennell D, Ray W, Johnson B, Sebastianelli W. Functional abnormalities in normally appearing athletes following mild traumatic brain injury: a functional MRI study. Exp Brain Res. 2010;202(2):341–54.
78. Zhang K, Johnson B, Pennell D, Ray W, Sebastianelli W, Slobounov S. Are functional deficits in concussed individuals consistent with white matter structural alterations: combined FMRI & DTI study. Exp Brain Res. 2010;204(1):57–70.
79. Mayer AR, Mannell MV, Ling J, Elgie R, Gasparovic C, Phillips JP, et al. Auditory orienting and inhibition of return in mild traumatic brain injury: a FMRI study. Hum Brain Mapp. 2009;30(12):4152–66.
80. Witt ST, Lovejoy DW, Pearlson GD, Stevens MC. Decreased prefrontal cortex activity in mild traumatic brain injury during performance of an auditory oddball task. Brain Imaging Behav. 2010;4(3–4):232–47.
81. Mayer AR, Yang Z, Yeo RA, Pena A, Ling JM, Mannell MV, et al. A functional MRI study of multimodal selective attention following mild traumatic brain injury. Brain Imaging Behav. 2012;6(2):343–54.
82. Lovell MR, Pardini JE, Welling J, Collins MW, Bakal J, Lazar N, et al. Functional brain abnormalities are related to clinical recovery and time to return-to-play in athletes. Neurosurgery. 2007;61(2):352–60.
83. Dettwiler A, Murugavel M, Putukian M, Cubon V, Furtado J, Osherson D. Persistent differences in patterns of brain activation after sports-related concussion: a longitudinal functional magnetic resonance imaging study. J Neurotrauma. 2014;31(2):180–8.
84. Hammeke TA, McCrea M, Coats SM, Verber MD, Durgerian S, Flora K, et al. Acute and subacute changes in neural activation during the recovery from sport-related concussion. J Int Neuropsychol Soc. 2013;19(8):863.
85. Johnson B, Hallett M, Slobounov S. Follow-up evaluation of oculomotor performance with fMRI in the subacute phase of concussion. Neurology. 2015;85(13):1163–6.
86. Elbin R, Covassin T, Hakun J, Kontos AP, Berger K, Pfeiffer K, et al. Do brain activation changes persist in athletes with a history of multiple concussions who are asymptomatic? Brain Inj. 2012;26(10):1217–25.
87. Terry DP, Faraco CC, Smith D, Diddams MJ, Puente AN, Miller LS. Lack of long-term fMRI differences after multiple sports-related concussions. Brain Inj. 2012;26(13–14): 1684–96.
88. Chen J-K, Johnston KM, Petrides M, Ptito A. Neural substrates of symptoms of depression following concussion in male athletes with persisting postconcussion symptoms. Arch Gen Psychiatry. 2008;65(1):81–9.
89. Chen J-K, Johnston K, Frey S, Petrides M, Worsley K, Ptito A. Functional abnormalities in symptomatic concussed athletes: an fMRI study. NeuroImage. 2004;22(1):68–82.
90. Chen J-K, Johnston KM, Collie A, McCrory P, Ptito A. A validation of the post concussion symptom scale in the assessment of complex concussion using cognitive testing and functional MRI. J Neurol Neurosurg Psychiatry. 2007;78(11):1231–8.
91. Chen J-K, Johnston KM, Petrides M, Ptito A. Recovery from mild head injury in sports: evidence from serial functional magnetic resonance imaging studies in male athletes. Clin J Sport Med. 2008;18(3):241–7.
92. Smits M, Dippel DW, Houston GC, Wielopolski PA, Koudstaal PJ, Hunink MM, et al. Postconcussion syndrome after minor head injury: brain activation of working memory and attention. Hum Brain Mapp. 2009;30(9):2789–803.
93. Badaut J, Bix G. Vascular neural network phenotypic transformation after traumatic injury: potential role in long-term sequelae. Transl Stroke Res. 2014;5(3):394–406.
94. Hoge RD. Calibrated fMRI. NeuroImage. 2012;62(2):930–7.
95. Gosselin N, Bottari C, Chen J-K, Petrides M, Tinawi S, de Guise E, et al. Electrophysiology and functional MRI in post-acute mild traumatic brain injury. J Neurotrauma. 2011;28(3):329–41.
96. Bergsneider M, Hovda DA, McArthur DL, Etchepare M, Huang S-C, Sehati N, et al. Metabolic recovery following human traumatic brain injury based on FDG-PET: time course and relationship to neurological disability. J Head Trauma Rehabil. 2001;16(2):135–48.

97. Biswal B, Zerrin Yetkin F, Haughton VM, Hyde JS. Functional connectivity in the motor cortex of resting human brain using echo-planar MRI. Magn Reson Med. 1995;34(4):537–41.
98. Laird AR, Fox PM, Eickhoff SB, Turner JA, Ray KL, McKay DR, et al. Behavioral interpretations of intrinsic connectivity networks. J Cogn Neurosci. 2011;23(12):4022–37.
99. Smith SM, Fox PT, Miller KL, Glahn DC, Fox PM, Mackay CE, et al. Correspondence of the brain's functional architecture during activation and rest. Proc Natl Acad Sci. 2009;106(31):13040–5.
100. Tagliazucchi E, Laufs H. Decoding wakefulness levels from typical fMRI resting-state data reveals reliable drifts between wakefulness and sleep. Neuron. 2014;82(3):695–708.
101. Sharp DJ, Beckmann CF, Greenwood R, Kinnunen KM, Bonnelle V, De Boissezon X, et al. Default mode network functional and structural connectivity after traumatic brain injury. Brain. 2011;134(8):2233–47.
102. Hillary FG, Rajtmajer SM, Roman CA, Medaglia JD, Slocomb-Dluzen JE, Calhoun VD, et al. The rich get richer: brain injury elicits hyperconnectivity in core subnetworks. PLoS One. 2014;9(8):e104021.
103. Broyd SJ, Demanuele C, Debener S, Helps SK, James CJ, Sonuga-Barke EJ. Default-mode brain dysfunction in mental disorders: a systematic review. Neurosci Biobehav Rev. 2009;33(3):279–96.
104. Iraji A, Benson RR, Welch RD, O'Neil BJ, Woodard JL, Ayaz SI, et al. Resting state functional connectivity in mild traumatic brain injury at the acute stage: independent component and seed-based analyses. J Neurotrauma. 2015;32(14):1031–45.
105. Johnson B, Zhang K, Gay M, Horovitz S, Hallett M, Sebastianelli W, et al. Alteration of brain default network in subacute phase of injury in concussed individuals: resting-state fMRI study. NeuroImage. 2012;59(1):511–8.
106. Zhu D, Covassin T, Nogle S, Doyle S, Russell D, Pearson R, et al. A potential biomarker in sports-related concussion: brain functional connectivity alteration of the default-mode network measured with longitudinal resting-State fMRI over thirty days. J Neurotrauma. 2015;32(5):327–41.
107. Hillary FG, Grafman JH. Injured brains and adaptive networks: the benefits and costs of hyperconnectivity. Trends Cogn Sci. 2017;21(5):385–401.
108. McCuddy WT, España LY, Nelson LD, Birn RM, Mayer AR, Meier TB. Association of acute depressive symptoms and functional connectivity of emotional processing regions following sport-related concussion. Neuroimage Clin. 2018;19:434–42.
109. Newsome MR, Li X, Lin X, Wilde EA, Ott S, Biekman B, et al. Functional connectivity is altered in concussed adolescent athletes despite medical clearance to return to play: a preliminary report. Front Neurol. 2016;7:116.
110. Czerniak SM, Sikoglu EM, Navarro AAL, McCafferty J, Eisenstock J, Stevenson JH, et al. A resting state functional magnetic resonance imaging study of concussion in collegiate athletes. Brain Imaging Behav. 2015;9(2):323–32.
111. Churchill N, Hutchison MG, Leung G, Graham S, Schweizer TA. Changes in functional connectivity of the brain associated with a history of sport concussion: a preliminary investigation. Brain Inj. 2017;31(1):39–48.
112. Zhou Y, Milham MP, Lui YW, Miles L, Reaume J, Sodickson DK, et al. Default-mode network disruption in mild traumatic brain injury. Radiology. 2012;265(3):882–92.
113. Mayer AR, Mannell MV, Ling J, Gasparovic C, Yeo RA. Functional connectivity in mild traumatic brain injury. Hum Brain Mapp. 2011;32(11):1825–35.
114. Churchill NW, Hutchison MG, Graham SJ, Schweizer TA. Connectomic markers of symptom severity in sport-related concussion: whole-brain analysis of resting-state fMRI. Neuroimage Clin. 2018;18:518–26.
115. van der Horn HJ, Liemburg EJ, Aleman A, Spikman JM, van der Naalt J. Brain networks subserving emotion regulation and adaptation after mild traumatic brain injury. J Neurotrauma. 2016;33(1):1–9.

116. Messé A, Caplain S, Pélégrini-Issac M, Blancho S, Lévy R, Aghakhani N, et al. Specific and evolving resting-state network alterations in post-concussion syndrome following mild traumatic brain injury. PLoS One. 2013;8(6):e65470.
117. Sours C, Zhuo J, Janowich J, Aarabi B, Shanmuganathan K, Gullapalli RP. Default mode network interference in mild traumatic brain injury–a pilot resting state study. Brain Res. 2013;1537:201–15.
118. Rubinov M, Sporns O. Complex network measures of brain connectivity: uses and interpretations. NeuroImage. 2010;52(3):1059–69.
119. Han K, Mac Donald CL, Johnson AM, Barnes Y, Wierzechowski L, Zonies D, et al. Disrupted modular organization of resting-state cortical functional connectivity in US military personnel following concussive 'mild' blast-related traumatic brain injury. NeuroImage. 2014;84:76–96.
120. Pandit AS, Expert P, Lambiotte R, Bonnelle V, Leech R, Turkheimer FE, et al. Traumatic brain injury impairs small-world topology. Neurology. 2013;80(20):1826–33.
121. van der Horn HJ, Liemburg EJ, Scheenen ME, de Koning ME, Spikman JM, van der Naalt J. Graph analysis of functional brain networks in patients with mild traumatic brain injury. PLoS One. 2017;12(1):e0171031.
122. Alsop DC, Detre JA, Golay X, Günther M, Hendrikse J, Hernandez-Garcia L, et al. Recommended implementation of arterial spin-labeled perfusion MRI for clinical applications: a consensus of the ISMRM perfusion study group and the European consortium for ASL in dementia. Magn Reson Med. 2015;73(1):102–16.
123. Yamakami I, McIntosh TK. Effects of traumatic brain injury on regional cerebral blood flow in rats as measured with radiolabeled microspheres. J Cereb Blood Flow Metab. 1989;9(1):117–24.
124. Yuan X-Q, Prough DS, Smith TL, DeWitt DS. The effects of traumatic brain injury on regional cerebral blood flow in rats. J Neurotrauma. 1988;5(4):289–301.
125. Ginsberg M, Zhao W, Alonso O, Loor-Estades J, Dietrich WD, Busto R. Uncoupling of local cerebral glucose metabolism and blood flow after acute fluid-percussion injury in rats. Am J Phys Heart Circ Phys. 1997;272(6):H2859–H68.
126. Wang CX, Shuaib A. Critical role of microvasculature basal lamina in ischemic brain injury. Prog Neurobiol. 2007;83(3):140–8.
127. Abdul-Muneer P, Schuetz H, Wang F, Skotak M, Jones J, Gorantla S, et al. Induction of oxidative and nitrosative damage leads to cerebrovascular inflammation in an animal model of mild traumatic brain injury induced by primary blast. Free Radic Biol Med. 2013;60:282–91.
128. Abdul-Muneer P, Chandra N, Haorah J. Interactions of oxidative stress and neurovascular inflammation in the pathogenesis of traumatic brain injury. Mol Neurobiol. 2015;51(3):966–79.
129. Menon DK. Brain ischaemia after traumatic brain injury: lessons from 15O2 positron emission tomography. Curr Opin Crit Care. 2006;12(2):85–9.
130. Meier TB, Bellgowan PS, Singh R, Kuplicki R, Polanski DW, Mayer AR. Recovery of cerebral blood flow following sports-related concussion. JAMA Neurol. 2015;72(5):530–8.
131. Wang Y, Nelson LD, LaRoche AA, Pfaller AY, Nencka AS, Koch KM, et al. Cerebral blood flow alterations in acute sport-related concussion. J Neurotrauma. 2016;33(13):1227–36.
132. Truettner JS, Alonso OF, Dietrich WD. Influence of therapeutic hypothermia on matrix metalloproteinase activity after traumatic brain injury in rats. J Cereb Blood Flow Metab. 2005;25(11):1505–16.
133. Jünger EC, Newell DW, Grant GA, Avellino AM, Ghatan S, Douville CM, et al. Cerebral autoregulation following minor head injury. J Neurosurg. 1997;86(3):425–32.
134. Strebel S, Lam AM, Matta BF, Newell DW. Impaired cerebral autoregulation after mild brain injury. Surg Neurol. 1997;47(2):128–31.
135. Wang Y, West JD, Bailey JN, Westfall DR, Xiao H, Arnold TW, et al. Decreased cerebral blood flow in chronic pediatric mild TBI: an MRI perfusion study. Dev Neuropsychol. 2015;40(1):40–4.

136. Ge Y, Patel MB, Chen Q, Grossman EJ, Zhang K, Miles L, et al. Assessment of thalamic perfusion in patients with mild traumatic brain injury by true FISP arterial spin labelling MR imaging at 3T. Brain Inj. 2009;23(7–8):666–74.
137. Critchley HD. Neural mechanisms of autonomic, affective, and cognitive integration. J Comp Neurol. 2005;493(1):154–66.
138. Churchill NW, Hutchison MG, Graham SJ, Schweizer TA. Symptom correlates of cerebral blood flow following acute concussion. Neuroimage Clin. 2017;16:234–9.
139. Lin C-M, Tseng Y-C, Hsu H-L, Chen C-J, Chen DY-T, Yan F-X, et al. Arterial spin labeling perfusion study in the patients with subacute mild traumatic brain injury. PLoS One. 2016;11(2):e0149109.
140. Werner C, Engelhard K. Pathophysiology of traumatic brain injury. Br J Anaesth. 2007;99(1):4–9.
141. Len TK, Neary JP, Asmundson GJ, Goodman DG, Bjornson B, Bhambhani YN. Cerebrovascular reactivity impairment after sport-induced concussion. Med Sci Sports Exerc. 2011;43(12):2241–8.
142. Kety SS, Schmidt CF. The effects of active and passive hyperventilation on cerebral blood flow, cerebral oxygen consumption, cardiac output, and blood pressure of normal young men. J Clin Invest. 1946;25(1):107–19.
143. Kassner A, Roberts TP. Beyond perfusion: cerebral vascular reactivity and assessment of microvascular permeability. Top Magn Reson Imaging. 2004;15(1):58–65.
144. Urback AL, MacIntosh BJ, Goldstein BI. Cerebrovascular reactivity measured by functional magnetic resonance imaging during breath-hold challenge: a systematic review. Neurosci Biobehav Rev. 2017;79:27–47.
145. Spano VR, Mandell DM, Poublanc J, Sam K, Battisti-Charbonney A, Pucci O, et al. CO2 blood oxygen level–dependent MR mapping of cerebrovascular reserve in a clinical population: safety, tolerability, and technical feasibility. Radiology. 2013;266(2):592–8.
146. Sobczyk O, Battisti-Charbonney A, Poublanc J, Crawley AP, Sam K, Fierstra J, et al. Assessing cerebrovascular reactivity abnormality by comparison to a reference atlas. J Cereb Blood Flow Metab. 2015;35(2):213–20.
147. Churchill NW, Hutchison MG, Graham SJ, Schweizer TA. Evaluating cerebrovascular reactivity during the early symptomatic phase of sport concussion. J Neurotrauma. 2019;36:1518–25.
148. Churchill N, Hutchison M, Graham S, Schweizer T. Cerebrovascular reactivity after sport concussion: from acute injury to one year after medical clearance. Front Neurol. 2020;11:558. https://doi.org/10.3389/fneur.2020.00558.
149. Moffett JR, Arun P, Ariyannur PS, Namboodiri AM. N-Acetylaspartate reductions in brain injury: impact on post-injury neuroenergetics, lipid synthesis, and protein acetylation. Front Neuroenerg. 2013;5:11.
150. Wyss M, Kaddurah-Daouk R. Creatine and creatinine metabolism. Physiol Rev. 2000;80(3):1107–213.
151. Miller BL, Changl L, Booth R, Ernst T, Cornford M, Nikas D, et al. In vivo 1H MRS choline: correlation with in vitro chemistry/histology. Life Sci. 1996;58(22):1929–35.
152. Haris M, Cai K, Singh A, Hariharan H, Reddy R. In vivo mapping of brain myo-inositol. NeuroImage. 2011;54(3):2079–85.
153. Garnett MR, Blamire AM, Corkill RG, Cadoux-Hudson TA, Rajagopalan B, Styles P. Early proton magnetic resonance spectroscopy in normal-appearing brain correlates with outcome in patients following traumatic brain injury. Brain. 2000;123(10):2046–54.
154. Vagnozzi R, Signoretti S, Tavazzi B, Floris R, Ludovici A, Marziali S, et al. Temporal window of metabolic brain vulnerability to concussion: a pilot 1H-magnetic resonance spectroscopic study in concussed athletes—Part III. Neurosurgery. 2008;62(6):1286–96.
155. Vagnozzi R, Tavazzi B, Signoretti S, Amorini AM, Belli A, Cimatti M, et al. Temporal window of metabolic brain vulnerability to concussions: mitochondrial-related impairment—Part I. Neurosurgery. 2007;61(2):379–89.

156. Henry LC, Tremblay S, Leclerc S, Khiat A, Boulanger Y, Ellemberg D, et al. Metabolic changes in concussed American football players during the acute and chronic post-injury phases. BMC Neurol. 2011;11(1):105.
157. Chamard E, Théoret H, Skopelja EN, Forwell LA, Johnson AM, Echlin PS. A prospective study of physician-observed concussion during a varsity university hockey season: metabolic changes in ice hockey players. Part 4 of 4. Neurosurg Focus. 2012;33(6):E4.
158. Churchill NW, Hutchison MG, Graham SJ, Schweizer TA. Neurometabolites and sport-related concussion: from acute injury to one year after medical clearance. Neuroimage Clin. 2020;27:102258.
159. Gasparovic C, Yeo R, Mannell M, Ling J, Elgie R, Phillips J, et al. Neurometabolite concentrations in gray and white matter in mild traumatic brain injury: an 1H–magnetic resonance spectroscopy study. J Neurotrauma. 2009;26(10):1635–43.
160. Lin A, Tran T, Bluml S, Merugumala S, Liao H-J, Ross BD. Guidelines for acquiring and reporting clinical neurospectroscopy. Semin Neurol. 2012;32(05):557–8.
161. Vink R, Golding EM, Williams JP, McIntosh TK. Blood glucose concentration does not affect outcome in brain trauma: a 31P MRS study. J Cereb Blood Flow Metab. 1997;17(1):50–3.
162. Vink R, McIntosh TK, Weiner MW, Faden AI. Effects of traumatic brain injury on cerebral high-energy phosphates and pH: a 31P magnetic resonance spectroscopy study. J Cereb Blood Flow Metab. 1987;7(5):563–71.
163. Kroshus E, Baugh CM, Stein CJ, Austin SB, Calzo JP. Concussion reporting, sex, and conformity to traditional gender norms in young adults. J Adolesc. 2017;54:110–9.
164. Foster CA, D'Lauro C, Johnson BR. Pilots and athletes: different concerns, similar concussion non-disclosure. PLoS One. 2019;14(5):e0215030.
165. Rawlins MLW, Johnson BR, Register-Mihalik JK, DeAngelis K, Schmidt JD, D'Lauro CJ. United States Air Force Academy cadets' perceived costs of concussion disclosure. Mil Med. 2020;185(1–2):e269–e75.
166. Leddy JJ, Sandhu H, Sodhi V, Baker JG, Willer B. Rehabilitation of concussion and post-concussion syndrome. Sports Health. 2012;4(2):147–54.
167. Kamins J, Bigler E, Covassin T, Henry L, Kemp S, Leddy JJ, et al. What is the physiological time to recovery after concussion? A systematic review. Br J Sports Med. 2017;51(12):935–40.
168. Button KS, Ioannidis JP, Mokrysz C, Nosek BA, Flint J, Robinson ES, et al. Power failure: why small sample size undermines the reliability of neuroscience. Nat Rev Neurosci. 2013;14(5):365–76.
169. Yue JK, Vassar MJ, Lingsma HF, Cooper SR, Okonkwo DO, Valadka AB, et al. Transforming research and clinical knowledge in traumatic brain injury pilot: multicenter implementation of the common data elements for traumatic brain injury. J Neurotrauma. 2013;30(22):1831–44.
170. Maas AI, Menon DK, Steyerberg EW, Citerio G, Lecky F, Manley GT, et al. Collaborative European NeuroTrauma effectiveness research in traumatic brain injury (CENTER-TBI) a prospective longitudinal observational study. Neurosurgery. 2015;76(1):67–80.
171. Broglio SP, McCrea M, McAllister T, Harezlak J, Katz B, Hack D, et al. A national study on the effects of concussion in collegiate athletes and US military service academy members: the NCAA–DoD concussion assessment, research and education (CARE) consortium structure and methods. Sports Med. 2017;47(7):1437–51.
172. Rosenbaum SB, Lipton ML. Embracing chaos: the scope and importance of clinical and pathological heterogeneity in mTBI. Brain Imaging Behav. 2012;6(2):255–82.
173. Hsu H-L, Chen DY-T, Tseng Y-C, Kuo Y-S, Huang Y-L, Chiu W-T, et al. Sex differences in working memory after mild traumatic brain injury: a functional MR imaging study. Radiology. 2015;276(3):828–35.
174. Wilke S, Prehn K, Taud B, List J, Flöel A. Multimodal assessment of recurrent mTBI across the lifespan. J Clin Med. 2018;7(5):95.
175. Jantzen KJ, Anderson B, Steinberg FL, Kelso JS. A prospective functional MR imaging study of mild traumatic brain injury in college football players. Am J Neuroradiol. 2004;25(5):738–45.

176. Pasternak O, Koerte IK, Bouix S, Fredman E, Sasaki T, Mayinger M, et al. Hockey Concussion Education Project, Part 2. Microstructural white matter alterations in acutely concussed ice hockey players: a longitudinal free-water MRI study. J Neurosurg. 2014;120(4):873–81.
177. Wright AD, Jarrett M, Vavasour I, Shahinfard E, Kolind S, van Donkelaar P, et al. Myelin water fraction is transiently reduced after a single mild traumatic brain injury–A prospective cohort study in collegiate hockey players. PLoS One. 2016;11(2):e0150215.
178. Churchill NW, Hutchison MG, Graham SJ, Schweizer TA. Baseline vs. cross-sectional MRI of concussion: distinct brain patterns in white matter and cerebral blood flow. Sci Rep. 2020;10(1):1–13.
179. Poldrack RA, Baker CI, Durnez J, Gorgolewski KJ, Matthews PM, Munafò MR, et al. Scanning the horizon: towards transparent and reproducible neuroimaging research. Nat Rev Neurosci. 2017;18(2):115.

Chapter 4
The Path Toward a Blood Test for Concussion: A Review of Biofluid Biomarkers for Concussive and Subconcussive Brain Trauma

Linda Papa

Introduction

Path Toward a Blood Test for Concussion

Currently, concussion (which is also known as mild traumatic brain injury) is largely a clinical diagnosis based on injury history, neurologic examination, neuropsychological testing, and, sometimes, neuroimaging. Brain-specific biomarkers measured through a simple blood test could complement the clinical evaluation of concussion and potentially guide management decisions [1–5]. The discovery of biomarkers that are specific for traumatic brain injury (TBI) is a quest that has been ongoing for several decades. The pursuit of these elusive markers has been most intense over the last 10–20 years [6–8]. Early human trials examined only moderate-to-severe TBI but are now expanding to include injuries on the milder end of the TBI spectrum, such as concussion and subconcussive injuries. Unlike other organ-based diseases, where rapid diagnostic tests of serum biomarkers are clinically essential to guide diagnosis and treatment, such as for myocardial ischemia or kidney dysfunction, TBI biomarker tests are just beginning to be validated and FDA-approved and are not currently part of clinical management.

Key features that would make a TBI biomarker clinically useful include (1) a high sensitivity (produced in the brain) and specificity (low or undetectable in blood in noninjury states) for brain injury; (2) the ability to stratify patients by severity of injury (concentration of the biomarker should increase with worsening injury); (3) the timely appearance in accessible biological fluid such as serum, saliva, or urine; (4) a well-defined time course; (5) the ability to monitor injury and response to

L. Papa (✉)
Department of Emergency Medicine, Orlando Health, Orlando Regional Medical Center, Orlando, FL, USA

T. A. Schweizer, A. J. Baker (eds.), *Tackling the Concussion Epidemic*,
https://doi.org/10.1007/978-3-030-93813-0_4

treatment; (6) the ability to predict functional outcome; and (7) be easily measured [8, 9].

The release of substances and potential biomarkers after an injury is not a static process. Understanding the biokinetic and temporal profile of a biomarker is critical to understanding the optimal time for measurement. The temporal profile may be affected by factors such as source of the sample (CSF, blood, urine, saliva), lesion type (mass lesion such as a hematoma versus more diffuse injury), location of injury, presence of concomitant intracranial and extracranial injuries (fractures, solid organ injuries), secondary insults (hypotension, hypoxia), and individual patient physiology. Ongoing secondary insults may contribute to secondary elevations in a marker, thereby altering its typical time course. For markers measured in blood, the level of a biomarker may also reflect the extent of blood–brain barrier disruption, rather than biomarker expression itself. It may also reflect release from extracranial sources that are traumatized as part of the injury event. In such a case, caution must be taken when interpreting levels that do not accurately reflect a brain injury, such as with S100β [10, 11]. Furthermore, individual physiology and pre-existing disease states may alter the metabolism or clearance of the biomarker, as with kidney or liver disease.

The following sections will review the most widely studied proteomic biomarkers for mild TBI and concussion in humans. Proteomic biomarkers are often represented by their neuroanatomic location in the central nervous system including astroglia (GFAP, S100β) and neuronal cells, with specific areas of the neuron such as the cell body (UCH-L1) and axon (tau, neurofilament) (Fig. 4.1). Furthermore, a novel group of promising transcriptomic biomarkers called microRNAs will also be discussed.

Proteomic Biomarkers

Glial Fibrillary Acidic Protein (GFAP) and Ubiquitin C-Terminal Hydrolase (UCH-L1)

GFAP and UCH-L1 for Mild-to-Moderate Traumatic Brain Injury

Glial fibrillary acidic protein (GFAP) is a protein found in the astroglial skeleton of both white and gray brain matter and has been used as a histological marker for glial cells. Ubiquitin C-terminal hydrolase-L1 (UCH-L1) is a protein in neurons that are involved in the addition and removal of ubiquitin from proteins that are destined to be metabolized and has been used as a histological marker for neurons [12, 13].

Several studies have looked at the specificity of GFAP and UCH-L1 to brain injury. GFAP and UCH-L1 have been shown to distinguish mild and moderate TBI patients from orthopedic controls and motor vehicle crash controls as well as from those TBI patients with negative CT scans [11, 14, 15]. In these studies, trauma control patients were exposed to significant trauma including the

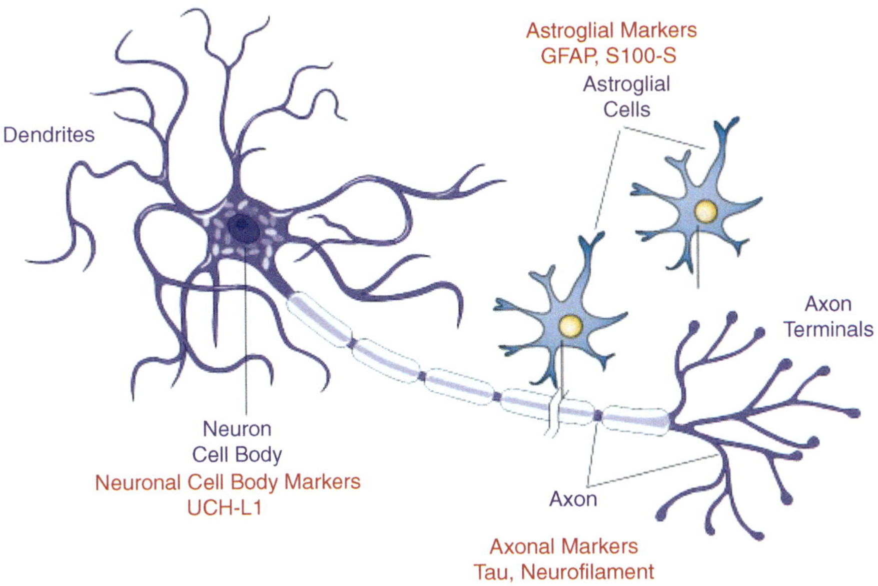

Fig. 4.1 This figure shows the TBI biomarkers reviewed in this chapter represented by their neuroanatomic location in the central nervous system including astroglial (GFAP, S100β) and neuronal cells, with specific areas of the neuron such as the cell body (UCH-L1) and axon (tau, neurofilament)

acceleration–deceleration vectors of motor vehicle collisions and substantial falls. Both GFAP and UCH-L1 showed a graded response to severity of injury from uninjured to orthopedic trauma, to mild and moderate TBI. However, GFAP appears to be most brain-specific in the setting of polytrauma with substantial extracranial injuries and fractures [2, 3, 10, 11, 14–16].

The temporal profiles of GFAP and UCH-L1 have also been evaluated in a large cohort of emergency department trauma patients. GFAP was performed consistently over 7 days in identifying concussion, detecting traumatic intracranial lesions on head computed tomography (CT), and predicting neurosurgical intervention [15]. GFAP was detectible in serum within an hour of concussion and remained elevated for several days after, rendering it a promising contender for clinical use for concussion diagnosis within a week of injury [15]. In contrast, UCH-L1 rose rapidly within 30 minutes of injury and peaked at 8 hours after injury and decreased steadily over 48 hours with small peaks and toughs over 7 days—making UCH-L1 a very early marker of concussion [15].

Both GFAP and UCH-L1 have now been evaluated in several studies to determine the need for CT scan and neurosurgical intervention in adults with mild-to-moderate TBI [11, 14, 15, 17–20] and more recently in children [2, 3, 21]. In early 2018, GFAP and UCH-L1 were FDA-approved for clinical use in adult patients with mild-to-moderate TBI to help determine the need for a CT scan within 12 hours of injury [22]. The approval was based on the ability to find lesions on CT scans but was not approved to diagnose a concussion. Moreover, it was not approved for use in children.

GFAP and UCH-L1 for Concussion and Subconcussive Brain Injury

CT is the standard imaging modality for assessing damage in TBI during the acute phase of injury. A CT scan can detect macroscopic traumatic lesions such as skull fractures, intracranial hematomas, contusions, subarachnoid hemorrhages, and swelling. However, more subtle injuries associated with mild TBI are often not detectable by this imaging modality. This discrepancy is evidenced by the lack of CT abnormalities in most patients with cognitive, physical, and behavioral dysfunction following a mild TBI. This group of TBI patients therefore represents the greatest challenge to accurate diagnosis and outcome prediction. Metting et al. found that mild TBI patients with axonal injury on anatomical MRI, but not CT, had elevated GFAP levels. Similarly, Yue et al. assessed GFAP in mild TBI patients with a negative CT scan and found GFAP was able to detect anatomical MRI lesions with an area under the curve of 0.78 despite the CT scan not showing any lesions [23]. UCH-L1 was not examined in either of these studies.

A significantly understudied group in whom biomarkers are rarely examined consists of individuals who experience head trauma without any symptoms of concussion. They are often classified as having "no injury" when, in fact, they may represent milder forms of concussion that do not elicit the typical signs or symptoms associated with concussion. Such injuries have been referred to as subconcussive injuries. Emerging data have demonstrated that significant alterations in brain function can occur in the absence of clinically obvious symptoms following even a single head trauma [24–26]. The issue of subconcussive trauma has been a particular concern in groups at high risk of exposure including military personnel [27] and athletes, as repetitive subconcussive impacts have the potential for long-term deleterious effects [26, 28, 29]. To address this deficiency, a recent study evaluated GFAP and UCH-L1 in subconcussive trauma in a large cohort of children and adult trauma patients presenting to three level I trauma centers with a Glasgow Coma Scale (GCS) score of 15 and a normal mental status. The biomarkers were measured at 20 distinct time points in patients with concussive, subconcussive, and nonconcussive trauma (Fig. 4.2). Although blood levels of both GFAP and UCH-L1 showed incremental increases from body trauma (lowest levels) to head trauma without concussion (higher levels than body trauma), to concussion (highest levels), GFAP was much better in distinguishing between the groups than UCH-L1. UCH-L1 was expressed at much higher levels than GFAP in those with nonconcussive trauma, particularly in children, suggesting that UCH-L1 is either not completely brain-specific or ultrasensitive to subtle impacts [30]. In athletes, UCH-L1 has shown elevations in both concussive [31] and subconcussive trauma [32]. However, these results are not significant in all studies [33, 34].

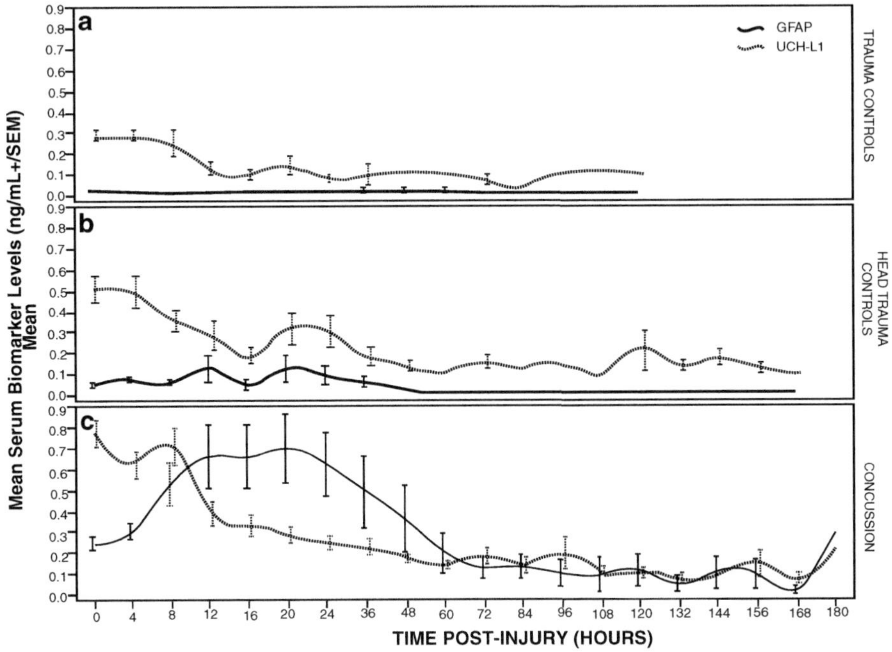

Fig. 4.2 Temporal profile of GFAP and UCH-L1 in three groups of trauma patients. (**a**) Temporal profile of GFAP and UCH-L1 in body trauma control patients. Means with error bars representing SEM. (**b**) Temporal profile of GFAP and UCH-L1 in head trauma control patients. Means with error bars representing SEM. (**c**) Temporal profile of GFAP and UCH-L1 in trauma patients with concussion. Means with error bars representing SEM. GFAP, glial fibrillary acidic protein; UCH-L1, ubiquitin C-terminal hydrolase. (Reprinted from Papa et al. [30]. With permission from BMJ Publishing Group Ltd.)

Other Potential Proteomic Biofluid Biomarkers of Concussion

A recent systematic review of biomarkers in sports-related concussion showed that there have been at least 11 different biomarkers assessed in athletes [5]. Besides GFAP and UCH-L1, other potential biomarkers include S100β, neuron-specific enolase, tau, neurofilament, amyloid beta, and brain-derived neurotrophic factor. Some correlate with number of hits to the head (soccer), acceleration–deceleration forces (jumps, collisions, and falls), post-concussive symptoms, trauma to the body versus the head, and dynamics of injury [5]. Some of these and other novel markers are discussed below.

S100β

S100β is expressed in astrocytes and helps to regulate intracellular levels of calcium. It is considered a marker of astrocyte injury or death. It can also be found in cells that are not neuronal such as adipocytes, chondrocytes, and melanoma cells, and, therefore, it is not brain-specific [35, 36]. Despite this, S100β is one of the most extensively studied biomarkers for TBI [4, 5, 37].

Several studies have found correlations between elevated serum levels of S100β and CT abnormalities in adults and children with TBI [4, 38]. Elevated concentrations of S100β in serum have been associated with an increased incidence of post-concussive symptoms, problems with cognition, and traumatic abnormalities on MRI [39–43]. However, there are also a number of studies contradicting these findings [16, 44–46]. Similarly, several studies have shown that serum S100β increases after concussive [31, 47, 48] and subconcussive brain injury [34, 49, 50]. However, a few studies have shown poor association with other prognostic parameters [33, 46, 51]. Peripheral sources of S100β complicate its use as a brain-specific marker, particularly in the setting of polytrauma. S100β has been shown to be elevated in injured patients with peripheral trauma who have had no direct head trauma [2, 11, 52]. Since many of these results have been inconsistently reproduced, the clinical value of S100β in TBI, particularly mild TBI and concussion, is still controversial.

Tau Protein

Tau is an intracellular, microtubule-associated protein that is amplified in axons and is involved with assembling axonal microtubule bundles and participating in anterograde axoplasmic transport [53]. Tau lesions are apparently related to axonal disruption such as in trauma or hypoxia [54, 55]. After release, it is proteolytically cleaved at the N and C terminals. In a study by Shaw et al., an elevated level of C-tau was associated with a poor outcome at hospital discharge in patients with different severities of closed head injury and with an increased likelihood of an intracranial injury, as identified on head CT [56]. However, these findings were not reproducible when C-tau was measured in peripheral blood in those strictly with mild TBI [57]. Two additional studies showed that C-tau was a poor predictor of the presence of any CT lesions and a poor predictor of post-concussive syndrome [44, 58]. Similarly, Bulut et al. found total tau (T-tau) differentiated patients with intracranial injury from those without intracranial injury [59]. However, they were unable to distinguish between healthy controls and mild TBI patients without intracranial lesions.

In contrast with the hospital setting, T-tau studies using newer assays in the sport setting have shown greater promise for detecting mild TBI. In 2014, a study of professional hockey players showed that serum T-tau out-performed S-100B and NSE in detecting concussion at 1-hour after injury and that levels were significantly higher in post-concussion samples at all times compared with preseason levels [48]. T-tau at 1 hour after concussion also correlated with the number of days it took for concussion symptoms to resolve. Accordingly, T-tau remained significantly elevated at 144 hours in players with post-concussive symptoms (PCS) lasting more than 6 days versus players with PCS for less than 6 days [48].

Phosphorylated tau (P-tau) has also been examined as a potential biomarker for brain trauma. Following TBI, axonal injury is coupled to tau hyperphosphorylation, leading to microtubule instability and tau-mediated neurodegeneration [60]. P-tau has been shown to outperform T-tau in distinguishing CT-positive from CT-negative cases and identifying patients with poor outcome [61]. Moreover, several months after TBI, P-tau has been shown to be elevated in TBI patients compared to healthy controls. The ratio between P-tau and T-tau has shown similar results [61]. High levels of total and phosphorylated tau have been found in postmortem samples of TBI patients and athletes [62, 63]. Further study is needed to elucidate the role of T-tau and P-tau in detecting chronic encephalopathy.

Neurofilaments

Neurofilaments are heteropolymeric components of the neuron cytoskeleton that consist of a 68 kDa light neurofilament subunit (NF-L) backbone with either 160 kDa medium (NF-M) or 200 kDa heavy subunit (NF-H) sidearms [64]. Following TBI, calcium influx into the cell contributes to a cascade of events that activates calcineurin, a calcium-dependent phosphatase that dephosphorylates neurofilament sidearms, presumably contributing to axonal injury [65]. Phosphorylated NF-H has been found to be elevated in the cerebrospinal fluid (CSF) of adults and children with severe TBI [66, 67]. It remains significantly elevated after a few days in children with poor outcome and diffuse axonal injury (DAI) on initial CT scan [67]. Similarly, in a study by Vajtr et al., serum NF-H was much higher in patients with DAI over 10 days after admission, with highest levels from day 4 to day 10 [68].

In a cohort of professional hockey players who underwent blood biomarker assessment at 1, 12, 36, and 144 hours after concussion and at return to play, serum NF-L increased over time and returned to normal at return to play. Also, serum NF-L levels were higher in players with prolonged post-concussive symptoms [69]. In a group of amateur boxers, serum NF-L concentrations showed elevations 7–10 days after a bout and subsequently decreased following 3 months of rest from boxing. Levels were also significantly correlated with the number of hits to the head [69]. Moreover, NF-L has been shown to increase in adult soccer players following repetitive subconcussive head impacts compared to baseline levels, however only after 24 hours post-impacts [70]. In contrast, NF-L levels in blood taken at baseline and at 6 and 14 days post-concussion in contact sport athletes showed no differences between any of the pre–post-time points [71].

Transcriptomic Biomarkers

MicroRNAs as the Next Generation of Biomarkers for Concussion

The earliest studies of TBI biomarkers were conducted using animal models. These models have been very helpful in providing histologic and pathophysiologic information on potential biomarkers. As a result, the selection of TBI biomarkers has been based on neuroanatomic location and on mechanisms of injury induced by trauma, such as neuroinflammation and ischemia.

A novel set of biomarkers, called microRNAs (miRNA), are now being studied as the next generation of biomarkers for many diseases and disorders such as cancer, and cardiovascular and neurodegenerative diseases [72]. miRNAs are small (19–28 nucleotides) endogenous RNA molecules that regulate protein synthesis at the post-transcriptional level. MiRNAs can be detected in serum and can be an indicator of disease pathology in neuronal cells. MiRNAs are relatively abundant in biofluids such as CSF, serum, and urine, are relatively stable at variable pH conditions, and are resistant to repeated freeze/thaw cycles and enzymatic degradation. Due to these properties, miRNA has advantages over protein-based markers. The utility of miRNAs as diagnostic markers of mild TBI or concussion has recently been explored [73–77]. In 2016, Bhomia et al. identified specific and sensitive miRNA-based biomarkers for mild and moderate TBI using real-time PCR methodology [74]. Samples from human subjects with mild-to-severe TBI were compared to trauma and normal controls and identified 10 miRNA signatures miR-151-5p, miR-328, miR-362-3p, miR-486, miR-505*, miR-451, miR-30d, miR-20a, miR-195, and miR-92a. Moreover, Johnson et al. identified 6 salivary miRNAs with overlapping CSF alterations (miR-182-5p, miR-221-3p, miR-26b-5p, miR-320c, miR-29c-3p, and miR-30e-5p) that distinguished children with TBI from healthy controls [75]. In a study by the same group, 52 children with concussion had 5 salivary miRNAs (miR-320c, miR-133a-5p, miR-769-5p, let-7a-3c, and miR-1307-3p) that were associated with prolonged post-concussive symptoms [76].

More recently, studies have been evaluating the role of microRNA in sports-related concussion. In one recent study, microRNA biomarkers measured pre- and post-season in collegiate football players were associated with worsening neurocognitive functioning over the course of a season in those with no concussions [78]. The study found significant elevations in circulating miRNA measured before the athletic season began and prior to any contact practices. All the players had significantly elevated levels compared to non-athlete controls ($p < 0.001$) indicating the presence of residual circulating miRNA biomarkers from prior concussive and sub-concussive impacts [78, 79]. Preseason miRNA levels predicted baseline SAC scores with very good areas under the curve, the highest being miR-195 (0.90), miR-20a (0.89), miR-151-5p (0.86) miR-505* (0.85), and miR-9-3p (0.77). Athletes who demonstrated worsening neurocognitive function from pre to post-season showed elevations in concentrations of miRNAs over the same period. The miRNAs with the most significant increases over the course of the season were miR-505*, miR-362-3p, miR-30d, miR-92a, and miR-486. Similarly, a study of saliva miRNA levels from 32 rugby players detected 5 miRNAs (miR-27b-3p, miR-142-3p, let-7i, miR-107, and miR-135b-5p) at 48–72 hours after sports-related concussion that correlated with reaction time on ImPACT testing and predicted concussion better than other protein biomarkers [80].

Based on the important function of miRNA in neurons in the central nervous system and their association with measures of brain injury in adults and children, specifically in sports-related injury, this class of biomarkers should be examined with more vigor in larger cohorts of patients.

Conclusion

TBI biomarkers measured through a simple blood test have the potential to provide invaluable information for the management of concussion by facilitating diagnosis and risk stratification, offering timely information about the pathophysiology of injury; monitoring recovery; and furnishing opportunities for drug target identification and surrogate measures for future clinical trials. In light of their timeliness, accuracy, and risk stratification potential, biofluid biomarkers with reliable sensitivity and specificity would be welcomed tools in treating concussion. This is especially so in settings limited by acute care resources such as in rural settings and non-hospital environments such as the playing field, battlefield, and primary care practices. With the first set of TBI biomarkers now FDA-approved to detect CT lesions in adult patients with mild-to-moderate the potential impact on clinical care will soon be seen. Additionally, more TBI biomarkers and devices are in queue for approval. Will these discoveries be incorporated into clinical algorithms and revolutionize clinical care as troponin did for cardiac disease? The final choice will eventually rest on the level of evidence, accessibility to testing, ease of use, the clinician's comfort with using and interpreting the results, and shared decision-making with patients and their families.

References

1. Papa L. Potential blood-based biomarkers for concussion. Sports Med Arthrosc Rev. 2016;24:108–15.
2. Papa L, Mittal MK, Ramirez J, Ramia M, Kirby S, Silvestri S, Giordano P, Weber K, Braga CF, Tan CN, Ameli NJ, Lopez M, Zonfrillo M. In children and youth with mild and moderate traumatic brain injury, glial fibrillary acidic protein out-performs S100beta in detecting traumatic intracranial lesions on computed tomography. J Neurotrauma. 2016;33:58–64.
3. Papa L, Zonfrillo MR, Ramirez J, Silvestri S, Giordano P, Braga CF, Tan CN, Ameli NJ, Lopez M, Mittal MK. Performance of glial fibrillary acidic protein in detecting traumatic intracranial lesions on computed tomography in children and youth with mild head trauma. Acad Emerg Med. 2015;22:1274–82.
4. Papa L, Ramia MM, Kelly JM, Burks SS, Pawlowicz A, Berger RP. Systematic review of clinical research on biomarkers for pediatric traumatic brain injury. J Neurotrauma. 2013;30:324–38.
5. Papa L, Ramia MM, Edwards D, Johnson BD, Slobounov SM. Systematic review of clinical studies examining biomarkers of brain injury in athletes after sports-related concussion. J Neurotrauma. 2015;32:661–73.
6. Papa L. Exploring the role of biomarkers for the diagnosis and management of traumatic brain injury patients. In: Man TK, Flores RJ, editors. Poteomics – human diseases and protein functions. 1st ed. London, UK: In Tech Open Access Publisher; 2012.
7. Papa L, Edwards D, Ramia M. Exploring serum biomarkers for mild traumatic brain injury. In: Kobeissy FH, editor. Brain neurotrauma: molecular, neuropsychological, and rehabilitation aspects. Boca Raton: CRC Press/Taylor & Francis; 2015. p. 301–8.
8. Papa L, Wang KKW. Raising the bar for traumatic brain injury biomarker research: methods make a difference. J Neurotrauma. 2017;34:2187–9.

9. Papa L, Robinson G, Oli M, Pineda J, Demery J, Brophy G, Robicsek SA, Gabrielli A, Robertson CS, Wang KW, Hayes RL. Use of biomarkers for diagnosis and management of traumatic brain injury patients. Expert Opin Med Diagn. 2008;2:937–45.
10. Pelinka LE, Kroepfl A, Leixnering M, Buchinger W, Raabe A, Redl H. GFAP versus S100B in serum after traumatic brain injury: relationship to brain damage and outcome. J Neurotrauma. 2004;21:1553–61.
11. Papa L, Silvestri S, Brophy GM, Giordano P, Falk JL, Braga CF, Tan CN, Ameli NJ, Demery JA, Dixit NK, Mendes ME, Hayes RL, Wang KK, Robertson CS. GFAP out-performs S100beta in detecting traumatic intracranial lesions on computed tomography in trauma patients with mild traumatic brain injury and those with extracranial lesions. J Neurotrauma. 2014;31:1815–22.
12. Tongaonkar P, Chen L, Lambertson D, Ko B, Madura K. Evidence for an interaction between ubiquitin-conjugating enzymes and the 26S proteasome. Mol Cell Biol. 2000;20:4691–8.
13. Gong B, Leznik E. The role of ubiquitin C-terminal hydrolase L1 in neurodegenerative disorders. Drug News Perspect. 2007;20:365–70.
14. Papa L, Lewis LM, Falk JL, Zhang Z, Silvestri S, Giordano P, Brophy GM, Demery JA, Dixit NK, Ferguson I, Liu MC, Mo J, Akinyi L, Schmid K, Mondello S, Robertson CS, Tortella FC, Hayes RL, Wang KK. Elevated levels of serum glial fibrillary acidic protein breakdown products in mild and moderate traumatic brain injury are associated with intracranial lesions and neurosurgical intervention. Ann Emerg Med. 2012;59:471–83.
15. Papa L, Brophy GM, Welch RD, Lewis LM, Braga CF, Tan CN, Ameli NJ, Lopez MA, Haeussler CA, Mendez Giordano DI, Silvestri S, Giordano P, Weber KD, Hill-Pryor C, Hack DC. Time course and diagnostic accuracy of glial and neuronal blood biomarkers GFAP and UCH-L1 in a large cohort of trauma patients with and without mild traumatic brain injury. JAMA Neurol. 2016;73:551–60.
16. Metting Z, Wilczak N, Rodiger LA, Schaaf JM, Van Der Naalt J. GFAP and S100B in the acute phase of mild traumatic brain injury. Neurology. 2012;78:1428–33.
17. Papa L, Lewis LM, Silvestri S, Falk JL, Giordano P, Brophy GM, Demery JA, Liu MC, Mo J, Akinyi L, Mondello S, Schmid K, Robertson CS, Tortella FC, Hayes RL, Wang KK. Serum levels of ubiquitin C-terminal hydrolase distinguish mild traumatic brain injury from trauma controls and are elevated in mild and moderate traumatic brain injury patients with intracranial lesions and neurosurgical intervention. J Trauma Acute Care Surg. 2012;72:1335–44.
18. Welch RD, Ellis M, Lewis LM, Ayaz SI, Mika VH, Millis S, Papa L. Modeling the kinetics of serum glial fibrillary acidic protein, ubiquitin carboxyl-terminal hydrolase-L1, and S100B concentrations in patients with traumatic brain injury. J Neurotrauma. 2017;34:1957–71.
19. Lewis LM, Schloemann DT, Papa L, Fucetola RP, Bazarian J, Lindburg M, Welch RD. Utility of serum biomarkers in the diagnosis and stratification of mild traumatic brain injury. Acad Emerg Med. 2017;24:710–20.
20. Bazarian JJ, Biberthaler P, Welch RD, Lewis LM, Barzo P, Bogner-Flatz V, Gunnar Brolinson P, Buki A, Chen JY, Christenson RH, Hack D, Huff JS, Johar S, Jordan JD, Leidel BA, Lindner T, Ludington E, Okonkwo DO, Ornato J, Peacock WF, Schmidt K, Tyndall JA, Vossough A, Jagoda AS. Serum GFAP and UCH-L1 for prediction of absence of intracranial injuries on head CT (ALERT-TBI): a multicentre observational study. Lancet Neurol. 2018;17:782–9.
21. Papa L, Mittal MK, Ramirez J, Silvestri S, Giordano P, Braga CF, Tan CN, Ameli NJ, Lopez MA, Haeussler CA, Mendez Giordano D, Zonfrillo MR. Neuronal biomarker ubiquitin C-terminal hydrolase detects traumatic intracranial lesions on computed tomography in children and youth with mild traumatic brain injury. J Neurotrauma. 2017;34:2132–40.
22. FDA authorizes marketing of first blood test to aid in the evaluation of concussion in adults. [Online]. Silver Springs: US Food & Drug Administration; 2018. Accessed 2 July 2018.
23. Yue JK, Yuh EL, Korley FK, Winkler EA, Sun X, Puffer RC, Deng H, Choy W, Chandra A, Taylor SR, Ferguson AR, Huie JR, Rabinowitz M, Puccio AM, Mukherjee P, Vassar MJ, Wang KKW, Diaz-Arrastia R, Okonkwo DO, Jain S, Manley GT, Investigators, T.-T. Association between plasma GFAP concentrations and MRI abnormalities in patients with CT-negative

traumatic brain injury in the TRACK-TBI cohort: a prospective multicentre study. Lancet Neurol. 2019;18:953–61.
24. Zhou Y, Kierans A, Kenul D, Ge Y, Rath J, Reaume J, Grossman RI, Lui YW. Mild traumatic brain injury: longitudinal regional brain volume changes. Radiology. 2013;267:880–90.
25. Bailes JE, Petraglia AL, Omalu BI, Nauman E, Talavage T. Role of subconcussion in repetitive mild traumatic brain injury. J Neurosurg. 2013;119:1235–45.
26. Bailes JE, Dashnaw ML, Petraglia AL, Turner RC. Cumulative effects of repetitive mild traumatic brain injury. Prog Neurol Surg. 2014;28:50–62.
27. Tate CM, Wang KK, Eonta S, Zhang Y, Carr W, Tortella FC, Hayes RL, Kamimori GH. Serum brain biomarker level, neurocognitive performance, and self-reported symptom changes in soldiers repeatedly exposed to low-level blast: a breacher pilot study. J Neurotrauma. 2013;30:1620–30.
28. Gavett BE, Stern RA, Mckee AC. Chronic traumatic encephalopathy: a potential late effect of sport-related concussive and subconcussive head trauma. Clin Sports Med. 2011;30:179–88.
29. Huber BR, Alosco ML, Stein TD, Mckee AC. Potential long-term consequences of concussive and subconcussive injury. Phys Med Rehabil Clin N Am. 2016;27:503–11.
30. Papa L, Zonfrillo MR, Welch RD, Lewis LM, Braga CF, Tan CN, Ameli NJ, Lopez MA, Haeussler CA, Mendez Giordano D, Giordano PA, Ramirez J, Mittal MK. Evaluating glial and neuronal blood biomarkers GFAP and UCH-L1 as gradients of brain injury in concussive, subconcussive and non-concussive trauma: a prospective cohort study. BMJ Paediatr Open. 2019;3:e000473.
31. Meier TB, Nelson LD, Huber DL, Bazarian JJ, Hayes RL, Mccrea MA. Prospective assessment of acute blood markers of brain injury in sport-related concussion. J Neurotrauma. 2017;34:3134–42.
32. Joseph JR, Swallow JS, Willsey K, Lapointe AP, Khalatbari S, Korley FK, Oppenlander ME, Park P, Szerlip NJ, Broglio SP. Elevated markers of brain injury as a result of clinically asymptomatic high-acceleration head impacts in high-school football athletes. J Neurosurg. 2018;130(5):1–7.
33. Asken BM, Bauer RM, Dekosky ST, Svingos AM, Hromas G, Boone JK, Dubose DN, Hayes RL, Clugston JR. Concussion BASICS III: serum biomarker changes following sport-related concussion. Neurology. 2018;91:e2133–43.
34. Puvenna V, Brennan C, Shaw G, Yang C, Marchi N, Bazarian JJ, Merchant-Borna K, Janigro D. Significance of ubiquitin carboxy-terminal hydrolase L1 elevations in athletes after subconcussive head hits. PLoS One. 2014;9:e96296.
35. Zimmer DB, Cornwall EH, Landar A, Song W. The S100 protein family: history, function, and expression. Brain Res Bull. 1995;37:417–29.
36. Olsson B, Zetterberg H, Hampel H, Blennow K. Biomarker-based dissection of neurodegenerative diseases. Prog Neurobiol. 2011;95:520–34.
37. Schulte S, Podlog LW, Hamson-Utley JJ, Strathmann FG, Struder HK. A systematic review of the biomarker S100B: implications for sport-related concussion management. J Athl Train. 2014;49:830–50.
38. Heidari K, Vafaee A, Rastekenari AM, Taghizadeh M, Shad EG, Eley R, Sinnott M, Asadollahi S. S100B protein as a screening tool for computed tomography findings after mild traumatic brain injury: systematic review and meta-analysis. Brain Inj. 2015;29:1146–57.
39. Ingebrigtsen T, Romner B. Management of minor head injuries in hospitals in Norway. Acta Neurol Scand. 1997;95:51–5.
40. Waterloo K, Ingebrigtsen T, Romner B. Neuropsychological function in patients with increased serum levels of protein S-100 after minor head injury. Acta Neurochir. 1997;139:26–31; discussion 31–2.
41. Ingebrigtsen T, Romner B. Serial S-100 protein serum measurements related to early magnetic resonance imaging after minor head injury. Case report. J Neurosurg. 1996;85:945–8.

42. Ingebrigtsen T, Waterloo K, Jacobsen EA, Langbakk B, Romner B. Traumatic brain damage in minor head injury: relation of serum S-100 protein measurements to magnetic resonance imaging and neurobehavioral outcome. Neurosurgery. 1999;45:468–75; discussion 475–6.
43. Heidari K, Asadollahi S, Jamshidian M, Abrishamchi SN, Nouroozi M. Prediction of neuropsychological outcome after mild traumatic brain injury using clinical parameters, serum S100B protein and findings on computed tomography. Brain Inj. 2015;29:33–40.
44. Bazarian JJ, Zemlan FP, Mookerjee S, Stigbrand T. Serum S-100B and cleaved-tau are poor predictors of long-term outcome after mild traumatic brain injury. Brain Inj. 2006;20:759–65.
45. Lima DP, Simao Filho C, Abib Sde C, De Figueiredo LF. Quality of life and neuropsychological changes in mild head trauma. Late analysis and correlation with S100B protein and cranial CT scan performed at hospital admission. Injury. 2008;39:604–11.
46. Dorminy M, Hoogeveen A, Tierney RT, Higgins M, Mcdevitt JK, Kretzschmar J. Effect of soccer heading ball speed on S100B, sideline concussion assessments and head impact kinematics. Brain Inj. 2015;29:1158–64.
47. Kiechle K, Bazarian JJ, Merchant-Borna K, Stoecklein V, Rozen E, Blyth B, Huang JH, Dayawansa S, Kanz K, Biberthaler P. Subject-specific increases in serum S-100B distinguish sports-related concussion from sports-related exertion. PLoS One. 2014;9:e84977.
48. Shahim P, Tegner Y, Wilson DH, Randall J, Skillback T, Pazooki D, Kallberg B, Blennow K, Zetterberg H. Blood biomarkers for brain injury in concussed professional ice hockey players. JAMA Neurol. 2014;71:684–92.
49. Kawata K, Rubin LH, Takahagi M, Lee JH, Sim T, Szwanki V, Bellamy A, Tierney R, Langford D. Subconcussive impact-dependent increase in plasma s100beta levels in collegiate football players. J Neurotrauma. 2017;34:2254–60.
50. Zonner SW, Ejima K, Bevilacqua ZW, Huibregtse ME, Charleston C, Fulgar C, Kawata K. Association of increased serum s100b levels with high school football subconcussive head impacts. Front Neurol. 2019;10:327.
51. Babcock L, Byczkowski T, Wade SL, Ho M, Bazarian JJ. Inability of S100B to predict post-concussion syndrome in children who present to the emergency department with mild traumatic brain injury: a brief report. Pediatr Emerg Care. 2013;29:458–61.
52. Pelinka LE, Kroepfl A, Schmidhammer R, Krenn M, Buchinger W, Redl H, Raabe A. Glial fibrillary acidic protein in serum after traumatic brain injury and multiple trauma. J Trauma. 2004;57:1006–12.
53. Teunissen CE, Dijkstra C, Polman C. Biological markers in CSF and blood for axonal degeneration in multiple sclerosis. Lancet Neurol. 2005;4:32–41.
54. Kosik KS, Finch EA. MAP2 and tau segregate into dendritic and axonal domains after the elaboration of morphologically distinct neurites: an immunocytochemical study of cultured rat cerebrum. J Neurosci. 1987;7:3142–53.
55. Higuchi M, Lee VM, Trojanowski JQ. Tau and axonopathy in neurodegenerative disorders. NeuroMolecular Med. 2002;2:131–50.
56. Shaw GJ, Jauch EC, Zemlan FP. Serum cleaved tau protein levels and clinical outcome in adult patients with closed head injury. Ann Emerg Med. 2002;39:254–7.
57. Chatfield DA, Zemlan FP, Day DJ, Menon DK. Discordant temporal patterns of S100beta and cleaved tau protein elevation after head injury: a pilot study. Br J Neurosurg. 2002;16:471–6.
58. Ma M, Lindsell CJ, Rosenberry CM, Shaw GJ, Zemlan FP. Serum cleaved tau does not predict postconcussion syndrome after mild traumatic brain injury. Am J Emerg Med. 2008;26:763–8.
59. Bulut M, Koksal O, Dogan S, Bolca N, Ozguc H, Korfali E, Ilcol YO, Parklak M. Tau protein as a serum marker of brain damage in mild traumatic brain injury: preliminary results. Adv Ther. 2006;23:12–22.
60. Rubenstein R, Chang B, Davies P, Wagner AK, Robertson CS, Wang KK. A novel, ultrasensitive assay for tau: potential for assessing traumatic brain injury in tissues and biofluids. J Neurotrauma. 2015;32:342–52.
61. Rubenstein R, Chang B, Yue JK, Chiu A, Winkler EA, Puccio AM, Diaz-Arrastia R, Yuh EL, Mukherjee P, Valadka AB, Gordon WA, Okonkwo DO, Davies P, Agarwal S, Lin F, Sarkis G,

Yadikar H, Yang Z, Manley GT, Wang KKW, The, T.-T. B. I. I, Cooper SR, Dams-O'connor K, Borrasso AJ, Inoue T, Maas AIR, Menon DK, Schnyer DM, Vassar MJ. Comparing Plasma Phospho Tau, Total Tau, and Phospho Tau-Total Tau ratio as acute and chronic traumatic brain injury biomarkers. JAMA Neurol. 2017;74:1063–72.
62. Puvenna V, Engeler M, Banjara M, Brennan C, Schreiber P, Dadas A, Bahrami A, Solanki J, Bandyopadhyay A, Morris JK, Bernick C, Ghosh C, Rapp E, Bazarian JJ, Janigro D. Is phosphorylated tau unique to chronic traumatic encephalopathy? Phosphorylated tau in epileptic brain and chronic traumatic encephalopathy. Brain Res. 2016;1630:225–40.
63. Alosco ML, Tripodis Y, Fritts NG, Heslegrave A, Baugh CM, Conneely S, Mariani M, Martin BM, Frank S, Mez J, Stein TD, Cantu RC, Mckee AC, Shaw LM, Trojanowski JQ, Blennow K, Zetterberg H, Stern RA. Cerebrospinal fluid tau, Abeta, and sTREM2 in Former National Football League Players: modeling the relationship between repetitive head impacts, microglial activation, and neurodegeneration. Alzheimers Dement. 2018;14:1159–70.
64. Julien JP, Mushynski WE. Neurofilaments in health and disease. Prog Nucleic Acid Res Mol Biol. 1998;61:1–23.
65. Buki A, Povlishock JT. All roads lead to disconnection?--Traumatic axonal injury revisited. Acta Neurochir. 2006;148:181–93; discussion 193–4.
66. Siman R, Toraskar N, Dang A, Mcneil E, Mcgarvey M, Plaum J, Maloney E, Grady MS. A panel of neuron-enriched proteins as markers for traumatic brain injury in humans. J Neurotrauma. 2009;26:1867–77.
67. Zurek J, Bartlova L, Fedora M. Hyperphosphorylated neurofilament NF-H as a predictor of mortality after brain injury in children. Brain Inj. 2012;25:221–6.
68. Vajtr D, Benada O, Linzer P, Samal F, Springer D, Strejc P, Beran M, Prusa R, Zima T. Immunohistochemistry and serum values of S-100B, glial fibrillary acidic protein, and hyperphosphorylated neurofilaments in brain injuries. Soud Lek. 2013;57:7–12.
69. Shahim P, Zetterberg H, Tegner Y, Blennow K. Serum neurofilament light as a biomarker for mild traumatic brain injury in contact sports. Neurology. 2017;88:1788–94.
70. Wirsching A, Chen Z, Bevilacqua ZW, Huibregtse ME, Kawata K. Association of acute increase in plasma neurofilament light with repetitive subconcussive head impacts: a pilot randomized control trial. J Neurotrauma. 2019;36:548–53.
71. Wallace C, Zetterberg H, Blennow K, Van Donkelaar P. No change in plasma tau and serum neurofilament light concentrations in adolescent athletes following sport-related concussion. PLoS One. 2018;13:e0206466.
72. Jin XF, Wu N, Wang L, Li J. Circulating microRNAs: a novel class of potential biomarkers for diagnosing and prognosing central nervous system diseases. Cell Mol Neurobiol. 2013;33:601–13.
73. Balakathiresan N, Bhomia M, Chandran R, Chavko M, Mccarron RM, Maheshwari RK. MicroRNA let-7i is a promising serum biomarker for blast-induced traumatic brain injury. J Neurotrauma. 2012;29:1379–87.
74. Bhomia M, Balakathiresan NS, Wang KK, Papa L, Maheshwari RK. A panel of serum MiRNA biomarkers for the diagnosis of severe to mild traumatic brain injury in humans. Sci Rep. 2016;6:28148.
75. Hicks SD, Johnson J, Carney MC, Bramley H, Olympia RP, Loeffert AC, Thomas NJ. Overlapping MicroRNA expression in saliva and cerebrospinal fluid accurately identifies pediatric traumatic brain injury. J Neurotrauma. 2018;35:64–72.
76. Johnson JJ, Loeffert AC, Stokes J, Olympia RP, Bramley H, Hicks SD. Association of salivary MicroRNA changes with prolonged concussion symptoms. JAMA Pediatr. 2018;172:65–73.
77. Mitra B, Rau TF, Surendran N, Brennan JH, Thaveenthiran P, Sorich E, Fitzgerald MC, Rosenfeld JV, Patel SA. Plasma micro-RNA biomarkers for diagnosis and prognosis after traumatic brain injury: a pilot study. J Clin Neurosci. 2017;38:37–42.
78. Papa L, Slobounov SM, Breiter HC, Walter A, Bream T, Seidenberg P, Bailes JE, Bravo S, Johnson B, Kaufman D, Molfese DL, Talavage TM, Zhu DC, Knollmann-Ritschel B, Bhomia M. Elevations in MicroRNA biomarkers in serum are associated with measures of concus-

sion, neurocognitive function, and subconcussive trauma over a Single National Collegiate Athletic Association Division I Season in Collegiate Football Players. J Neurotrauma. 2019;36:1343–51.
79. Abbas K, Shenk TE, Poole VN, Breedlove EL, Leverenz LJ, Nauman EA, Talavage TM, Robinson ME. Alteration of default mode network in high school football athletes due to repetitive subconcussive mild traumatic brain injury: a resting-state functional magnetic resonance imaging study. Brain Connect. 2015;5:91–101.
80. Di Pietro V, Porto E, Ragusa M, Barbagallo C, Davies D, Forcione M, Logan A, Di Pietro C, Purrello M, Grey M, Hammond D, Sawlani V, Barbey AK, Belli A. Salivary MicroRNAs: diagnostic markers of mild traumatic brain injury in contact-sport. Front Mol Neurosci. 2018;11:290.

Chapter 5
Sport-Related Concussion Guideline Development: Acute Management to Return to Activity

Michael G. Hutchison

Introduction

Concussion is recognized as a difficult injury to diagnose and manage, despite focused research efforts to understand its pathophysiology. Fundamentally, concussion is a clinical syndrome represented by the onset of signs and/or symptoms following direct or indirect head contact, which are caused by the significant rotational and translational forces exerted on brain tissue during the collision event. One of the early hallmarks of concussion is that the neurological signs and symptoms from the trauma's biochemical forces do not result in macroscopic neural damage. However, the effects of concussion are not negligible, as they perturb many facets of cellular and/or physiological function including, but not limited to ionic shifts, metabolic changes, and impaired neurotransmission [1]. While the basic neurobiology of concussion/mild traumatic brain injury (mTBI) was initially elucidated in animal models, we have observed an exponential growth of human studies over the last 20 years corroborating these early findings. Most recently, concussion has been described as a neurometabolic cascade of events involving bioenergetic challenges, cytoskeletal and axonal alterations, impairments in neurotransmission, and vulnerability to delayed cell death and chronic dysfunction [1]. The contributions by researchers and corresponding knowledge translation activities have motivated healthcare professionals to adopt a more conservative approach to the diagnosis and management of concussion (e.g., "when in doubt, sit them out"), as well as more refined strategies to return an individual recovering from a concussion back to full activities.

The purpose of this chapter is to review the advancements in the acute management and return-to-activity principles for an individual sustaining a sport-related concussion (SRC). I start the chapter with the historical context of concussion, followed by describing the development of return-to-activity protocols and key issues,

M. G. Hutchison (✉)
Faculty of Kinesiology and Physical Education, University of Toronto, Toronto, ON, Canada
e-mail: michael.hutchison@utoronto.ca

T. A. Schweizer, A. J. Baker (eds.), *Tackling the Concussion Epidemic*,
https://doi.org/10.1007/978-3-030-93813-0_5

along with specific considerations when applying in the sport context. I also acknowledge that return to activity following SRC often includes school and work considerations and discusses the development of guidelines specific to school and work. I conclude the chapter by highlighting that an interdisciplinary team with concussion expertise is warranted in certain situations irrespective of return-to-activity goals.

Historical Evolution of Guidelines

The clinical manifestations of concussion in the absence of gross bleeding or focal parenchymal damage have been known since the tenth century AD when the Arabian physician Rhazes first defined the condition [2]. Although concussion was first documented centuries ago, the formulation and guidance of concussion in the sport setting are in its relative infancy, having been first articulated approximately 50 years ago.

Concussion is a subset of TBI, and the operational definition for diagnosis includes (i) induced by direct or indirect trauma, (ii) physiological disruption of brain function, and (iii) behavioral changes (i.e., signs), and/or symptoms. Also, concussion may include (i) a loss of consciousness (LOC) or (ii) period of post-traumatic amnesia (PTA). Numerous authors and organizations have proposed injury severity scales and return-to-sport recommendations for the management of concussions that occurred during sport participation. Table 5.1 provides examples of the more

Table 5.1 Summary of frequently referenced sport-related concussion severity scales

	Cantu [6]	Colorado Medical Society Guidelines [7]	American Academy of Neurology [9]	Kelly and Rosenburg [10]
Severe Grade 3	LOC lasting at least 5 minutes	LOC	Any LOC either brief (seconds) or prolonged (minutes)	Any LOC either brief (seconds) or prolonged (minutes)
Moderate Grade 2	LOC lasting less than 5 minutes	Confusion with amnesia No LOC (Amnesia could be delayed)	Transient confusion No LOC Concussion symptoms or mental status abnormalities on examination last more than 15 minutes	Transient confusion No LOC Concussion symptoms or mental status abnormalities on examination last more than 15 minutes
Mild Grade 1	No LOC but suffers from impaired intellectual function especially in short-term memory and interpreting new information	Confusion without amnesia (for at least 20 minutes); no LOC	Transient confusion No LOC Concussion symptoms or mental status abnormalities on examination resolve in less than 15 minutes	Transient confusion No LOC Concussion symptoms or mental status abnormalities on examination resolve in less than 15 minutes

frequently cited SRC severity scales; detailed reviews of scales have been published elsewhere (see Johnston and colleagues [3] or McCrory [4]). Early severity scales had neurosurgical underpinnings, as evidenced by their composition being primarily driven by the duration of LOC and post-traumatic amnesia (PTA) (e.g., Maroon, Steele, and Berlin [5]; Cantu [6]; Kelly and colleagues [7]). While LOC and PTA duration are significant outcome predictors in severe brain injury, clinical evidence has indicated these are not typical features of concussion, and their prognostic value is inconclusive [8]. Attempts were later made to combine neurosurgical concepts with symptoms (e.g., American Academy of Neurology [9]; Kelly and Rosenberg [10]). Still, the subcategories of severity were loosely defined and arbitrary, which made interpretation and application difficult in practice and with little prognostic value.

In 2001, significant progress was made with the establishment of the Concussion In Sport Group (CISG) during the initial consensus meeting in Vienna. One of the key outcomes from the meeting was the first consensus statement, where an attempt was made to unify the field by providing a consensus definition of concussion with specified clinical features [11]. An additional key outcome from the initial consensus meeting was the acknowledgment of the limitations and lack of scientific rigor associated with published scales to date; thus, the CISGs recommended that a multimodality assessment of an individual determines safe return to play rather than reliance on predetermined scales. The CISG has held an additional four meetings with corresponding consensus documents [12–15]. Concurrently, various other institutional [16], sport-specific [17, 18], and professional organization meetings [19–24] have taken place that have informed clinical practice and prioritized future SRC research initiatives.

Acute Evaluation

The first attempt to standardize a concussion evaluation within the sporting context was seen with the Sport Concussion Assessment Tool (SCAT). The first iteration of the SCAT was developed by a consensus panel using and modifying existing concussion assessment tools [i.e., Standardized Assessment of Concussion [25] and Balance Error Scoring System (BESS) [26, 27]] designed to assess an individual's immediate awareness and orientation, cognition, balance, and symptoms. More recent versions of the SCAT have been created and disseminated, which have been globally adopted. Also, there is a version for the acute phase of injury in children 5–12 years old who are suspected of having an SRC [28], as well as the Concussion Recognition Tool 5 [29], developed for use by non-medically trained individuals to assist in identifying and providing basic management of information for suspected SRC. The SCAT global adoption is likely due to the international composition of the CISG, diversity of sports and medical professions represented, and support by many sports organizations such as Hockey Federation (IIHF), Fédération Internationale de Football Association (FIFA), and the International Olympic Committee (IOC) Medical Commission.

Although tools exist to provide health professionals a framework for concussion evaluation, in an acute sport setting, the priority of a first responder is to confirm whether the individual is medically stable (e.g., airway, breathing, and cardiac functioning) and to rule out more serious injuries (e.g., spinal injury or serious form of TBI). Once this is established, the health professional needs to determine whether the individual meets the diagnostic criteria of concussion: (1) sufficient mechanism (i.e., acceleration/deceleration forces to produce brain trauma) and (2) symptoms, signs (e.g., loss of consciousness, vacant look, motor incoordination, or balance problems), or alteration in mental status. This apparent simple task is arguably one of the most complicated decisions made by the sports medicine team. Certain sports have a limited number of substitutions per game, and thus, evaluations are time-limited. Also, empirical investigations of the multimodal SCAT in the acute setting have highlighted that symptoms are the best discriminatory element for concussion [30–32], however, a motivated athlete may minimize symptoms in hopes of returning to play, or the clinician may have difficulty differentiating exertional-related symptoms from suspected concussion. Furthermore, a well-documented phenomenon is that symptoms are not always immediately apparent and take minutes to hours to appear [33]. However, the evaluation of a suspected concussion in the days or even weeks following the event has its own set of challenges for clinicians. Confirming a sufficient mechanism without video evidence of the event and time course of symptoms and behavioral changes suffers from recall bias and incomplete information at times. Additionally, it has been recognized that the diagnostic utility of the SCAT declines significantly at 3–5 days following injury [34].

There are unique complexities and issues to the clinical setting for a suspected concussion; however, the general guidance for modern concussion management is the oft-quoted maxim: "*when in doubt, sit them out.*" There are sound, evidence-based reasons to support a conservative approach for a suspected concussion. The first line of evidence is that continuing to be active ("play through," "work through symptoms") in the early stages of the neurometabolic cascade (minutes to hours after trauma) can prolong recovery time [35]. Additionally, there is a growing body of literature indicating that even after symptoms have resolved, athletes may experience subtle cognitive deficits [36–39] or physiological perturbations [40]. It is also recognized that an athlete who sustains a concussion is at increased risk of suffering future concussions [41] with potentially more severe physiological consequences [42, 43]. Finally, there is literature suggesting that the effects of concussion in some individuals may be deleterious to cognition and mood dysregulation later in life [44–46], a process that might be related to the frequency of concussive injuries over time. Collectively, conservative concussion management is well-informed and prudent, and in practical terms, this means that once an individual is diagnosed with a concussion, they are not medically "cleared" or "recovered from concussion" (i.e., able to return to high-risk practice or gameplay) until several clinical thresholds are met. Present-day conservative concussion management is a significant departure from some earlier guidelines, as athletes diagnosed with a concussion were previously permitted to return to the same game or practice under certain conditions.

Development of Return-to-Activity Protocols

It is now recognized that determination of "recovery" is *process*-driven, where an individual is advised to move through a gradual progression of functional capacity, concluding with a final medical evaluation confirming the successful completion of all stages. Currently, there is no validated stepwise exercise protocol, but the accepted essential components of return-to-sport protocols (RTSp) include progressively increasing physical demands with added components of head acceleration, sensory stimulation, and cognitive burden simulating sport-specific game participation (and controlled body contact if required). In the first CISG consensus statement [11], we find the first iteration of a structured and supervised concussion rehabilitation protocol conducive to optimal injury recovery and safe and successful RTSp. The following stages were identified the following:

1. No activity, complete rest. Once asymptomatic, proceed to next level.
2. Light aerobic exercise such as walking or stationary cycling.
3. Sport-specific training—for example, skating in hockey, running in soccer.
4. Non-contact training drills.
5. Full contact training after medical clearance.
6. Gameplay.

With this stepwise progression, the athlete should proceed to the next level if asymptomatic at their current level. It was recommended that if any symptoms occur following an SRC, however, the patient should drop back to the previous asymptomatic level and try to progress again after 24 h. This means that the most basic process of concussion recovery can take a minimum of several days. Since the original graded exertional protocol was proposed almost 20 years ago, empirical evidence and changes in consensus views have led to the evolution of RTSp strategies. Several organizations have published consensus statements recommending return-to-activity protocols: American Academy of Pediatrics [22, 47], American Medical Society for Sports Medicine [21, 22], American Academic of Neurology [23], and the National Athletic Trainers' Association [24]. The most cited RTSp strategies are associated with the CISG consensus documents [11–13, 15, 48]. There have been three key aspects of RTSp guidelines that have evolved substantially over time: defining and implementing *"rest,"* the importance of being *"asymptomatic,"* and the recommended ways to re-engage and progress in physical activity.

Rest

Historically, the clinical management of concussion has focused on complete *rest* following concussion. The objective was physical and cognitive rest until the athlete was symptom-free, in order to ease discomfort from post-concussive symptoms and possibly promote recovery by decreasing energy demands on the brain. Although well-intentioned, this language was interpreted by many as the recommendation of a deleterious management approach by advocating isolation or "bedroom jail." Observational studies have consistently documented the negative consequences of inactivity in various pathological conditions [49]. Furthermore, sedentary behavior following injury/illness is the most consistent risk factor for chronic disability [50], and prescribed rest may begin to adversely affect the cardiopulmonary and musculoskeletal systems in healthy people within just 3 days [51]. In the larger body of mTBI literature, it has been noted that inactivity may also exacerbate and/or prolong recovery in commonly identified comorbidities such as vestibular disorders, mental health and pain disorders, and chronic fatigue syndrome [49]. Athletes in particular are vulnerable to the deleterious effects of prolonged rest with respect to physical deconditioning, as well as the secondary symptoms arising from fatigue and reactive depression [52]. The ambiguity of rest was addressed in the most recent consensus statement, which explicitly noted that there is currently insufficient evidence to prescribe complete rest; instead, the advice is to encourage patients to become gradually and progressively more active "after a brief period of rest during the acute phase (24–48 h) after injury" [13].

Asymptomatic

Similar to the concept of *rest,* the term "asymptomatic" has created significant confusion and variability in interpretation among clinicians, patients, and researchers. The term "asymptomatic" was a key component of many earlier RTSp guidelines and a required milestone prior to moving to the next level of recovery. The term *asymptomatic* is clearly noted in the frequently cited Cantu [6] and American Academy of Neurology [9] guidelines; furthermore, many of the earlier iterations of the CISG RTSp process incorporated the term asymptomatic in their protocol. Determination of *asymptomatic* status is generally based on self-reported symptoms of the individual acquired using a standardized "checklist." A systematic review of SRC scales/checklists identified several tools to capture and describe symptoms [53]. Scales have differed in their terminology used to describe symptoms, as well as their approach to measurement of specific symptoms (e.g., dichotomously as "Yes" or "No," Likert scale) [53]. The description of symptoms following SRC has been primarily influenced by prior research on the non-sporting head injury population [53]. However, *asymptomatic* status was never operationally defined, and many health professionals and patients interpreted the term in a literal sense, whereby there was a requirement for individuals to be "symptom-free" or report an "absence of symptoms." Unfortunately, this resulted in many individuals remaining in a "complete rest" stage for extended periods. Some individuals were also advised not to progress due to symptoms following exercise, not because of exacerbation of concussion symptoms but due to exertional-related health status changes. This became apparent in the literature when researchers reported exertional-related symptoms among healthy athletic samples following sport participation [53, 54] tend to vary with the time and day of measurement, sleep status, emotional status/depressive symptoms, attitude, motivation, honesty, and the willingness of the individual [55–64]. Presently, the measurement of symptoms by standardized checklists remains a common practice. There remains no gold standard in measuring symptoms following concussion; however, the Post-concussion Symptom Scale (PCSS) included in the SCAT-5 is used globally and frequently cited in the SRC literature. In an attempt to determine an acceptable level of symptoms to be considered asymptomatic, normative scores have been established to assist management decisions following SRC [65]. Furthermore, pre-injury or baseline symptom status may be acquired during preseason medicals or before competitive play when possible in some high-performance settings. In these baseline assessments, the most recent SCAT-5 has moved away from "how do you feel right now" to asking athletes to rate their symptoms based on "how he/she typically feels" [34].

Presently, following an initial period of rest (24–48 h), patients can be encouraged to become gradually and progressively more active while staying below their cognitive and physical symptom exacerbation thresholds (i.e., activity level should not bring on or worsen their symptoms) [34]. Similar to determining asymptomatic status, clinical judgment is required to determine the threshold for exacerbation and whether the symptoms are specific to concussive trauma or comorbid injuries (e.g., neck) or the individual's general health status at the time of assessment. Thus, there has been a movement away from employing a single metric of symptom severity, with consideration given to additional approaches that have a patient quantify their current state as a percentage out of 100 (i.e., normal, healthy state), and then tracking their main clinical complaints (e.g., top three symptoms), or even broad symptom categories/subtypes (e.g., somatic, cognitive, affective, oculomotor).

Defining the Do's and Don'ts of Reengaging in Activity

Early versions of RTSp protocols were vague when referring to the re-introduction of exercise following the initial period of rest. In the initial Vienna statement [11], it was suggested that re-introduction of exercises (i.e., Stage 2) be "light aerobic exercise such as walking or stationary cycling"; however, such guidance lacked specifics to common exercise intervention principles such as frequency, intensity, type, and duration of activity. In 2016, the re-introduction of exercises was further refined to "walking, swimming, or stationary cycling, keeping intensity less than 70% maximum permitted heart rate" [13]. At the University of Toronto, we recommend patients initially engage in aerobic exercise using a stationary cycle because it ensures minimal head movement with little burden to the cognitive, oculomotor, and vestibular systems. Although a pragmatic approach, future studies should identify whether specific modalities of exercise may be better than others, and ultimately, these findings should be used to inform healthcare professionals.

The optimal period of rest post-SRC remains unknown, but it is now well-recognized that prolonged rest can be associated with amplified symptoms and delayed recovery [50, 66]. In recent years, we have seen the emergence of research highlighting the beneficial effects of an early introduction of aerobic exercise. Initially, the utility of aerobic exercise was examined in those with persistent systems, whereby researchers found a reduction in symptom burden and recovery time following the use of a standardized sub-symptom exacerbation aerobic exercise intervention [52]. The biological basis for aerobic exercise in the treatment of concussion includes its ability to improve cortical connectivity and activation [67], facilitate neuroplasticity [68, 69], and enhance cerebral blood flow and cerebrovascular reactivity [52]. Since the initial work in those with persistent symptoms, we have seen a consistent increase in studies investigating the use of aerobic exercise early following injury. In addition, a retrospective covariate balancing propensity study of acute SRC found early initiation aerobic exercise was associated with faster full return to sport and school or work [70]. More recently, findings from randomized controlled trials have emerged [71–73]. Leddy and colleagues conducted the first RCT with a large sample size and showed that individualized sub-symptom threshold aerobic exercise treatment prescribed to adolescents with concussion symptoms during the first week following SRC increases the speed of recovery and may reduce the incidence of delayed recovery [74]. This is a promising finding in the treatment of concussion and could have widespread impact because aerobic exercise is easily implementable and does not require specialized equipment. Notwithstanding, most of the evidence has been limited to adolescents with SRC; therefore, continued research is required to ensure the beneficial effects across different age groups and populations.

Evolution of Return to Activity: Sport

The decision to return an athlete to sport is significant due to the potential risk of sustaining another concussion or even more serious injury. As noted, the consensus is that RTSp should occur gradually, initiating with aerobic activity and concluding with participation in a restricted practice setting (i.e., minimizing risk of head trauma) [13]; this process should not occur on the same day of injury.

The initial Vienna document was especially notable for incorporating a stepwise RTSp approach [11]. This framework was consistent in the subsequent consensus statement (i.e., Prague 2004); however, it was noted "progressive addition of resistance training at steps 3 or 4" [75]. In 2008, the Zurich statement [48] revised Step 2 with the additional guidance of intensity, suggesting aerobic exercise should be <70% max HR. Also, greater details were provided around functional exercises at each stage of the RTS strategy and their objectives (e.g., increase heart rate, head movement, and coordination) [48]. The 2008 version also specified that progressive resistance training begins in Step 4 [48]. In 2012, Step 1 was changed from "no activity/physical and cognitive rest" to "symptom-limited activity/daily activities that do not provoke symptoms" [15]. There was also added emphasis for the recommendation that there should be at least 24 h for each step. Furthermore, in concussion cases where clinical recovery extended beyond the expected window (i.e., 10 days), concussions should be managed in a multidisciplinary manner by healthcare providers with SRC experience. In the most recent 2016 consensus document [13], there were no changes to the steps of the progression; however, the time window of concussion cases warranting an evaluation by an expert in concussion was stratified by age group (i.e., >10–14 days in adults and >4 weeks in children). Finally, language was added that is specific to resistance training, specifying that it should only be added in the later stages (stage 3 or 4 at the earliest).

Since the implementation of the Vienna consensus RTSp protocol, several organizations have since modified or developed additional RTSp protocols. Although most medical professionals have accepted the graded return-to-activity progression as the standard of care for returning athletes to sport participation, the evidence justifying the number and components of various levels and steps is lacking. Therefore, validation of the existing RTSp progressions is still needed. We have observed a transition from waiting until athletes are asymptomatic, to encouraging an earlier introduction to aerobic activity, and there is evidence that the latter approach facilitates recovery [70, 71, 74]. Future research is encouraged to provide greater details on the type and intensity of exercises throughout the RTSp progression. For example, it is generally accepted that various steps of the progression include adding head acceleration, intensity, and cognitive load elements. However, the specific tasks and exercises that address these various components are not delineated. There is emerging evidence that the introduction of a dual task—the simultaneous measurement of motor and cognitive performance—can be altered following concussion [76–78]. Future studies should explore how this can be incorporated into an RTSp strategy.

Return to Activity: School

Importantly, RTSp is only one component of the concussion recovery process. Being able to concentrate and take notes in class, take tests and term examinations, complete assignments, and remain social are also crucial elements of a student's daily life. Although this is an accepted consideration in evaluating many individuals recovering from an SRC, educational needs (or modifications) have only recently been recognized. In 2010, the CDC published its initial *Heads Up* toolkit for schools (https://www.cdc.gov/headsup/schools), providing an overview of the issues that schools might face and the types of problems and associated supports that they might need to provide the returning student [79]. Since that time, several clinicians and researchers have highlighted the importance of academic support as an essential component of concussion recovery [80–82]. In 2013 and 2014, we saw medical organizations such as the American Medical Society for Sports Medicine and the Canadian Pediatric Society note the importance of a formalized return to school (RTSc) process [22, 83]. At the 2016 Berlin International Concussion in Sport Group meeting, a pediatric subgroup was tasked with "What factors must be considered in 'return to school' following concussion and what strategy or accommodations should be followed?" [84]. The CISG's first return to school strategy was recommended in the corresponding consensus statement [13].

Similar to RTSp protocols, despite the lack of evidence-informed guidelines, there exist several documented protocols that advise a gradual return to school [81, 82, 85–87]. Numerous outstanding questions regarding return to school exist, including when, at what capacity, and how many levels of progression are advised in a RTSc protocol. At the University of Toronto, we have adopted a similar approach to sport for school re-integration. We can describe an individualized return to school strategy within the following domains of stress: physical, cognitive, and sensory. The *physical* domain refers to attendance and duration (e.g., one-class versus half-day versus full-day) and considerations for participating in activity-based classes (e.g., physical education/gym classes, field trips, or laboratories). The *cognitive* domain considers the individual's level of cognitive participation or workload, such as note-taking, assignments, homework, and test-taking. Finally, the *sensory* domain considers a student's ability to tolerate visual and auditory stimuli; therefore, modifications to screen time, group work, music classes, etc., may be necessary and gradually re-introduced. Further research is needed to validate appropriate recommendations for academic support; however, key components of all return to school strategies include the following:

- *School support personnel*: Depending on the school's resources and size, one or many individuals should be assigned to efficiently and effectively manage the return and support processes. This is the initial touch point for the student upon reporting the concussion, and they allow for initiating and managing an individualized return to school plan.
- *Communication*: A key facet in a successful return to school plan is clear communication among the healthcare provider, family, and school. This often takes the form of a written note or the completion of a predefined form by the healthcare provider to parents and the school. These recommendations should be considered by the school, although they are not obligatory until the educational expert(s) examine their relevance and capacity for implementation [79]. It is also necessary to educate parents and patients about symptoms and management plans and make all teachers interacting with the student aware of the individualized return to school plan.
- *Individualized Re-integration Plan*: The return to school plan should be individually adapted and guided by the injured student's medical direction and needs. The individualized plan is a practical educational roadmap to be implemented by a coordinated school personnel team.
- *Monitoring*: Recovery from concussion may take weeks; therefore, periodic evaluations and updates by a healthcare provider will be required, and often, it is the responsibility of parents to communicate or provide documentation to school personnel for consideration and to appropriately adjust the RTSc plan.

The RTSc is a central task in caring for an individual recovering from a concussion. Active communication among the healthcare, family, school, and student often creates an environment wherein the vast majority of cases will demonstrate a gradual symptom resolution and recovery with positive movement toward full academic participation. In situations where the recovery process is longer than projected, the personnel and resources should be in place to make targeted adjustments and accommodate the student. Although RTSc guidelines are structured and logical, there is still need for further research to provide more precise guidance than currently is recommended in the recent Berlin pediatric concussion statement [84] and other protocols [85, 86, 88–90].

Return to Activity: Work

The high frequency of concussions in the sport context warrants the development of RTSp and RTSs protocols. Accordingly, the literature and evidence of best practices reviewed and summarized in this chapter have focused on youth through university athlete populations and professional adults, where school and/or sport would be considered their main activities. However, for many adults who sustain a sport-related concussion during recreational sport or physical activity, return-to-work (RTW) is their primary concern.

Many researchers and clinicians view the acute management principles for concussion to be the same, regardless of their level of sport participation [15, 91]. However, there is very little information in the sport medicine literature specific to RTW protocols; thus, one must look to the larger mTBI literature for recommendations. General recommendations for progressive return to activities within the workplace environment include returning to work for half days, then part time, and finally full time, with progressive adjustment in work tasks according to the cognitive and sensory demands required to accomplish them. A review of the literature suggests more than half of patients with mTBI return to work by 1 month following injury, and more than 80% by 6 months [92]. Thus, in addition to general accommodation principles, there exist published guidelines for health professionals in applying the best practices with those showing persistent post-concussion symptoms. The Ontario Neurotrauma Foundation provides an exemplary example of such guidelines [93].

Interdisciplinary Team

The heterogeneity of recovery trajectories following concussion is well-documented and consistent in both the SRC and general mTBI literature and is evident in both youth and older adult cohorts. We have seen the CISG consensus documents evolve, with the most recent version highlighting that the management of individuals considered to have persistent symptoms (i.e., >10–14 days in adults and >4 weeks in children) should involve a multidisciplinary team of healthcare providers with experience in treating SRC. We have also observed the development of dedicated guidelines (e.g., Guideline for Concussion/Mild Traumatic Brain Injury and Prolonged Symptoms: 3rd Edition for Adults 18+ years of age [93]) for dealing with individuals with persistent symptoms. What is emphasized in those with persistent symptoms is that these individuals do not necessarily reflect a single etiology but rather a constellation of nonspecific post-traumatic symptoms that may be linked to coexisting and/or confounding factors, or ongoing physiological injury to the brain.

Given that concussion represents an all-encompassing term for a spectrum of conditions that have a common primary injury mechanism, we have observed the emergence of concussion subtypes, such as cognitive, ocular motor, headache/migraine, vestibular, and anxiety/mood, as well as concussion-associated conditions such as sleep disturbance and cervical strain [94, 95]. The subtyping approach has advantages in monitoring and determining symptom exacerbation and provides directed intervention opportunities. For example, closely monitored active rehabilitation programs involving controlled sub-symptom threshold using submaximal exercise are safe and are beneficial to facilitate recovery, particularly for those with autonomic instability or physical deconditioning [52]. Also, targeted physical therapy may be recommended for patients with cervical spine or vestibular dysfunction, while cognitive behavioral therapy and/or pharmacological approaches can be employed to deal with individuals dealing with persistent mood or behavioral issues. To date, subtypes have been determined by expert panel/opinion or derived by existing checklists that are potentially skewed toward certain symptom types due to their prevalence (i.e., cognitive and somatic symptoms); therefore, it has been recognized that expanding empirical investigations to include other potential subtypes (e.g., autonomic, endocrine, and auditory dysfunction) and applying well designed methodological approaches such as cluster analysis would be advantageous [96] and may inform future rehabilitation strategies.

Summary

The diagnosis, management, and ultimate decision for medical clearance following SRC are complex and multifactorial. We have seen the necessity for creating a clinical framework for evaluation, and empirical evidence suggests some of the components of the SCAT offer clinical utility in the acute period of concussion (<24 h). However, given the heterogeneity of clinical presentations and recovery trajectories, identifying a unified clinical tool that provides all relevant information remains elusive. We have observed progressions in operationalizing and standardizing the process for the return-to-activity protocol for sport. More recently, individualized pathways have been created from adjustments and accommodations for both school and work. Regardless of the activity (sport, school, or work), the literature is consistent in that prolonged concussion recoveries would benefit from an interdisciplinary team with experience dealing with SRC to help inform the management process and decision-making. Although concussion management has progressed logically and systematically to best assist the patient, school, and family, the field is still in need of further evidence to provide more precise, evidence-based guidance and recommendations.

References

1. Giza CC, Hovda DA. The new neurometabolic cascade of concussion. Neurosurgery. 2014;75 Suppl 4:S24–33. https://doi.org/10.1227/NEU.0000000000000505.
2. McCrory PR, Berkovic SF. Concussion: the history of clinical and pathophysiological concepts and misconceptions. Neurology. 2001;57:2283–9. https://doi.org/10.1212/wnl.57.12.2283.
3. Johnston KM, McCrory P, Mohtadi NG, Meeuwisse W. Evidence-based review of sport-related concussion: clinical science. Clin J Sport Med. 2001;11:150–9. https://doi.org/10.1097/00042752-200107000-00005.
4. McCrory P. Concussion revisited: a historical perspective. How has the focus on concussions evolved over the years? In: Gagnon I, Ptito A, editors. Sports concussions. Boca Raton: CRC; 2018.
5. Maroon JC, Steele PB, Berlin R. Football head and neck injuries--an update. Clin Neurosurg. 1980;27:414–29.
6. Cantu RC. Head injuries in sport. Br J Sports Med. 1996;30:289–96. https://doi.org/10.1136/bjsm.30.4.289.
7. Kelly JP, et al. Concussion in sports. Guidelines for the prevention of catastrophic outcome. JAMA. 1991;266:2867–9. https://doi.org/10.1001/jama.266.20.2867.
8. Iverson GL, et al. Predictors of clinical recovery from concussion: a systematic review. Br J Sports Med. 2017;51:941–8. https://doi.org/10.1136/bjsports-2017-097729.
9. Practice parameter: the management of concussion in sports (summary statement). Report of the Quality Standards Subcommittee. Neurology. 1997;48:581–5. https://doi.org/10.1212/wnl.48.3.581.
10. Kelly JP, Rosenberg JH. The development of guidelines for the management of concussion in sports. J Head Trauma Rehabil. 1998;13:53–65. https://doi.org/10.1097/00001199-199804000-00008.
11. Aubry M, et al. Summary and agreement statement of the First International Conference on Concussion in Sport, Vienna 2001. Recommendations for the improvement of safety and health of athletes who may suffer concussive injuries. Br J Sports Med. 2002;36:6–10. https://doi.org/10.1136/bjsm.36.1.6.
12. McCrory P, et al. Summary and agreement statement of the 2nd International Conference on Concussion in Sport, Prague 2004. Br J Sports Med. 2005;39:196–204. https://doi.org/10.1136/bjsm.2005.018614.
13. McCrory P, et al. Consensus statement on concussion in sport-the 5th international conference on concussion in sport held in Berlin, October 2016. Br J Sport Med. 2017;51:838–47. https://doi.org/10.1136/bjsports-2017-097699.
14. McCrory P, et al. Consensus statement on Concussion in Sport 3rd International Conference on Concussion in Sport held in Zurich, November 2008. Clin J Sport Med. 2009;19:185–200. https://doi.org/10.1097/JSM.0b013e3181a501db.
15. McCrory P, et al. Consensus statement on concussion in sport: the 4th International Conference on Concussion in Sport held in Zurich, November 2012. Br J Sports Med. 2013;47:250–8. https://doi.org/10.1136/bjsports-2013-092313.
16. Collins MW, et al. Statements of Agreement From the Targeted Evaluation and Active Management (TEAM) approaches to treating concussion meeting held in Pittsburgh, October 15–16, 2015. Neurosurgery. 2016;79:912–29. https://doi.org/10.1227/NEU.0000000000001447.

17. Smith AM, et al. Ice Hockey Summit II: zero tolerance for head hits and fighting. Clin J Sport Med. 2015;25:78–87. https://doi.org/10.1097/JSM.0000000000000195.
18. Smith AM, et al. Proceedings from the ice hockey summit on concussion: a call to action. Clin J Sport Med. 2011;21:281–7. https://doi.org/10.1097/jsm.0b013e318225bc15.
19. Guskiewicz KM, et al. National Athletic Trainers' Association position statement: management of sport-related concussion. J Athl Train. 2004;39:280–97.
20. Herring SA, et al. Concussion (mild traumatic brain injury) and the team physician: a consensus statement--2011 update. Med Sci Sports Exerc. 2011;43:2412–22. https://doi.org/10.1249/MSS.0b013e3182342e64.
21. Harmon KG, et al. American Medical Society for Sports Medicine position statement on concussion in sport. Br J Sports Med. 2019;53:213–25. https://doi.org/10.1136/bjsports-2018-100338.
22. Harmon KG, et al. American Medical Society for Sports Medicine position statement: concussion in sport. Br J Sports Med. 2013;47:15–26. https://doi.org/10.1136/bjsports-2012-091941.
23. Giza CC, et al. Summary of evidence-based guideline update: evaluation and management of concussion in sports: report of the Guideline Development Subcommittee of the American Academy of Neurology. Neurology. 2013;80:2250–7. https://doi.org/10.1212/WNL.0b013e31828d57dd.
24. Broglio SP, et al. National Athletic Trainers' Association position statement: management of sport concussion. J Athl Train. 2014;49:245–65. https://doi.org/10.4085/1062-6050-49.1.07.
25. McCrea M, et al. Standardized assessment of concussion (SAC): on-site mental status evaluation of the athlete. J Head Trauma Rehabil. 1998;13:27–35. https://doi.org/10.1097/00001199-199804000-00005.
26. Guskiewicz KM, Ross SE, Marshall SW. Postural stability and neuropsychological deficits after concussion in collegiate athletes. J Athl Train. 2001;36:263–73.
27. Riemann BL, Guskiewicz KM. Effects of mild head injury on postural stability as measured through clinical balance testing. J Athl Train. 2000;35:19–25.
28. Davis GA, et al. The Child Sport Concussion Assessment Tool 5th Edition (Child SCAT5): background and rationale. Br J Sports Med. 2017;51:859–61. https://doi.org/10.1136/bjsports-2017-097492.
29. Echemendia RJ, et al. The Concussion Recognition Tool 5th Edition (CRT5): background and rationale. Br J Sports Med. 2017;51:870–1. https://doi.org/10.1136/bjsports-2017-097508.
30. Carson JD, et al. Premature return to play and return to learn after a sport-related concussion: physician's chart review. Can Fam Physician. 2014;60(e310):e312–5.
31. Downey RI, Hutchison MG, Comper P. Determining sensitivity and specificity of the Sport Concussion Assessment Tool 3 (SCAT3) components in university athletes. Brain Inj. 2018;32:1345–52. https://doi.org/10.1080/02699052.2018.1484166.
32. Putukian M, et al. Prospective clinical assessment using Sideline Concussion Assessment Tool-2 testing in the evaluation of sport-related concussion in college athletes. Clin J Sport Med. 2015;25:36–42. https://doi.org/10.1097/JSM.0000000000000102.
33. Olson A, Ellis MJ, Selci E, Russell K. Delayed symptom onset following pediatric sport-related concussion. Front Neurol. 2020;11:220. https://doi.org/10.3389/fneur.2020.00220.
34. Echemendia RJ, et al. The Sport Concussion Assessment Tool 5th Edition (SCAT5): background and rationale. Br J Sports Med. 2017;51:848–50. https://doi.org/10.1136/bjsports-2017-097506.
35. Asken BM, et al. "Playing through it": delayed reporting and removal from athletic activity after concussion predicts prolonged recovery. J Athl Train. 2016;51:329–35. https://doi.org/10.4085/1062-6050-51.5.02.
36. Munia TT, et al. Preliminary results of residual deficits observed in athletes with concussion history: combined EEG and cognitive study. Annu Int Conf IEEE Eng Med Biol Soc. 2016;2016:41–4. https://doi.org/10.1109/EMBC.2016.7590635.
37. Brown JA, Dalecki M, Hughes C, Macpherson AK, Sergio LE. Cognitive-motor integration deficits in young adult athletes following concussion. BMC Sports Sci Med Rehabil. 2015;7:25. https://doi.org/10.1186/s13102-015-0019-4.

38. Comper P, Hutchison M, Magrys S, Mainwaring L, Richards D. Evaluating the methodological quality of sports neuropsychology concussion research: a systematic review. Brain Inj. 2010;24:1257–71. https://doi.org/10.3109/02699052.2010.506854.
39. Echemendia RJ, et al. Testing the hybrid battery approach to evaluating sports-related concussion in the National Hockey League: a factor analytic study. Clin Neuropsychol. 2019:1–20. https://doi.org/10.1080/13854046.2019.1690051.
40. Kamins J, et al. What is the physiological time to recovery after concussion? A systematic review. Br J Sports Med. 2017;51:935–40. https://doi.org/10.1136/bjsports-2016-097464.
41. Abrahams S, Fie SM, Patricios J, Posthumus M, September AV. Risk factors for sports concussion: an evidence-based systematic review. Br J Sports Med. 2014;48:91–7. https://doi.org/10.1136/bjsports-2013-092734.
42. Signoretti S, Lazzarino G, Tavazzi B, Vagnozzi R. The pathophysiology of concussion. PM R. 2011;3:S359–68. https://doi.org/10.1016/j.pmrj.2011.07.018.
43. Vagnozzi R, et al. Temporal window of metabolic brain vulnerability to concussion: a pilot 1H-magnetic resonance spectroscopic study in concussed athletes--part III. Neurosurgery. 2008;62:1286–95. https://doi.org/10.1227/01.neu.0000333300.34189.74; discussion 1295–1286.
44. Broglio SP, Eckner JT, Paulson HL, Kutcher JS. Cognitive decline and aging: the role of concussive and subconcussive impacts. Exerc Sport Sci Rev. 2012;40:138–44. https://doi.org/10.1097/JES.0b013e3182524273.
45. Gallo V, et al. Concussion and long-term cognitive impairment among professional or elite sport-persons: a systematic review. J Neurol Neurosurg Psychiatry. 2020;91:455–68. https://doi.org/10.1136/jnnp-2019-321170.
46. Hutchison MG, Di Battista AP, McCoskey J, Watling SE. Systematic review of mental health measures associated with concussive and subconcussive head trauma in former athletes. Int J Psychophysiol. 2018;132:55–61. https://doi.org/10.1016/j.ijpsycho.2017.11.006.
47. Halstead ME, Walter KD, Council on Sports, M. & Fitness. American Academy of Pediatrics. Clinical report--sport-related concussion in children and adolescents. Pediatrics. 2010;126:597–615. https://doi.org/10.1542/peds.2010-2005.
48. McCrory P, et al. Consensus statement on concussion in sport: the 3rd International Conference on Concussion in Sport held in Zurich, November 2008. Br J Sports Med. 2009;43 Suppl 1:i76–90. https://doi.org/10.1136/bjsm.2009.058248.
49. Hanisch UK, Kettenmann H. Microglia: active sensor and versatile effector cells in the normal and pathologic brain. Nat Neurosci. 2007;10:1387–94. https://doi.org/10.1038/nn1997.
50. Silverberg ND, Iverson GL. Is rest after concussion "the best medicine?": recommendations for activity resumption following concussion in athletes, civilians, and military service members. J Head Trauma Rehabil. 2013;28:250–9. https://doi.org/10.1097/HTR.0b013e31825ad658.
51. Winkelman C. Bed rest in health and critical illness: a body systems approach. AACN Adv Crit Care. 2009;20:254–66. https://doi.org/10.1097/NCI.0b013e3181ac838d.
52. Leddy J, Hinds A, Sirica D, Willer B. The role of controlled exercise in concussion management. PM R. 2016;8:S91–S100. https://doi.org/10.1016/j.pmrj.2015.10.017.
53. Alla S, Sullivan SJ, McCrory P, Schneiders AG, Handcock P. Does exercise evoke neurological symptoms in healthy subjects? J Sci Med Sport. 2010;13:24–6. https://doi.org/10.1016/j.jsams.2008.12.629.
54. Gaetz MB, Iverson GL. Sex differences in self-reported symptoms after aerobic exercise in non-injured athletes: implications for concussion management programmes. Br J Sports Med. 2009;43:508–13. https://doi.org/10.1136/bjsm.2008.051748.
55. Iverson GL, Lange RT. Examination of "postconcussion-like" symptoms in a healthy sample. Appl Neuropsychol. 2003;10:137–44. https://doi.org/10.1207/S15324826AN1003_02.
56. Cook NE, et al. Baseline cognitive test performance and concussion-like symptoms among adolescent athletes with ADHD: examining differences based on medication use. Clin Neuropsychol. 2017;31:1341–52. https://doi.org/10.1080/13854046.2017.1317031.
57. Cottle JE, Hall EE, Patel K, Barnes KP, Ketcham CJ. Concussion baseline testing: preexisting factors, symptoms, and neurocognitive performance. J Athl Train. 2017;52:77–81. https://doi.org/10.4085/1062-6050-51.12.21.

58. Covassin T, Elbin RJ 3rd, Larson E, Kontos AP. Sex and age differences in depression and baseline sport-related concussion neurocognitive performance and symptoms. Clin J Sport Med. 2012;22:98–104. https://doi.org/10.1097/JSM.0b013e31823403d2.
59. Covassin T, et al. Sex differences in baseline neuropsychological function and concussion symptoms of collegiate athletes. Br J Sports Med. 2006;40:923–7. https://doi.org/10.1136/bjsm.2006.029496; discussion 927.
60. Hunt AW, Paniccia M, Reed N, Keightley M. Concussion-like symptoms in child and youth athletes at baseline: what is "typical"? J Athl Train. 2016;51:749–57. https://doi.org/10.4085/1062-6050-51.11.12.
61. Hurtubise JM, Hughes CE, Sergio LE, Macpherson AK. Comparison of baseline and post-concussion SCAT3 scores and symptoms in varsity athletes: an investigation into differences by sex and history of concussion. BMJ Open Sport Exerc Med. 2018;4:e000312. https://doi.org/10.1136/bmjsem-2017-000312.
62. Mihalik JP, et al. The effects of sleep quality and sleep quantity on concussion baseline assessment. Clin J Sport Med. 2013;23:343–8. https://doi.org/10.1097/JSM.0b013e318295a834.
63. Mrazik M, Naidu D, Lebrun C, Game A, Matthews-White J. Does an individual's fitness level affect baseline concussion symptoms? J Athl Train. 2013;48:654–8. https://doi.org/10.4085/1062-6050-48.3.19.
64. Silverberg ND, Berkner PD, Atkins JE, Zafonte R, Iverson GL. Relationship between short sleep duration and preseason concussion testing. Clin J Sport Med. 2016;26:226–31. https://doi.org/10.1097/JSM.0000000000000241.
65. Kontos AP, et al. A revised factor structure for the post-concussion symptom scale: baseline and postconcussion factors. Am J Sports Med. 2012;40:2375–84. https://doi.org/10.1177/0363546512455400.
66. Schneider KJ, et al. Rest and treatment/rehabilitation following sport-related concussion: a systematic review. Br J Sports Med. 2017;51:930–4. https://doi.org/10.1136/bjsports-2016-097475.
67. Colcombe SJ, Kramer AF, McAuley E, Erickson KI, Scalf P. Neurocognitive aging and cardiovascular fitness: recent findings and future directions. J Mol Neurosci. 2004;24:9–14. https://doi.org/10.1385/JMN:24:1:009.
68. Griesbach GS, Hovda DA, Molteni R, Wu A, Gomez-Pinilla F. Voluntary exercise following traumatic brain injury: brain-derived neurotrophic factor upregulation and recovery of function. Neuroscience. 2004;125:129–39. https://doi.org/10.1016/j.neuroscience.2004.01.030.
69. Neeper SA, Gomez-Pinilla F, Choi J, Cotman C. Exercise and brain neurotrophins. Nature. 1995;373:109. https://doi.org/10.1038/373109a0.
70. Lawrence DW, Richards D, Comper P, Hutchison MG. Earlier time to aerobic exercise is associated with faster recovery following acute sport concussion. PLoS One. 2018;13:e0196062. https://doi.org/10.1371/journal.pone.0196062.
71. Leddy JJ, Haider MN, Hinds AL, Darling S, Willer BS. A preliminary study of the effect of early aerobic exercise treatment for sport-related concussion in males. Clin J Sport Med. 2019;29:353–60. https://doi.org/10.1097/JSM.0000000000000663.
72. Willer BS, et al. Comparison of rest to aerobic exercise and placebo-like treatment of acute sport-related concussion in male and female adolescents. Arch Phys Med Rehabil. 2019;100(12):2267–75. https://doi.org/10.1016/j.apmr.2019.07.003.
73. Micay R, Richards D, Hutchison MG. Feasibility of a postacute structured aerobic exercise intervention following sport concussion in symptomatic adolescents: a randomised controlled study. BMJ Open Sport Exerc Med. 2018;4:e000404. https://doi.org/10.1136/bmjsem-2018-000404.
74. Leddy JJ, et al. Early subthreshold aerobic exercise for sport-related concussion: a randomized clinical trial. JAMA Pediatr. 2019;173:319–25. https://doi.org/10.1001/jamapediatrics.2018.4397.
75. McCrory P. Summary and agreement statement of the 2nd International Conference on Concussion in Sport, Prague 2004. Br J Sport Med. 2005;39:i78–86. https://doi.org/10.1136/bjsm.2005.018614.

76. Buttner F, et al. Concussed athletes walk slower than non-concussed athletes during cognitive-motor dual-task assessments but not during single-task assessments 2 months after sports concussion: a systematic review and meta-analysis using individual participant data. Br J Sports Med. 2020;54:94–101. https://doi.org/10.1136/bjsports-2018-100164.
77. Dorman JC, et al. Tracking postural stability of young concussion patients using dual-task interference. J Sci Med Sport. 2015;18:2–7. https://doi.org/10.1016/j.jsams.2013.11.010.
78. Fino PC, et al. Detecting gait abnormalities after concussion or mild traumatic brain injury: a systematic review of single-task, dual-task, and complex gait. Gait Posture. 2018;62:157–66. https://doi.org/10.1016/j.gaitpost.2018.03.021.
79. Gioia G. Return to school: when and how should return to school be organized after a concussion? In: Gagnon I, Ptito A, editors. Sports concussions. Boca Raton: CRC; 2018.
80. Baker JG, Willer BS, Leddy JJ. Integrating neuropsychology services in a multidisciplinary concussion clinic. J Head Trauma Rehabil. 2019;34:419–24. https://doi.org/10.1097/HTR.0000000000000541.
81. McGrath N. Supporting the student-athlete's return to the classroom after a sport-related concussion. J Athl Train. 2010;45:492–8. https://doi.org/10.4085/1062-6050-45.5.492.
82. Sady MD, Vaughan CG, Gioia GA. School and the concussed youth: recommendations for concussion education and management. Phys Med Rehabil Clin N Am. 2011;22:701–19. https://doi.org/10.1016/j.pmr.2011.08.008, ix.
83. Purcell L, Kissick J, Rizos J, Canadian Concussion, C. Concussion. CMAJ. 2013;185:981. https://doi.org/10.1503/cmaj.120511.
84. Davis GA, et al. What is the difference in concussion management in children as compared with adults? A systematic review. Br J Sports Med. 2017;51:949–57. https://doi.org/10.1136/bjsports-2016-097415.
85. Gioia GA. Medical-School partnership in guiding return to school following mild traumatic brain injury in youth. J Child Neurol. 2016;31:93–108. https://doi.org/10.1177/0883073814555604.
86. Iverson GL, Gioia GA. Returning to school following sport-related concussion. Phys Med Rehabil Clin N Am. 2016;27:429–36. https://doi.org/10.1016/j.pmr.2015.12.002.
87. DeMatteo C, et al. Post-concussion return to play and return to school guidelines for children and youth: a scoping methodology. Disabil Rehabil. 2015;37:1107–12. https://doi.org/10.3109/09638288.2014.952452.
88. Choe MC, et al. A multicenter look at multidisciplinary youth concussion/mild traumatic brain injury programs: the Four Corners Youth Consortium (4CYC). Pediatr Neurol. 2020;107:84–5. https://doi.org/10.1016/j.pediatrneurol.2020.01.008.
89. Gioia GA. Multimodal evaluation and management of children with concussion: using our heads and available evidence. Brain Inj. 2015;29:195–206. https://doi.org/10.3109/02699052.2014.965210.
90. Purcell LK, Davis GA, Gioia GA. What factors must be considered in 'return to school' following concussion and what strategies or accommodations should be followed? A systematic review. Br J Sports Med. 2019;53:250. https://doi.org/10.1136/bjsports-2017-097853.
91. Sojka P. "Sport" and "non-sport" concussions. CMAJ. 2011;183:887–8. https://doi.org/10.1503/cmaj.110504.
92. Bloom B, et al. A systematic review and meta-analysis of return to work after mild traumatic brain injury. Brain Inj. 2018;32:1623–36. https://doi.org/10.1080/02699052.2018.1532111.
93. Guideline for concussion/mild traumatic brain injury and prolonged symptoms. 3rd ed. (for Adults 18+ years of age). Ontario: Neurotrauma Foundation; 2021. Retrieved from: https://braininjuryguidelines.org/concussion/.
94. Lumba-Brown A, et al. Representation of concussion subtypes in common postconcussion symptom-rating scales. Concussion. 2019;4:CNC65. https://doi.org/10.2217/cnc-2019-0005.
95. Lumba-Brown A, et al. Concussion guidelines step 2: evidence for subtype classification. Neurosurgery. 2020;86:2–13. https://doi.org/10.1093/neuros/nyz332.
96. Davis G, Commentary A. Concussion guidelines step 2: evidence for subtype classification. Neurosurgery. 2020;86:E222–3. https://doi.org/10.1093/neuros/nyz364.

Chapter 6
Concussion Evaluation and Neurocognitive Outcomes

McKyla McIntyre and Mark T. Bayley

Introduction

There are many purposes for conducting concussion assessments, including the recognition of a potential concussion, confirming diagnosis, ruling out life-threatening traumatic brain injury, establishing prognosis, developing a focused treatment plan, and deciding about return to play or usual activity.

Recognition of concussion has classically been a challenge in the sport context, in part due to conflicting demands including the desire of the athlete (and sometimes coaches) to return to play, as well as failure to recognize the significance of their symptoms. In the context of the general population, emergency personnel may also overlook concussion and mild traumatic brain injury because of lack of training or the presence of polytrauma, where more obvious orthopedic and life-threatening injuries may take priority. For these reasons among others, there is a need for a standardized objective tools that can reliably detect subtle symptoms.

Concussion remains a clinical diagnosis made by an appropriately trained clinician, as there are no validated objective assessments that conclusively make the diagnosis, and conventional neuroimaging is typically normal in the absence of superimposed structural pathology. That being said, the focus initially should include ruling out life-threatening intracranial processes, including subdural or epidural hematoma, skull and neck fractures, and raised intracranial pressure. To do this, the clinician must establish the mechanism of injury and the progression of

M. McIntyre
Physical Medicine and Rehabilitation, Toronto Rehabilitation Institute, Toronto, ON, Canada
e-mail: mckyla.mcintyre@uhn.ca

M. T. Bayley (✉)
Division of Physical Medicine and Rehabilitation, Department of Medicine, Toronto Rehabilitation Institute-University Health Network and University of Toronto, Toronto, ON, Canada
e-mail: Mark.bayley@uhn.ca

T. A. Schweizer, A. J. Baker (eds.), *Tackling the Concussion Epidemic*,
https://doi.org/10.1007/978-3-030-93813-0_6

early symptoms and establish whether urgent neuroimaging is needed using standard rules such as the Canadian CT head rule.

The next focus, post-diagnosis, is on establishing the prognosis for recovery and developing an early symptom-focused treatment plan, which requires a detailed assessment of the various functions that are affected by concussion. This also has been historically controversial, as many researchers have sought to identify discrete phenotypes of concussion, yet research to date has not yielded validated, reproducible, or discrete categories of concussion that are associated with a typical prognosis. Furthermore, the clinician must contend with the fact that pre-concussion comorbidities and characteristics, including mental health and previous concussions, may have a greater impact on the prognosis than the severity of the most recent concussion itself. Finally, the clinician is faced with safety concerns surrounding return to activity and, specifically in contact sports, return to play, which may expose the individual with concussion to significant risks of re-injury and rare but serious complications such as second-impact syndrome.

The purpose of this chapter is to provide an overview of the assessment of the individual at risk of concussion pre-exposure, recognition of the person with suspected concussion, the diagnostic approach to concussion and its associated symptoms, and, finally, the neurocognitive/neurophysical examination to determine readiness to return to usual activities.

The chapter will also present a brief overview of the approach to assessment in special controversial situations, such as when litigation or compensation is involved, when severe concomitant pain or mental health issues are present or when there is evidence of cognitive deterioration.

Baseline or Pre-Exposure Assessment and Testing

It is important to note that pre-exposure assessment or testing is not usually available in the general population who sustain a concussion. However, the purpose of pre-exposure or pre-participation assessments when relevant is twofold: first, to establish the safety for participation for those who had previous concussions and those who have a pre-existing neurological, musculoskeletal, or other disability that may increase their vulnerability to the effects of a concussion; and second, to establish a baseline among individuals in high-risk activities such as military theater or contact sports, such that in the event that a concussion occurs, the treating team can determine when recovery has occurred. There are currently no evidence-based guidelines for pre-participation assessment. Concussion is an injury that is specific to an individual patient and their pre-injury health status, with often only subtle findings. Baseline or pre-exposure testing has been proposed as a better way to facilitate recognition and diagnosis; however, as we will discuss, there remains a number of confounding factors that may undermine the value of this approach.

Pre-Participation History and Examination

A number of sports, military, and other industries require an assessment by a trained clinician prior to participation in play or work. The components of this assessment should include a history of previous illness and injuries, a neurological and musculoskeletal examination, and, if needed, neuroimaging, cognitive, and/or psychological testing [1]. A useful worksheet is available with this reference [2].

Pre-participation History should address the following questions:

(a) Has the player had a previous head injury or concussion (with dates), or history of increasing recovery time and lower injury threshold? The circumstances of previous injury should be noted including any medical documentation if possible, and any neuroimaging findings were completed.
(b) In the event of prior concussion(s), are there residual symptoms, for example, in the domains of physical (dizziness, headaches, and vision), sleep, emotional (irritability, changes in mood), and cognitive symptom categories?
(c) Is there residual neck pain from prior injuries, or any other musculoskeletal injuries?
(d) Is there a history of any relevant neurological or mental health concerns, such as learning and communication deficits, migraine headaches, attention deficit hyperactivity disorder, physical disabilities, and/or mental health conditions)?

N.B. *Establishing the number of previous concussions may not be straightforward.* Kerr and colleagues [3, 4] noted that there are multiple levels of influence that may affect disclosure of sport-related concussion and concussion symptoms in general. These may include (a) personal factors such as lack of knowledge, internal pressure, sex, and gender considerations; (b) interpersonal factors such as others' knowledge and attitudes toward concussion, external pressures, and supports; and (c) environmental factors, such as access to concussion prevention materials or sports culture and policy considerations, such as concussion-related legislation. Among adolescents, the nature and phrasing of questions may also affect disclosure of previous concussions. One group found that these three questions did not yield the same estimates of previous number of concussions: "Have you had a concussion or head injury?" "Have you been knocked out?" and "Have you had your bell rung or been dinged?" Indeed, more adolescents endorsed the term having your "bell rung" or being "dinged" than if they were asked if they had had a concussion [5]. Other authors have highlighted the inconsistency of prior reporting of concussions, particularly among young males with a greater number of baseline concussions and those with attention deficit hyperactivity disorder, although they did find 80% consistency among high school students without any of those conditions [6]. Among professional football players, about 50% of athletes had not reported concussions [7]. There have been attempts to standardize the approach to identifying concussions, and the Ohio State University TBI Identification Method short form has been used primarily in adults [8]. Overall, these study findings suggest that clinicians must be cautious and comprehensive when interviewing these populations and use multiple terms to describe concussion that might elucidate subtle concussion events by encouraging open and comprehensive reporting.

Pre-participation examination should include a neurological examination, particularly focused on vision including extraocular movements, vestibular examination, balance, muscle tone bulk, strength, and reflexes.

Pre-participation Neuroimaging may be required in those with a history of multiple concussions with residual symptoms and neurological signs; however, it is unnecessary on a routine basis as there are no widely agreed-upon imaging signs or clinical interpretations thereof. See Chap. 3 for further details.

Role of Baseline Neurocognitive Testing

A number of authors have proposed baseline neurocognitive testing as a method of improving detection of residual symptoms and signs of concussion, as well as a strategy for preventing premature return to contact sports, work, or military service. Several neuropsychological assessments have been identified using pen and paper type tasks; however, more recently because of the increasing number of participants in sports, computerized testing has come into favor [9]. These computerized tests allow for the ability to (a) test multiple athletes simultaneously; (b) administer and interpret tests in the absence of neuropsychologists; (c) ensure maximally standardized components of test administration; (d) readily use alternative test forms via randomized presentation of stimuli; (e) quantify reaction time; and (f) take advantage of centralized data repositories for easy comparison with normative data and secure storage [10]. It is fair to say that there has been controversy about the benefit of computerized testing, because of the possibility of basing return to play decisions solely on cognitive testing when athletes are still symptomatic, as well as the intrinsic variability in cognitive testing due to factors such as sleep, time of day, and presence of pain. Furthermore, when used in the context of professional or collegiate athletes to make decisions, reports of intentionally reducing effort or "feigning worse" on baseline testing so that if a concussion occurs performance will look normal exist. These factors should give clinicians pause to the widespread implementation of these tools. In addition, developers of these tools have a financial incentive to promote the benefits and de-emphasize the weaknesses of such testing. Therefore, the reader is cautioned to keep in mind the available evidence for baseline testing, which remains limited at this point as will be outlined below.

What Is the Evidence for Baseline Computerized Testing?

Table 6.1 reviews the four most common computerized tests below [9]. A closer look at the psychometric properties of these tools demonstrates that they have only adequate test–retest reliability, with the Pearson coefficients greater than 0.6. At a

Table 6.1 Commonly used pre-participation and post-concussion neurocognitive computerized testing

Computerized test battery	Domains studied	Age range	Reliability	Validity and sensitivity to change
ANAM—Automated Neuropsychological Assessment Metrics tests	8 tasks—simple reaction time, code substitution learning, procedural reaction time, math processing, matching to sample, code substitution-delay simple reaction time, and go/no-go	9 to adult	Test–retest Pearson coefficient = 0.79 at 7 days and 0.71 at 198 days	Sensitivity to change of 6 to 24% at 24 hours. Lower at 8 days. False-positive rates of 16.5%
Axon Sports/Cogstate Sports	Four tasks-processing speed [simple reaction time], attention [choice reaction time] learning including visual recognition memory and working memory	10 to adult	Test–retest Pearson = 0.6 overall. Highly variable only 5 tasks with Pearson >0.6 and up to 0.94 for some tasks.	Sensitivity to change of 7 to 48% at 24 hours but diminished by eight days with positive predictive value (PPV) of 24%. False-positive rate 28.2%
ImPACT—Immediate Post-Concussion Assessment and Aognitive Testing	Six tasks—word memory, design memory, Xs and Os, symbol match, color match, and three letters which result in verbal memory, visual memory, visual–motor speed, reaction time, and impulse control composite scores	5 to adult	Test–retest Pearson coefficient 0.23–0.39 over 45 days but >0.6 beyond this window	Sensitivity to change of 24–40% at 24 hours and worse at 8 days with PPV of only 50%. Significant false-positive rates
Concussion Vital Signs (CNS-VS)	Seven subtests—verbal memory, visual memory, finger tapping, symbol digit coding, Stroop test, shifting attention test, and continuous performance test	10–70	Test-retest reliability "comparable" to other measures though Pearson not provided	Several studies involving ages 7 to 19 years. Good discriminant validity between mild and severe TBI [13]

group level, the tests reveal moderate to large changes, but only during the first 5–7 days post-injury [11, 12]. In a prospective head-to-head study of the three computerized tests below, Nelson and colleagues found that there were statistical differences evident in the concussed versus control group at 24 hours, but most of these differences were not statistically significant after eight days [10]. Similarly, sensitivity to those who were diagnosed with concussion ranged from 6 to 23.8% for ANAMs subtests, 7% to 48.6% for Cogstate, and 24 to 39.5% for ImPACT at 24 hours and these diminished at eight days. The overall sensitivity of the IMPACT instrument was 67.8% at 24 hours. The authors of these papers note that modest reliability and validity may be explained by other factors: For example, sport-related concussion is associated with a rapid recovery within the first few hours. It is well known that cognitive impairment may persist further out from injury; however, these tests appeared to lack the sensitivity to detect any neurocognitive abnormality. The authors concluded that the clinical utility of these tools was highest in the first 24 hours.

Another study assessed recovery from concussion in active-duty military service members. The sample was divided into those active military personnel classified as concussed or non-concussed using the ANAM tool [14]. Although the magnitude of the change did seem to be greater in those considered concussed by medical personnel, the authors cautioned against the use of the tool in isolation and suggested that a multidimensional approach for post-deployment screening was preferable.

Baseline and normative referenced approaches have been proposed as a method of interpreting the results of the neurocognitive testing outlined above. In other words, in the absence of baseline data, the question is as follows: Could the assessment scores from individuals with suspected concussion simply be compared to normal ranges in age-matched controls to support the diagnosis? If normative referenced approaches are as good as baseline referenced approaches, then individuals could simply be tested after their concussion and compared against normative data, without going to the trouble of baseline testing. Haran et al. [14] addressed this question in a sample of individuals in the military using the ANAM tool. The study found that there was no significant difference between the baseline and norm-referenced approaches. The two approaches correctly identified nearly the same percentage of injured service members that had lower post-injury scores. Conversely, a study using the ANAM found domain-specific advantages of each method; the baseline comparison method identified 2.6 times more impairment than the normative comparison method for simple reaction time, while the normative comparison method identified 7.6 times more impairment than the baseline method for mathematical processing [15]. No significant disagreements were observed in the comparison of concussed vs controls with either method. In contrast, using the COGsport test battery, another study found that the baseline method was slightly more sensitive than the normative method. Both methods had high specificity [16]. Studies have found differences in sensitivity using the ImPACT assessment tool among athletes classified as above average or already abnormal at baseline. They note that athletes with learning disability or attention deficit hyperactivity disorder (ADHD) may be useful to test at baseline.

Summarizing the literature and consensus, the Berlin statement regarding concussion in sport indicated that “Baseline or pre-season Neuropsychological testing was considered by the panel and was not felt to be required as a mandatory aspect of every assessment." [17].

There may be unique athlete populations where baseline testing may be considered [18]. These populations include the following: selecting youth athletes who have pre-existing conditions that may impact the interpretation of post-injury test results, such as a history of previous concussion or learning disorders, which may impact the interpretation of post-injury test results. Baseline testing is considered as an optional assessment within the comprehensive concussion protocol, as long as the medical teams caring for the athletes include experienced healthcare professionals who have competency-based training and clinical experience to allow them to correctly administer and interpret these tests.

Concussion Recognition on the Field or at the Location of Injury

There is a strong consensus that individuals suspected of having a concussion should be removed from the game, arena, or workplace immediately, to prevent further brain injury or complications. However, the rapid pace of sports, workplace machinery or blasts/explosion, the varying angles of impact to the body and head, and the difficulty of determining thresholds for concussion injury can make instant recognition very challenging. A number of tools have been developed that can be used by coaches, trainers, and emergency personnel to recognize that an individual has sustained a concussion.

Key Elements of Concussion Recognition Tools

Key elements of concussion recognition tools have been examined in a number of systematic reviews, and one author reviewed available tools and associated risk of bias, the objectivity of the reference standard, and the flow and timing of the tools [19]. It is important to note that the reference or "gold" standard for most of these recognition tools is the clinical diagnosis of concussion by a physician. The authors identified and classified tools into the following categories: elicited symptoms and signs, oculomotor tests, balance tests, and cognitive and multimodal tests. They used the Quadas-2 and Newcastle–Ottawa instruments to assess risk of bias and cautioned that the published research suggested most studies were at high risk of bias, and the overall strength of the evidence examining sideline screening tools remained low.

Despite these limitations, the authors had some important observations regarding the predictive value of certain symptoms and likelihood of a clinical diagnosis of concussion. They warn that estimates of sensitivity vary based on the sport (i.e., the pre-test probability of concussion) as some contact sports have higher prevalence of concussion. Thus, these estimates must be interpreted in the context of sports concussion. The presence of dizziness as a symptom had an estimated sensitivity of between 0.65 and 0.85 in sport-related concussion with a specificity of 0.96 [20]. Nausea on the other hand was less sensitive [0.22–0.65] but did have high specificity (0.93). Headache had estimates of sensitivity between 0.76 and 0.99 with a specificity of 0.82. Interestingly, the total number of symptoms on the Sport Concussion Assessment Tool Version 2 had a sensitivity of 0.84 and the symptom severity score had a sensitivity of 0.75 and was highly specific for concussion. The Maddocks questions were developed to test attention and awareness of game situation and have been incorporated into other concussion recognition tools. The Maddocks questions had a range of sensitivity estimates between 0.34–0.75, and the presence or absence of normal orientation had limited value with the sensitivity of only 0.21 [19].

Cognitive testing for early recognition has also been examined. The elements in the Standardized Assessment of Concussion (SAC) demonstrated a wide range of sensitivity values ranging from 0.20 to 0.94, evaluated across several studies suggesting they may not be consistently helpful [21].

Balance tests are another strategy for recognition of the presence of concussion. The balance error scoring system (BESS) had relatively low sensitivity ranging from 0.36 to 0.8 as did a modified version estimate at 0.25. Similarly, tandem stance and tandem gait observations had sensitivity of 0.28 and 0.83, respectively.

Oculomotor testing can also be used for concussion recognition, and the King-Devick assessment, which had estimates of sensitivity of 0.79–1.0 and high specificity, suggests that the presence of oculomotor abnormalities may be an important sign of concussion.

A number of efforts have been made to develop multimodal assessments that include both physical symptoms and cognitive testing. When combined, some of these combinations have yielded good sensitivity, including the Sport Concussion Assessment Tool (SCAT) or the BESS combined with the SAC. The authors highlighted that all these studies continue to have limitations including a high risk of bias.

These tools continue to evolve, and the most commonly used instruments and content covered are outlined in the table below (Table 6.2). The Concussion Recognition Tool 5 (CRT5) is the most recent revision of the Pocket Sport Concussion Assessment Tool 2 that was initially introduced by the Concussion in Sport Group in 2005. It remains the most used tool in sports [22]. Based on the evidence, symptom assessments in and of themselves are reasonable strategies for recognition at the sideline by non-professional trainers; however, the reader is cautioned that if there is any doubt about whether a concussion occurred, the individual should be removed from the activity as soon as possible.

Table 6.2 Key Elements of Concussion Recognition Tools

Recognition tool	Target functions	Elements	Validity
Symptoms	Specific symptoms	Dizziness Nausea Headache	Sensitivity (0.65–0.85) specificity 0.95 Sensitivity (0.22–0.65) specificity 0.93 Sensitivity (0.76–0.99) specificity of 0.82.
SCAT—Symptom Inventory	Cognitive, physical	Total number severity	Sensitivity of 0.84 specificity 0.95–1.0 Sensitivity of 0.75 specificity 0.95–1.0
Maddocks Questions	Awareness orientation	Knowledge of game and situation	Range of sensitivity estimates between 0.34 and 0.75
Concussion Recognition Tool (version 5)	Symptoms, cognition, orientation	Red flags, signs of balance, facial trauma, staring ahead, and lying motionless) memory/awareness	Sensitivity of 0.84 and highly specific 0.95–1.0
Standardized Assessment of Concussion SAC	Cognitive	Number of cognitive tasks for attention	Sensitivity of 0.20 to 0.94 and specificity of 0.96
King–Devick test	Oculomotor	Number scanning Eye movement task	Sensitivity of 0.79–1.0 and high specificity
Balance Error Scoring System	Balance	Balance errors with different challenges	Sensitivity 0.36 to 0.8 and specificity range of 0.71–1.0

Video Review for Recognition of Concussion

Video Review for Recognition of Concussion has been proposed as an option for recognizing concussive injuries. Video analysis has been used to evaluate suspected concussion in a number of professional sports [23–26]. A group of experts convened to develop consensus definitions identifying the following six video signs of concussion: lying motionless, motor incoordination, impact seizure, tonic posturing, and no protective action-floppy and blank/vacant look [27]. The authors provide definitions and observational characteristics for each of these signs. These signs have a variety of estimates of sensitivity among different types of sports. Among rugby players, the presence of blank/vacant look had a sensitivity of 54%, while the most specific sign was tonic posturing, with a specificity of 0.99. The rest of the signs had estimates of 2.8% [impact seizure] to motor incoordination (36.8%). If any of the six signs were present though, there was a 60.4% sensitivity rate. All of these signs though have a high specificity estimated at greater than 0.9.

In conclusion, the Concussion Recognition Tool 5 represents the most well-established multimodal instrument that includes symptoms, signs, and cognitive assessments. The addition of video review could potentially offer a promising approach to improving identification and evaluation of significant head impact events.

The Comprehensive Medical Assessment to Make a Diagnosis of Concussion

The History of the Injury—Key Elements to Establish the Diagnosis

Description of the Accident

There are many purposes to asking about the accident. The patient's description of the accident helps to establish how much they remember, the mechanism of the accident and the potential forces at play, initial symptoms and any period of disturbed consciousness, or alteration in mental state. Ideally, it will also establish if witnesses were present who could corroborate the story. Particularly in the medico-legal context, the description of the accident helps to establish whether the forces at play are likely to have been sufficient to create a concussion. The kinematics of injury likely play a role in concussion risk, as research suggests that adding rotational forces to the flexion and extension forces applied to the head increase the likelihood of axonal injury and concussion [28]. Blast forces associated with explosion in military theater also can exert forces on the brain and have a similar acute clinical presentation to other causes of concussion [29–31]. The examiner is encouraged to take an open-ended question approach that probes sequentially what happened. Important prompts include the following: How did you land or what struck you? Did you have any obvious scrapes, bleeding, or bruising and where were they? It is important also to establish whether the injury was psycho-traumatic in nature, for example, in the case of assaults.

Retrograde Amnesia represents the time from last memory until the impact that caused the concussion. Retrograde amnesia (RTA) is not present in all episodes of concussion as it rarely occurs without evidence of post-traumatic amnesia. In a Finnish study, about 24% of individuals with concussion had RTA but only 1.3% reported RTA without post-traumatic amnesia [32, 33]. It may be helpful in understanding the injury trajectory, though.

Post-traumatic Amnesia represents the period of time from the impact until the individual regains continuous memory for ongoing events, which has been proposed as an indicator of severity of brain injury. Establishing the duration of PTA can be challenging when not done immediately after the accident. Furthermore, some individuals assume that they were unconscious because they cannot remember when they were only in PTA. It is also possible that people may have some islands of memory in a sea of amnesia just after the concussion, so it is important to ask the patient sequentially what happened and who they remember, whether they remember the ambulance ride or emergency personnel until they truly regain memory functioning. The effects of sedative medication or substance use must be considered in this assessment as they may confound memory formation. Ideally, direct observations of orientation are most helpful. The abbreviated Westmead Post-Traumatic Amnesia Scale, the Maddocks Questions in Sports, and the Galveston Orientation and Amnesia tests all have established normative data for assessing whether an individual remains in PTA but have limited usefulness after the person has recovered [34–36].

Loss of Consciousness
It is important to note that loss of consciousness is not required to establish a diagnosis of concussion. Acute symptoms following head trauma may reflect the presence of acute physiological disruption of brain function even in the absence of loss of consciousness. Witnesses observing and reporting that the individual experienced a loss of consciousness is helpful in establishing the diagnosis particularly in the medical-legal context to confirm diagnosis of mild traumatic brain injury. However, a number of studies document that there were no differences in cognitive testing between those with and without loss of consciousness [37].

Early Symptoms and Signs
The most common symptom after concussion is headache; however, it can be induced by injury to the neck as well. Other important symptoms include dizziness, nausea, and/or visual distortion; although no individual symptom is diagnostic of concussion, it is important to assess for these as outlined below under the symptom inventory. It is also important to ask if witnesses observed any neurological signs such as seizure/convulsions, balance difficulties, or weakness.

Associated Injuries
It is important to ask about concurrent injuries, especially in the context of polytrauma, as they may establish the degree of forces involved, the proximity of forces to the head, and the risk of other injuries that may exacerbate concussion symptoms and severity. For example, a period of hypotension or hypoxia associated with polytrauma particularly involving chest injuries and hypo-perfusion due to severe bleeding can exacerbate a brain injury [38].

Ruling out Life-Threatening Brain Injury

Adult Assessment The critical priority of the early medical assessment is to rule out a life-threatening brain injury or intracranial complication requiring neurosurgical intervention. A number of clinical decision rules exist for this purpose, including the Canadian CT Head Rule and The New Orleans Criteria. The Canadian CT Head Rule has been validated extensively. The sensitivity for detecting clinically important brain injury and injuries requiring neurosurgical intervention was 100%. The

Canadian CT Head Rule for neuroimaging include GCS less than 15 at two hours post-injury, suspected open or depressed skull fracture, any signs of basal skull fracture, two or more episodes of vomiting, and age 65 years or older. Other criteria for brain injury detection include greater than 30 minutes of retrograde amnesia or dangerous mechanism of injury. Examples of the latter include pedestrian struck by a motor vehicle, an occupant ejected from a motor vehicle, or a fall from an elevation of three or more feet or five stairs [39].

Assessment of Children Several age-specific decision rules have been developed for children. The most commonly used assessments include the PECARN management algorithm for children after trauma, CATCH2 (Canadian Assessment of Tomography for Childhood Head injury), and the CHALICE (Children's Head injury Algorithm for the prediction of Important Clinical Events). These tools have high sensitivity, and they all attempt to reduce unnecessary use of head CT in the decision-making process as there remains a prominent concern as to the effects of radiation from CT scans, and the associated risk of cancer in children. These tools include considerations such as the Glasgow coma scale, signs of diminished arousal, and nausea and loss of consciousness as part of their decision rules. The specifics are available online and in references [40].

Evaluation of Pre-Existing Conditions and Risk Factors for Prolonged Concussion Symptoms

The purpose of identifying risk factors for prolonged symptoms is to identify those who may need either a different approach to management, or an earlier referral to specialized interprofessional care.

Adult risk factors have not been fully elucidated; however, systematic reviews have examined those that are at highest risk for prolonged recovery. One of the key predictors is the number and severity of early concussion symptoms. Older age and female gender are minimally modifiable risk factors. The clinician should inquire about pre-existing factors, i.e., previous concussions with associated duration of recovery for each, pre-existing attention deficit hyperactivity disorder, mental health disorders including anxiety, depression and substance use disorders, migraine headaches, and pain disorders [41–43]. While difficult to elicit in a traditional history, fear avoidance behavior and coping style are further implicated as predictors of recovery. Emerging assessment tools may be helpful in identifying those who may need more supportive approach to an earlier psychological intervention [44–46]. Recently, our group has developed a risk score for persistent concussion symptoms based on age, pre-existing health care use, and pre-existing mental health, available at https://kite-uhn.com/tricordrr [47]. Other contextual factors include injuries sustained in motor vehicle accident, and the potential influence of secondary gain related to litigation or compensation, while a lack of social supports and lower socioeconomic status appear to influence outcomes as well.

Pediatric Risk Factors The risk factors for prolonged symptoms among children have been elucidated and include age (increased risk with increased age), sex (female), duration of recovery from a previous concussion, high pre-injury symptom burden, high symptom burden at initial presentation, clinical evidence of vestibular or oculomotor dysfunction, personal and family history of mental health issues or migraines, history of learning or behavioral difficulties, and family socioeconomic status/education. The large-scale 5P study led to the development of a risk score calculator. This calculator is readily available online at http://www.5pconcussion.com/en/scorecalculator and utilizes age, sex, duration of previous concussion symptoms, history of migraine, slow response to questions, BESS Tandem stance balance testing errors, presence of a headache, sensitivity to noise, and increased fatigue in the emergency room [48].

Concussion Symptom Inventories

The development of a number of inventories of symptoms provides the clinician with a rapid means of delineating the different types of presentation of each concussion patient. These inventories can be completed online or using checklists. Clinicians can then use the inventory to guide their clinical assessment highlighted below. A number of inventories are embedded in instruments including the SCAT5, the ImPACT computerized testing, and others. Perhaps the two most frequently used are the SCAT5 Symptom Inventory and the RiverMead Post-Concussion Questionnaire (RPQ). Between these, there are differences in the way the symptom questions are asked and the Likert scale used. The SCAT consists of 22 questions with a 7-point Likert scale; it is reviewed and updated by the Concussion in Sport group every 4 years [17]. It is currently in the fifth iteration. The SCAT was developed for field assessments of sports-related concussions but is used for concussion in the general population as well. The RPQ debuted in 1995 and was the first measure of concussion severity. It has 16 questions and a 5-point Likert scale and also queries premorbid symptom state. The RPQ is more commonly used in neuropsychological research on concussions [49]. Our research suggests the two inventories yield very similar profiles and information and are highly correlated [50].

Clinical Assessment of Neurophysical Concussion Symptoms

The clinical assessment for concussion involves a thorough review of the neurological system, as well as an assessment for life-threatening brain injury or potential contributors to overall symptom burden including secondary or comorbid diagnoses. The general approach after recognition and diagnosis of the concussion is to assess the nature of the individual symptoms using a focused history and physical examination relevant to that symptom. Every concussion assessment should include the components of the neurological examination, namely cognition/mental status, cranial nerves, muscle bulk and tone, power, reflexes, sensory examination, coordination, and gait assessment with functional tests including balance assessment. The bulk of the neurological examination is typically normal in concussion, but a comprehensive approach provides reassurance to the examiner that concurrent neurological pathology is not missed. Further explanation of portions of the assessment, including pertinent history and physical examination, will be included in this section as they pertain to the concussion patient. Table 6.3 also summarizes key concussion symptoms with associated differential diagnoses, history taking points, pertinent physical examination maneuvers, and relevant standardized assessments.

Headaches

The International Headache Society has set out criteria for post-traumatic headache (PTH) attributed to mTBI/concussion. As part of the diagnosis, the headache should develop within 7 days of a head trauma that otherwise meets the definition of concussion. Post-traumatic headaches may be sub-classified into acute, where headache resolves within 3 months of head trauma, or chronic, where headaches persist beyond 3 months. In clinical practice, the focus of the headache history is to identify which headache phenotype that post-traumatic headache most resembles. In other words, which type of headache experienced in the general population most resembles the post-traumatic headache? It is also important to note prior history of headaches, as certain headache disorders are associated with prolonged recovery after concussion.

(a) *History*: The post-traumatic headache history should explore symptomatology of the various headache types, including migraine, tension type, neuralgic, or cervicogenic. Characteristic features of migraine-type PTH include (1) duration of 4–72 hours if untreated, (2) at least 2 symptoms out of unilateral, pulsating, moderate-severe intensity, and aggravation with physical activity, and (3) at least one of nausea and/or vomiting, or both photophobia and phonophobia. The migraine-type PTH may be associated with an aura consisting of at least one fully reversible visual, sensory, or speech symptom, and at least two of homonymous visual and/or unilateral sensory symptoms, timeline of symptoms developing gradually over 5 minutes or different symptoms occurring in succession over more than 5 minutes, and each symptom lasting between 5 and 60 minutes. Characteristic features of tension-type PTH include duration between 30 minutes and 7 days, at least 2 symptoms out of bilateral location, pressing/tightening quality, mild-to-moderate intensity, and no aggravation by routine physical activity, and both of no nausea or vomiting, and no more than one of photophobia or phonophobia. Cervicogenic headaches are classically moderate–severe in intensity, with non-throbbing, unilateral quality of pain without side shift [51]. They typically originate in the cervical region, but may

Table 6.3 Summary of common symptoms after concussion with associated differential diagnoses, history and physical findings, and standardized tools

Symptom	Differential diagnosis	Key history features	Key physical findings	Standardized tools
Headache	Tension type Migraine TMJ Cervicogenic Site of injury Occipital neuralgia	Location Quality Severity Photophobia or Phonophobia Nausea or vomiting	Fundoscopy Tenderness over greater or lesser occipital nerves Hematoma	Not applicable
Neck pain	Cervical sprain/strain injuries Radiculopathy Chronic pain syndrome	Location Quality Severity Focal neurological symptoms	C-spine range of motion Spurling's maneuver	Goniometric measurement of range of motion
Sleep dysfunction	Sleep apnea/ organic sleep dysfunction	Sleep initiation vs sleep interruption difficulty Energy Sleep efficiency (total hours in bed/total hours asleep) Snoring/apnea	Blood pressure Upper airway stridor	Epworth sleepiness scale insomnia severity index
Fatigue	Medical conditions (anemia, renal/liver dysfunction, viral, etc.) Fibromyalgia Hormone dysregulation	Constitutional symptoms	Conjunctival pallor Goiter Neck circumference Dehydration/malnutrition	Fatigue Severity Scale Multidimensional Fatigue Inventory Fatigue Impact Scale
Vision	Primary visual injury vs nonspecific Headache exacerbation	Acuity Diplopia	Extraocular movements Convergence Saccades	King-Devick test VOMS
Dizziness	Vestibular injury vs nonspecific Orthostatic hypotension/ autonomic dysfunction	Vertigo (spinning or lurching) vs wooziness	Dix–Hallpike Orthostatic BP	Electronystagmography

(continued)

Table 6.3 (continued)

Symptom	Differential diagnosis	Key history features	Key physical findings	Standardized tools
Balance	Peripheral vs central causes	Falls Sensory deficits Coordination deficits	Cerebellar signs Gait Romberg test Proprioception	Caloric testing BESS PSAP with force plate system
Symptom	Differential diagnosis	Key history features	Key examination findings	Standardized tools
Anxiety	Situational phobia Post-traumatic stress disorder	Anxiety, nightmares, flashbacks panic attacks Social anxiety Accident-related anxiety	Hyperarousal, startle response	GAD-7 (anxiety) PC-PTSD (PTSD) PHQ-9
Mood	Major depressive disorder Generalized anxiety disorder PTSD	Low mood, anhedonia, irritability, emotional lability, or suicidal ideation	Low affect General appearance (grooming, etc.)	PHQ-9, BDI-II (depression) BAI, Brief symptom inventory
Substance use disorders	Depression, anxiety, PTSD	Low mood, withdrawal symptoms, and signs for opiates and alcohol	Affect appearance, vital signs	CAGE Question and Substance Abuse Subtle Screening Inventory (SASSI-3)

refer to other regions of the head, and they may be exacerbated by cervical spine range of motion. Neuralgic headaches will typically fall over the distribution of the greater and/or lesser occipital nerves and may have associated scalp numbness or paresthesia.

(b) *Examination:* Examination of the head and neck should begin with general inspection for scalp wounds or hematomas that may have occurred during acute injury, as well as head and neck alignment. A complete cranial nerve examination should be performed, including fundoscopy, to exclude signs of increased intracranial pressure such as papilledema. Palpation of the head and neck should include territories of the greater and lesser occipital nerves with Tinel's test to assess for neuralgia that may contribute to headaches. Assessment of the temporomandibular joint (TMJ) through palpation and range of motion may also elicit pain as a headache contributor.

Concurrent Neck Pain

(a) *History:* In addition to the headache history above, neck pain/discomfort should be specifically assessed, including exacerbating and alleviating factors. Review of systems should include radicular pain and focal weakness, numbness, or paresthesia in the extremities.

(b) *Examination:* Prior to palpation, screening active and passive range of motion of the cervical spine should be performed to assess for concurrent cervical sprain or strain injuries, or cervicogenic etiology of headaches. Supine positioning may be utilized for passive range of motion to facilitate relaxation of the cervical muscles if the standing examination elicits guarding. Spurling's test can be performed if the history suggests cervical radiculopathy. Palpation of pertinent structures in the cervical spine should include the base of the occiput and the cervical spinous processes, as well as soft tissues such as the paraspinal tissues, suboccipital muscles, scalenes, trapezii, and rhomboids. These soft tissue structures are commonly tender in cervical sprain or strain injuries and may have associated myofascial trigger points.

Sleep Dysfunction

(a) *History*: A detailed history of premorbid sleep function may elicit clues as to the degree of impairment related to concussion. It is relevant to document average total time in bed and total time asleep, in order to calculate sleep efficiency. Lifestyle habits such as typical sleep and wake times, exercise, screen time, caffeine use, and sleep medications should be documented. The history should also include screening for organic sleep dysfunction such as sleep apnea and restless legs syndrome (RLS), as well as for triggers of sleep interruption such as pain, mood symptoms, or nocturia. Sleep scales that can be used for screening purposes include the Epworth Sleepiness Scale (ESS) or the Insomnia Severity Index [52].

(b) *Examination*: Physical examination for sleep is limited; review of a sleep diary may be a reasonable substitute for physical maneuvers. In addition, neck circumference can be measured as a sign of possible obstructive sleep apnea, and upper airway can be assessed for evidence of stridor.

Fatigue

(a) *History*: Aside from sleep difficulty, the differential diagnosis for fatigue after concussion includes hormonal dysregulation resulting from pituitary damage. In addition to fatigue, symptoms of hypothyroidism may include weight gain, temperature dysregulation, or changes to hair, skin, or nails; symptoms of hypocortisolism may include weight loss, hypotension, or lightheadedness. General medical conditions such as viral illnesses, malignancy, anemia, liver/renal dysfunction, or malnutrition/B12 deficiency may also contribute to fatigue. Lastly, fatigue may result from other concussion sequelae such as mood disorders, which are covered later in this section.

(b) *Examination*: Note should be made of patient's body habitus as a clue for overall nutritional status. Assessment for peripheral edema may elicit signs of liver dysfunction, heart failure, or malnutrition. Examination of the neck should include palpation for cervical lymphadenopathy as may be present in viral illnesses or malignancy, as well as palpation of the thyroid for goiter.

Vision

(a) *History*: Pertinent history features for vision include blurred vision/decreased acuity, or diplopia with rapid eye movements. Patients may notice symptom provocation with convergence/near focus, or exacerbation of headaches with focusing on screens or near objects. Due to their visual symptoms, patients may report general fogginess, difficulty concentrating, or impaired attention.

(b) *Examination*: Full visual examination should be performed including pupillary reaction, acuity, and visual fields. In addition to extraocular movements, which may elicit diplopia or nystagmus in the concussed patient, saccadic assessment should be completed. The King-Devick test can be used as a measure of reading-related eye tracking and saccades [53, 54]. The Vestibular Ocular Motor Screen (VOMS) can be administered to assess visual movements and the vestibular ocular reflex (VOR). Cold caloric testing can be performed as an additional test of the VOR.

Dizziness

(a) *History*: Key elements of the history include falls related to dizziness or lightheadedness, as well as orthostatic symptoms or presyncope/syncope events. Review of systems should also include screening for vertigo, as the injury that led to concussion may provoke otolith dysfunction causing benign paroxysmal positional vertigo (BPPV).
(b) *Examination*: For any patient presenting with vertigo, the Dix-Hallpike test should be completed, as BPPV can be diagnosed clinically; electronystagmogram (ENG) can also be performed by a qualified clinician. In the event of positional symptoms, baseline and orthostatic vital signs should be considered to rule out contribution of orthostatic hypotension, bradycardia, or other signs of dysautonomia [54, 55].

Balance Difficulty

(a) *History*: The history for balance difficulty overlaps with the history for dizziness as outlined above; visual symptoms such as blurred vision or diplopia may also exacerbate balance problems. Number of falls and related injuries should be asked about, as well as any compensatory gait aids or devices needed. Review of systems should include sensory deficits or coordination problems that may provide clues for central or peripheral nervous system contributors.
(b) *Examination*: Focused balance assessments include tandem gait, tandem stance, and single leg balance (20 seconds per side, per test). A standardized balance assessment such as the Balance Error Scoring System (BESS) may be utilized either as a sideline assessment or in the office [53]. Another option is the Postural Sway Assessment Protocol (PSAP) using a force plate system, which has been shown to have good test-retest reliability [56]. Coordination testing including finger-to-nose, heel-to-shin, and rapid alternating movements should be done to elicit contributing cerebellar dysfunction. The Romberg test should also be performed to elicit balance difficulty related to proprioceptive dysfunction. Sensory examination of bilateral feet should be performed including light touch, pinprick, and proprioception or vibration.

Assessment of Neurocognitive and Psychological Concussion Symptoms

Mood/Emotional Status

The mental health complications of concussion are extensively dealt with in Chap. 7, however, to highlight the importance, an overview of the assessment of low mood, and anxiety follows:

(a) *History*: The history should carefully elicit symptoms of depression including low mood, anhedonia, irritability, emotional lability, and suicidal ideation; underlying baseline or compensatory substance use is also important. In addition to general anxiety, screening for post-traumatic stress disorder (PTSD) should be administered, particularly if the injury occurred in a potentially life-threatening situation. It is important to note that cognitive assessments can be confounded by untreated mood disorders and that some of the symptoms such as poor concentration and focus may also occur due to these conditions.
(b) *Assessment tools*: The Patient Health Questionnaire 9 (PHQ-9) or Beck Depression Inventory II (BDI-II) can be used as screens for depression, while Beck Anxiety Inventory (BAI) or Generalized Anxiety Disorder 7-item scale (GAD-7) can be used for anxiety [54, 57]. The short Primary Care PTSD Screen (PC-PTSD) can be used for PTSD [58]. However, these questionnaires have not been specifically validated for the concussion population [54, 58]. The CAGE questionnaire (Cut Down, Annoyed, Feeling Guilty, Eye-Opener) and the Substance Abuse Subtle Screening Inventory (SASSI-3) have been shown to have good sensitivity >90% in people with TBI when compared to a detailed diagnostic interview [59, 60].

Neurocognitive Assessment

Cognitive assessment may take a number of forms, including administration of a cognitive screening tool, computerized standardized testing, or a full neuropsychological evaluation. The most common difficulties reported on clinical interview are attention and concentration, cognitive processing speed/efficiency, memory, and executive function [61]. On clinical examination, observation should be made of the patient's general appearance (grooming, affect, etc.), as important clues may be gleaned regarding overall function and mood. Any adaptive equipment utilized by the patient, such as sunglasses to compensate for bright lighting, should also be documented.

Cognitive Screening tools As mentioned above, there are several screening cognitive assessments that can be administered to assess orientation, attention, visuospatial/executive function, language, and memory. Standardized cognitive assessments that have been developed for the concussion population include the Immediate Post-Concussion Assessment and Cognitive Test (ImPACT), the Cogstate Computerized Cognitive Assessment Tool (CCAT), and the Standardized Assessment of Concussion (SAC) [53]. The ImPACT tool yielded higher sensitivity [79%] as compared to symptom reports, postural control data, or pencil and paper measures in the first 24 hours.

Neuropsychological examination is the most comprehensive form of cognitive assessment and is administered using standardized paper and pen tasks. *N.B. The purpose of this review is to provide the reader with an overview of the role of neuropsychological assessment, rather than to explain how to perform one, which requires extensive training and experience in interpreting testing*. Neuropsychologists are concerned with the assessment, diagnosis, treatment, and/or rehabilitation of brain psychopathology in the cognitive, behavioral, and emotional domains. Neuropsychologists will interview the individual, gathering medical history of developmental milestones, psychosocial history, and previous history of complaints. They will then coordinate the administration of standardized tests and questionnaires. Some neuropsychologists will also develop a hybrid model with use of the computerized testing discussed in Sect. 6.2.3 and Table 6.1 above, with additive paper and pencil tasks.

People with concussion are vulnerable to a number of psychological and emotional disturbances following injury, including but not limited to depression, anxiety, social isolation and loneliness, frustration, anger, and guilt. Whether concurrent psychological symptoms reflect a response to being injured and/or the pathophysiological consequences of concussions is not clear [62]. Behavioral questionnaires explore mood lability, irritability, insomnia, anxiety, and depression and reports of personality changes. There are a number of instruments that may be used. The Beck Depression Inventory (BDI-II) is a 21-question multiple-choice self-report inventory, one of the most widely used psychometric tests for measuring the severity of depression. Each answer is scored on a scale value of 0 to 3, where higher total scores indicate more severe depressive symptoms. The standardized cutoffs are available [63]. Other questionnaires include multidimensional fatigue inventories, anxiety measures, and personality inventories. Together, these provide a psychological picture of the person with concussion but may not provide insights as to causality of concussion in the genesis of symptoms.

Cognitive testing against normative ranges includes intelligence with estimates of premorbid functioning and academic achievement. Intelligence testing is divided into verbal and performance scores that may provide insights into injury to one particular area of the brain, or evidence of previous learning disability. The examination will then go on to test sensory perception, motor functions and hand dexterity, attention, memory, auditory and visual processing, language, problem-solving, planning, organization, and speed of processing [64].

This comprehensive testing provides important insights into the individual with concussion; however, the assessment can be lengthy and expensive and is therefore most appropriate when the differential diagnosis is unclear and/or the individual has pre-existing or comorbid conditions.

Special Concussion Assessment Considerations

Forensic Concussion Assessment

Forensic concussion assessment is sometimes required when questions of causal relationship exist between the injury and ongoing symptoms or disability. This occurs in a number of contexts including motor vehicle accidents, personal injury litigation, workplace injuries and compensation, eligibility for military compensation, and disability insurance disputes. One potentially significant contributing factor is symptom exaggeration [65]. When considering a diagnosis of post-concussion syndrome, clinicians must systematically evaluate and eliminate the possible contribution of many differential diagnoses, comorbidities, and factors that may cause, mimic, or maintain self-reported symptoms long after a concussion.

In the above context, the clinician must take a very careful clinical history looking for signs of inconsistency. Particular focus on the mechanism of injury and the availability of witnesses to validate the injury is critical. Any first responder medical information can be very helpful in providing observational corroboration. Clinicians must address the question of "does the resultant injury make sense?" Histories that suggest full recall of all events are not typical. Furthermore, the development of symptoms many days or weeks post-injury is also not typical. The loss of autobiographical memory is also highly unusual. On clinical examination, the observation of inconsistency between functioning under direct observation compared to when the individual is unaware of the observer can suggest inconsistent effort or embellishment of clinical signs.

The majority of neuropsychologists routinely use measures of effort [66], and leading neuropsychology organizations clearly state that measures of effort are an essential part of a neuropsychological evaluation [67, 68]. One example, the Test of Memory Malingering (TOMM) presents pairs of words for memorization. At first glance, the test appears daunting, but even random guessing achieves a 50% rate of recall on presentation of the words. As such, individuals falling well below the expected 50% threshold result were able to recognize the correct result and select the alternative suggesting an attempt to appear worse [69]. In a study by Lange et al., participants were divided into two groups based on TOMM performance (15 fail, 48 pass). Patients in the TOMM fail group scored higher than those in the TOMM pass group on measures of symptoms. Furthermore, there were significant main effects on cognition scores with patients in the TOMM fail group performed more poorly on the attention, memory, and executive functioning indexes [70].

These findings and others suggest that ideally a detailed interprofessional assessment be completed including neuropsychological assessment with effort testing.

Assessment of the Individual with Multiple Concussions or Long-Term Symptoms and Question of Chronic Traumatic Encephalopathy

Chronic traumatic encephalopathy (CTE) is a neurodegenerative disease that is thought to be caused by repeated exposure to brain trauma [71]. As noted previously, it is important to understand and estimate the number of previous concussions, as multiple concussions is likely a risk factor. Unfortunately, there are currently no universally accepted clinical diagnostic criteria for chronic traumatic encephalopathy (CTE) that can be applied when a patient is still alive. In its current form, CTE is considered to be a postmortem neuropathological diagnosis. Preliminary proposed research criteria for "traumatic encephalopathy syndrome (TES)" include three core features of CTE: (i) "cognitive," (ii) "behavioral" (i.e., anger dyscontrol), and (iii) "mood" (i.e., depression or hopelessness). These core features help to define diagnostic "subtypes" or "variants" according to the research criteria symptoms [72].

The following criteria have been used in the long-term cohort study called UNITE and are a further elucidation of the research criteria above [73]:

Required Features: Persistence of symptoms for longer than 2 years; no other neurologic disorder is more likely to account for all the clinical features; history of head trauma exposure, typically associated with history of concussion, although may be limited to subconcussive trauma; head trauma exposure is repetitive in nature; demonstrated progressive course; delayed symptom onset; and self-report or observer report of cognitive dysfunction, confirmed with objective cognitive decline documented by results of formal neuropsychological testing.

Supportive Features: Emotional dysregulation includes depression, anxiety, agitation, aggression, paranoid ideation, deterioration of interpersonal relationships, and suicidality; Behavioral change includes violence, poor impulse control, socially inappropriate behavior, avolition, apathy, change in personality, and comorbid substance abuse; Motor disturbance includes bradykinesia, tremor, rigidity, gait instability, dysarthria, dysphagia, ataxia, and gaze disturbance.

The patient should be observed for these features with concomitant inquiry of close family and friends. These criteria can help the clinician recognize this syndrome, and they are a good initial set to look for the diagnosis of this very challenging condition.

Conclusions

In summary, concussion remains a clinical diagnosis that is under-recognized and under-reported in sport and other populations. Although several concussion recognition tools exist, none have been validated to conclusively make the diagnosis, and conventional neuroimaging is typically normal in the absence of superimposed structural pathology. It is important to appropriately screen for concussion by obtaining details of the event in question, including the forces that were involved. Early diagnostic considerations include ruling out life-threatening sequelae of brain injury with neuroimaging, and decision tools such as the Canadian CT Head Rule are available to assist with screening.

The assessment for concussion includes a detailed symptom inventory that can be conveniently organized into physical, cognitive, and emotional symptoms. There are a number of concussion symptom inventories that can be used for serial monitoring of symptom recovery. The bulk of the neurological examination is typically normal in concussion, but a comprehensive approach provides reassurance to the examiner that concurrent neurological pathology is not missed. The physical examination can also rule out treatable alternative non-concussion causes of symptoms.

Neurocognitive testing at baseline is of limited value, with the exception of certain types of at-risk people who have a history of previous concussions, attention deficit disorder, or learning disorder; it should not be used as the sole determinant for diagnosing concussion nor for determining readiness for return to play. Similarly, post-concussion neuropsychological assessments can be lengthy and expensive and are therefore most appropriate when the differential diagnosis is unclear and/or the individual has pre-existing or comorbid conditions. The concussion assessment should be comprehensive and include effort testing when compensation or litigation is involved.

References

1. McCrory P. Pre-participation assessment for head injury. Clin J Sport Med. 2004;14(3):139–44.
2. McCrory P. Pre-participation assessment for head injury. Clin J Sport Med. 2004;14(3):139–44. Appendix: Suggested Baseline Assessment Form. Retrieved from: https://journals.lww.com/cjsportsmed/Fulltext/2004/05000/Preparticipation_Assessment_for_Head_Injury.00006.aspx#A1-6. Accessed 26 Jan, 2021.
3. Kerr ZY, Register-Mihalik JK, Marshall SW, Evenson KR, Mihalik JP, Guskiewicz KM. Disclosure and non-disclosure of concussion and concussion symptoms in athletes: review and application of the socio-ecological framework. Brain Inj. 2014;28(8):1009–21. https://doi.org/10.3109/02699052.2014.904049.
4. Kerr ZY, Register-Mihalik JK, Kroshus E, Baugh CM, Marshall SW. Motivations associated with nondisclosure of self-reported concussions in former collegiate athletes. Am J Sports Med. 2016;44:220–5.
5. McLeod TC, Bay RC, Heil J, McVeigh SD. Identification of sport and recreational activity concussion history through the pre-participation screening and a symptom survey in young athletes. Clin J Sport Med. 2008;18(3):235–40.
6. Wojtowicz M, Iverson GL, Silverberg ND, Mannix R, Zafonte R, Maxwell B, Berkner PD. Consistency of self-reported concussion history in adolescent athletes. J Neurotrauma. 2017;34(2):322–7.
7. Kerr ZY, Register-Mihalik JK, Kay MC, DeFreese JD, Marshall SW, Guskiewicz KM. Concussion nondisclosure during professional career among a cohort of former National Football League athletes. Am J Sports Med. 2018;46(1):22–9.
8. Corrigan JD, Bogner J. Initial reliability and validity of the Ohio State University TBI identification method. J Head Trauma Rehabil. 2007;22(6):318–29.
9. Lynch W. Options for evaluating and tracking pediatric concussion. J Head Trauma Rehabil. 2018;33(5):354–61.
10. Nelson LD, LaRoche AA, Pfaller AY, Lerner EB, Hammeke TA, Randolph C, Barr WB, Guskiewicz K, McCrea MA. Prospective, head-to-head study of three computerized neurocognitive assessment tools (CNTs): reliability and validity for the assessment of sport-related concussion. J Int Neuropsychol Soc. 2016;22(1):24–37.
11. Iverson GL, Brooks BL, Collins MW, Lovell MR. Tracking neuropsychological recovery following concussion in sport. Brain Inj. 2006;20:245–52. https://doi.org/10.1080/02699050500487910. (PubMed: 16537266).
12. Iverson GL, Lovell MR, Collins MW. Interpreting change on ImPACT following sport concussion. Clin Neuropsychol. 2003;17:460–7. https://doi.org/10.1076/clin.17.4.460.27934. [PubMed: 15168911].

13. Gualtieri CT, Johnson LG. Reliability and validity of a computerized neurocognitive test battery, CNS vital signs. Arch Clin Neuropsychol. 2006;21(7):623–43.
14. Haran FJ, Dretsch MN, Slaboda JC, Johnson DE, Adam OR, Tsao JW. Comparison of baseline-referenced versus norm-referenced analytical approaches for in-theatre assessment of mild traumatic brain injury neurocognitive impairment. Brain Inj. 2016;30(3):280–6.
15. Schmidt JD, Register-Mihalik JK, Mihalik JP, Kerr ZY, Guskiewicz KM. Identifying impairments after concussion: normative data versus individualized baselines. Med Sci Sports Exerc. 2012;44:1621–8.
16. Louey AG, Cromer JA, Schembri AJ, Darby DG, Maruff P, Makdissi M, Mccrory P. Detecting cognitive impairment after concussion: sensitivity of change from baseline and normative data methods using the CogSport/axon cognitive test battery. Arch Clin Neuropsychol. 2014;29(5):432–41.
17. McCrory P, Meeuwisse W, Dvorak J, Aubry M, Bailes J, Broglio S, Cantu RC, Cassidy D, Echemendia RJ, Castellani RJ, Davis GA. Consensus statement on concussion in sport—the 5th international conference on concussion in sport held in Berlin, October 2016. Br J Sports Med. 2017;51(11):838–47.
18. Statement on Concussion Baseline Testing in Canada. Parachute. Accessed 25 Jan, 2021 https://parachute.ca/wp-content/uploads/2019/06/BaselineTestingStatement-Parachute.pdf.
19. Patricios J, Fuller GW, Ellenbogen R, Herring S, Kutcher JS, Loosemore M, Makdissi M, McCrea M, Putukian M, Schneider KJ. What are the critical elements of sideline screening that can be used to establish the diagnosis of concussion? A systematic review. Br J Sports Med. 2017;51(11):888–94.
20. Maddocks DL, Dicker GD, Saling MM. The assessment of orientation following concussion in athletes. Clin J Sport Med. 1995;5:32–5.
21. Barr WB, McCrea M. Sensitivity and specificity of standardized neurocognitive testing immediately following sports concussion. J Int Neuropsychol Soc. 2001;7:693–702.
22. Echemendia RJ, Meeuwisse W, McCrory P, Davis GA, Putukian M, Leddy J, Makdissi M, Sullivan SJ, Broglio SP, Raftery M, Schneider K. The concussion recognition tool 5th edition (CRT5): background and rationale. Br J Sports Med. 2017;51(11):870–1.
23. Bruce JM, Echemendia RJ, Meeuwisse W, et al. Development of a risk prediction model among professional hockey players with visible signs of concussion. Br J Sports Med. 2018;52(17):1143–8.
24. Echemendia RJ, Bruce JM, Meeuwisse W, Hutchison MG, Comper P, Aubry M. Can visible signs predict concussion diagnosis in the National Hockey League? Br J Sports Med. 2018;52(17):1149–54.
25. Davis GA, Makdissi M, Bloomfield P, et al. International study of video review of concussion in professional sports. Br J Sports Med. 2019;53:1299–304.
26. McCrory PR, Berkovic SF. Video analysis of acute motor and convulsive manifestations in sport-related concussion. Neurology. 2000;54(7):1488–91.
27. Davis GA, Makdissi M, Bloomfield P, Clifton P, Echemendia RJ, Falvey ÉC, Fuller GW, Green G, Harcourt P, Hill T, McGuirk N. International consensus definitions of video signs of concussion in professional sports. Br J Sports Med. 2019;53(20):1264–7.
28. Gennarelli TA, Thibault LE, Adams JH, Graham DI, Thompson CJ, Marcincin RP. Diffuse axonal injury and traumatic coma in the primate. Annals of Neurology: Official Journal of the American Neurological Association and the Child Neurology Society. 1982;(6):564–74.
29. Luethcke CA, Bryan CJ, Morrow CE, Isler WC. Comparison of concussive symptoms, cognitive performance, and psychological symptoms between acute blast versus non-blast-induced mild traumatic brain injury. J Int Neuropsychol Soc. 2011;17(1):36–45.
30. Kontos AP, Elbin RJ, Kotwal RS, Lutz RH, Kane S, Benson PJ, et al. The effects of combat-related mild traumatic brain injury (mTBI): does blast mTBI history matter? J Trauma Acute Care Surg. 2015;79(4 Suppl 2):S146–51.
31. Dretsch MN, Kelly MP, Coldren RL, Parish RV, Russell ML. No significant acute and subacute differences between blast and blunt concussions across multiple neurocognitive measures and symptoms in deployed soldiers. J Neurotrauma. 2015;32(16):1217–22.

32. Paniak C, MacDonald J, Toller-Lobe G, Durand A, Nagy J. A preliminary normative profile of mild traumatic brain injury diagnostic criteria. J Clin Exp Neuropsychol. 1998;20(6):852–5.
33. Luoto TM, Iverson GL, Losoi H, Wäljas M, Tenovuo O, Kataja A, Brander A, Öhman J. Clinical correlates of retrograde amnesia in mild traumatic brain injury. Brain Inj. 2015;29(5):565–72.
34. Meares S, Shores EA, Smyth T, Batchelor J, Murphy M, Vukasovic M. Identifying posttraumatic amnesia in individuals with a Glasgow coma scale of 15 after mild traumatic brain injury. Arch Phys Med Rehabil. 2015;96(5):956–9.
35. Maddocks DL, Dicker GD, Saling MM. The assessment of orientation following concussion in athletes. Clinical Journal of Sport Medicine: Official Journal of the Canadian Academy of Sport Medicine. 1995;5(1):32–5.
36. Levin HS, O'Donnell VM, Grossman RG. The Galveston orientation and amnesia test: A practical scale to assess cognition after head injury. J Nerv Ment Dis. 1979;167(11):675–84. https://doi.org/10.1097/00005053-197911000-00004.
37. Lovell MR, Iverson GL, Collins MW, McKeag D, Maroon JC. Does loss of consciousness predict neuropsychological decrements after concussion? Clinical Journal of Sport Medicine: Official Journal of the Canadian Academy of Sport Medicine. 1999;9(4):193–8.
38. McDonald SJ, Sun M, Agoston DV, Shultz SR. The effect of concomitant peripheral injury on traumatic brain injury pathobiology and outcome. J Neuroinflammation. 2016;13(1):1–4.
39. Stiell IG, Clement CM, Rowe BH, Schull MJ, Brison R, Cass D, Eisenhauer MA, McKnight RD, Bandiera G, Holroyd B, Lee JS. Comparison of the Canadian CT head rule and the New Orleans criteria in patients with minor head injury. JAMA. 2005;294(12):1511–8.
40. Easter JS, Bakes K, Dhaliwal J, Miller M, Caruso E, Haukoos JS. Comparison of PECARN, CATCH, and CHALICE rules for children with minor head injury: a prospective cohort study. Ann Emerg Med. 2014;64(2):145–52.
41. van der Naalt J, Timmerman ME, de Koning ME, van der Horn HJ, Scheenen ME, Jacobs B, Hageman G, Yilmaz T, Roks G, Spikman JM. Early predictors of outcome after mild traumatic brain injury (UPFRONT): an observational cohort study. The Lancet Neurol. 2017;16(7):532–40.
42. Silverberg ND, Gardner AJ, Brubacher JR, Panenka WJ, Li JJ, Iverson GL. Systematic review of multivariable prognostic models for mild traumatic brain injury. J Neurotrauma. 2015;32(8):517–26.
43. Mikolić A, Polinder S, Steyerberg EW, Retel Helmrich IR, Giacino JT, Maas AI, van der Naalt J, Voormolen DC, Von Steinbüchel N, Wilson L, Lingsma HF. Prediction of global functional outcome and post-concussive symptoms after mild traumatic brain injury: external validation of prognostic models in the collaborative European NeuroTrauma effectiveness research in traumatic brain injury (CENTER-TBI) study. J Neurotrauma. 2021;38(2):196–209.
44. Silverberg ND, Iverson GL, Panenka W. Cogniphobia in mild traumatic brain injury. J Neurotrauma. 2017;34(13):2141–6.
45. Wijenberg ML, Hicks AJ, Downing MG, van Heugten CM, Stapert SZ, Ponsford JL. Relevance of the fear-avoidance model for chronic disability after traumatic brain injury. J Neurotrauma. 2020;37(24):2639–46.
46. Snell DL, Siegert RJ, Debert C, Cairncross M, Silverberg ND. Evaluation of the fear avoidance behavior after traumatic brain injury questionnaire. J Neurotrauma. 2020;37(13):1566–73.
47. Langer LK, Alavinia SM, Lawrence DW, Munce SE, Kam A, Tam A, Ruttan L, Comper P, Bayley MT. Prediction of risk of prolonged post-concussion symptoms: derivation and validation of the TRICORDRR (Toronto Rehabilitation Institute concussion outcome determination and rehab recommendations) score. PLoS Med. 2021;18(7):e1003652.
48. Zemek R, Barrowman N, Freedman SB, Gravel J, Gagnon I, McGahern C, Aglipay M, Sangha G, Boutis K, Beer D, Craig W. Clinical risk score for persistent postconcussion symptoms among children with acute concussion in the ED. JAMA. 2016;315(10):1014–25.
49. King NS, Crawford S, Wenden FJ, Moss NE, Wade DT. The Rivermead post concussion symptoms questionnaire: A measure of symptoms commonly experienced after head injury and its reliability. J Neurol. 1995;242:587–92.

50. Langer L, Comper P, Ruttan L, Inness EL, Kam A, Lawrence D, Tam A, Chandra T, Foster E, Bayley MT. Can Sport Concussion Assessment Tool (SCAT) symptom scores be converted to Rivermead Post-Concussion Questionnaire (RPQ) scores and vice versa? Findings from the Toronto Concussion Study. Frontiers in Sports and Active Living [In press, 2022].
51. Hall T, Briffa K, Hopper D. Clinical evaluation of cervicogenic headache: a clinical perspective. J Man Manip Ther. 2008;16(2):73–80. https://doi.org/10.1179/106698108790818422.
52. Mosti C, Spiers MV, Kloss JD. A practical guide to evaluating sleep disturbance in concussion patients. Neurol Clin Pract. 2016;6(2):129–37. https://doi.org/10.1212/CPJ.0000000000000225.
53. Broglio SP, Katz BP, Zhao S, McCrea M, McAllister T, CARE Consortium Investigators. Test-Retest Reliability and Interpretation of Common Concussion Assessment Tools: Findings from the NCAA-DoD CARE Consortium. Sports Med. 2018 May;48(5):1255–1268. https://doi.org/10.1007/s40279-017-0813-0. Erratum in: Sports Med. 2018 Mar 26;: PMID: 29138991; PMCID: PMC5889766.
54. Craton N, Ali H, Lenoski S. COACH CV: The seven clinical phenotypes of concussion. Brain Sci. 2017;7(9):119. Published 2017 Sep 16. https://doi.org/10.3390/brainsci7090119
55. Matuszak JM, McVige J, McPherson J, Willer B, Leddy J. A practical concussion physical examination toolbox: evidence-based physical examination for concussion. Sports Health. 2016;8:260–9.
56. Quatman-Yates CC, Lee A, Hugentobler JA, Kurowski BG, Myer GD, Riley MA. Test-retest consistency of a postural sway assessment protocol for adolescent athletes measured with a force plate. Int J Sports Phys Ther. 2013;8(6):741–8.
57. Sandel N, Reynolds E, Cohen PE, Gillie BL, Kontos AP. Anxiety and mood clinical profile following sport-related concussion: from risk factors to treatment. Sport Exerc Perform Psychol. 2017;6(3):304–23. https://doi.org/10.1037/spy0000098.
58. Guideline for Concussion/Mild Traumatic Brain Injury & Prolonged Symptoms, 3rd Edition, for Adults Over 18 Years of Age: Chapter 8 Mental Health Disorders https://braininjuryguidelines.org/concussion/. Ontario Neurotrauma Foundation. Accessed 8 Aug, 2021.
59. Ashman TA, Schwartz ME, Cantor JB, Hibbard MR, Gordon WA. Screening for substance abuse in individuals with traumatic brain injury. Brain Inj. 2004;18(2):191–202. https://doi.org/10.1080/0269905031000149506.
60. Ewing JA. Detecting alcoholism: the CAGE questionnaire. J Am Med Assoc. 1984;252:1905–7.
61. Covassin T, Elbin RJ. The cognitive effects and decrements following concussion. Open access. J Sports Med. 2010;1:55–61. https://doi.org/10.2147/oajsm.s6919, Published 2010 May 12.
62. Echemendia RJ, Iverson GL, McCrea M, Macciocchi SN, Gioia GA, Putukian M, Comper P. Advances in neuropsychological assessment of sport-related concussion. Br J Sports Med. 2013;47(5):294–8.
63. Beck AT, Steer RA, Ball R, Ranieri W. Comparison of Beck depression inventories -IA and -II in psychiatric outpatients. J Person Asses. 1996;67:588–97. https://doi.org/10.1207/s15327752jpa670313.
64. Iverson GL, Schatz P. Advanced topics in neuropsychological assessment following sport-related concussion. Brain Inj. 2015;29(2):263–75.
65. Nelson NW, Hoelzle JB, McGuire KA, Ferrier-Auerbach AG, Charlesworth MJ, Sponheim SR. Evaluation context impacts neuropsychological performance of OEF/OIF veterans with reported combat-related concussion. Arch Clin Neuropsychol. 2010;25(8):713–23.
66. Sharland MJ, Gfeller JD. A survey of neuropsychologists' beliefs and practices with respect to the assessment of effort. Arch Clin Neuropsychol. 2007;22(2):213–23.
67. Bush SS, Ruff RM, Tröster AI, Barth JT, Koffler SP, Pliskin NH, Reynolds CR, Silver CH. Symptom validity assessment: practice issues and medical necessity NAN policy & planning committee. Arch Clin Neuropsychol. 2005;20(4):419–26.
68. Heilbronner RL, Sweet JJ, Morgan JE, Larrabee GJ, Millis SR. And conference participants 1. American Academy of clinical neuropsychology consensus conference statement on the neuropsychological assessment of effort, response bias, and malingering. Clin Neuropsychol. 2009;23(7):1093–129.

69. Tombaugh TN. The test of memory malingering (TOMM): normative data from cognitively intact and cognitively impaired individuals. Psychol Assess. 1997;9(3):260.
70. Lange RT, Iverson GL, Brooks BL, Ashton Rennison VL. Influence of poor effort on self-reported symptoms and neurocognitive test performance following mild traumatic brain injury. J Clin Exp Neuropsychol. 2010;32(9):961–72.
71. McKee AC, Cantu RC, Nowinski CJ, et al. Chronic traumatic encephalopathy in athletes. J Neuropathol Exp Neurol. 2009;68:709–35.
72. Montenigro PH, Baugh CM, Daneshvar DH, Mez J, Budson AE, Au R, et al. Clinical subtypes of chronic traumatic encephalopathy: literature review and proposed research diagnostic criteria for traumatic encephalopathy syndrome. Alzheimers Res Ther. 2014;6:68. https://doi.org/10.1186/s13195-014-0068-z.
73. Reams N, Eckner JT, Almeida AA, Aagesen AL, Giordani B, Paulson H, Lorincz MT, Kutcher JS. A clinical approach to the diagnosis of traumatic encephalopathy syndrome: a review. JAMA Neurol. 2016;73(6):743–9.

Chapter 7
Psychological and Mental Health Sequelae of Concussion: Prevalence, Treatment Recommendations, Novel Biomarkers, and Diagnostic Challenges

Amanda K. Ceniti, Sakina J. Rizvi, and Sidney H. Kennedy

Mental Health Following Concussion

The physical and cognitive symptoms that may persist following a concussion are often compounded by psychiatric symptoms which are unfortunately common post-injury [1]. Several mental health concerns appear to have an increased prevalence following concussion: these include major depressive disorder (MDD), suicidality, anxiety disorders, and post-traumatic stress disorder (PTSD). Their increased prevalence post-concussion may be due to a combination of biological and psychosocial factors (Fig. 7.1). Biological factors include the impact of trauma on brain areas involved in the pathophysiology of mental illnesses, including disruption to brain networks and neurotransmitter systems. For example, disruptions to the prefrontal cortex are common in concussion and may be particularly relevant to the development of psychiatric disorders, as these brain areas are important for functioning that is impaired across mental illnesses including processing rewarding stimuli [2], decision-making, and cognition [3]. Psychological factors may also contribute, including the potential experience of trauma at the time of injury, as well as adaptation to changing roles or limitations due to persisting physical or cognitive deficits associated with the concussion; these role changes may further result in a loss of identity (e.g., in sport or at work). Finally, social factors may also play a role, including situational changes, isolation, and legal or interpersonal issues relating to the circumstances of the injury [4, 5].

This chapter will discuss the prevalence of and risk factors for various psychiatric disorders following concussion, highlight emerging biomarker research, summarize key considerations for interactions between these conditions, and outline evidence-based treatment recommendations available for depression and PTSD. The

A. K. Ceniti · S. J. Rizvi · S. H. Kennedy (✉)
Arthur Sommer Rotenberg Suicide & Depression Studies Program, St. Michael's Hospital, Toronto, ON, Canada
e-mail: Sidney.Kennedy@uhn.ca

T. A. Schweizer, A. J. Baker (eds.), *Tackling the Concussion Epidemic*,
https://doi.org/10.1007/978-3-030-93813-0_7

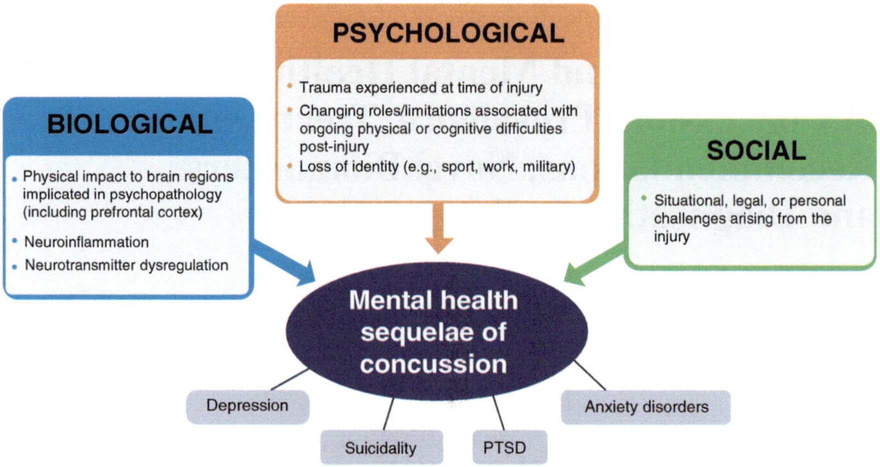

Fig. 7.1 Biological, psychological, and social factors that may contribute to the development of mental health sequelae of concussion

terms "concussion" and "mild traumatic brain injury (mTBI)" are frequently used interchangeably in the literature; when referencing external literature in this chapter, we have applied the terminology used by the original authors. We will focus on adult populations only; while the literature base has also identified an increased prevalence of mental health conditions post-concussion in children and adolescents [6], this is beyond the scope of this chapter.

Prevalence and Clinical Correlates

There is substantial evidence that the prevalence of psychiatric disorders increases following concussion; this may reflect recurrence of pre-injury disorders or emergent conditions [7, 8]. Conditions with the strongest evidence include mood and anxiety disorders, as well as PTSD, with limited evidence for increased risk of eating disorders, somatoform disorder, or psychotic disorders post-concussion [7, 8].

Depression

The prevalence of MDD following concussion ranges from 20 to 45% [9–12], which is much greater than the 12-month and lifetime rates seen in the general population (5% and 11%, respectively) [13]. The time course of MDD following injury is variable, occurring most commonly over the first 6–12 months, but with elevated risk persisting for decades following injury [14–16]. Some have proposed that the increased prevalence immediately following injury is more linked to biological changes such as the neurochemical cascade that occurs post-injury and has greater ties to lesion location, whereas MDD that develops months to years following injury may be more attributed to psychological and social factors [17]. The presence of depression post-concussion has been associated with decreased quality of life, cognitive impairments, increased post-concussive symptoms, and decreased work productivity [18–20].

Suicidality

In addition to the increased risk of depression following concussion, suicide risk also increases following brain injury. Though it is often associated with psychiatric disorders, there is an impetus to consider suicidality as an important independent factor, as it is a cross-cutting entity across multiple illnesses, has been associated with specific neurobiological and genetic factors, and may require targeted treatment [21, 22]. Furthermore, increased suicide risk post-concussion has been observed independently of psychiatric history [23].

Though the link between head injuries and suicide in relation to military, veteran, and athlete populations has attracted the most public attention, recent reports suggest that this elevated risk is present in community samples as well. A longitudinal cohort analysis of over 200,000 individuals who had sustained a concussion in Ontario reported a threefold increase in death by suicide over the long term as compared to the population norm [23]. This elevated risk was independent of psychiatric history or demographic variables. Similarly, a systematic review and meta-analysis of 17 studies identified an association between concussion and suicide risk in both military and non-military populations; a history of concussion was found to elevate risk across the spectrum of suicidality, ranging from suicidal ideation to suicide attempt and death by suicide [24]. Another longitudinal study reported that suicidal ideation was commonly endorsed at 3 months post-injury and was present at even higher rates at the 6-month follow-up [25]. Factors associated with ideation in that sample included depression, marital status, and mechanism of injury. In a separate study, severity of injury was found to play a role in suicide risk, as severe TBI was associated with higher suicide risk than mild TBI and concussion [26].

Increased risk of death by suicide has also been identified in military personnel and veterans with a history of TBI [27]. Preliminary findings among military personnel with mTBI suggest that increased anger and depression symptoms may be mediators of the association between head injury and suicide risk, highlighting these factors as potential therapeutic targets for reducing suicidality in this population [28]. Another study of military personnel reported that the number of previous TBIs increased suicide risk, after controlling for the effects of PTSD, depression, and TBI symptom severity [29]. The generalizability of these findings in military and veteran samples is limited by their predominantly male composition (90–93%), particularly given the known sex and gender differences in suicide risk, with higher rates of death by suicide in men [30], although ideation and attempts are more prevalent in women [31].

The relationship between concussion and suicide risk in professional athletes is inconclusive [32, 33]. These conflicting findings may exist because suicide risk is complex, with numerous risk and protective factors; mixed findings may be due to the relative contributions of these factors. For instance, connectedness and belongingness are largely thought to be protective factors for suicidality [34, 35], so the presence of these factors among the professional athletes in this study may have served to counteract the increased risk afforded by the injury.

Anxiety Disorders

There is limited evidence for increased risk of anxiety disorders following concussion, though results do suggest a subtle increase in risk. One report found a preliminary association between concussion history and anxiety symptoms among college students [36]. In a separate study of collegiate athletes, post-concussion depression and anxiety were predicted by pre-injury depression, emphasizing the importance of baseline mental health screening among athletes [37].

PTSD

PTSD following concussion has been studied primarily in military and veteran populations who have sustained a TBI. In a large longitudinal study involving more than 4000 US army personnel, there was an increased risk of having PTSD among those who had sustained a TBI: this relationship was observed both 3- and 9 months following their return to the US and remained after adjusting for pre-deployment psychiatric history [38]. Injury-related factors may play a role in the clinical symptom presentation of PTSD. For instance, loss of consciousness (LOC) has been associated with greater avoidance symptoms and poorer quality of life measures among veterans with PTSD [39]; LOC was also associated with an increased prevalence of PTSD following mTBI in both servicemember and civilian samples [40, 41]. In a separate study, servicemembers who experienced combat-related mTBI had a significantly higher prevalence of PTSD than those with non-combat mTBI; this association with injury type was not present with depression [42].

Risk for PTSD has also been shown to increase in non-military populations: for example, in one report, PTSD was identified 6 months after mTBI in 27% of individuals, and was associated with post-concussive symptoms, decreased life satisfaction, functional disability, and cognitive impairments; in this report, pre-injury psychiatric history was identified as a significant predictor [43]. Athletes who sustained a concussion were also found to report more PTSD symptoms than non-concussed athletes, despite not reporting these symptoms pre-injury [44].

The presence of PTSD post-concussion also appears to be associated with increased cognitive impairment [45], greater pain intensity [46], poorer psychosocial functioning [47], maladaptive coping [46], and persistent post-concussive symptoms [48].

Treatment Recommendations

Treatment of the psychiatric symptoms occurring post-concussion is an important component of long-term management. Though there is substantial variability across individuals, TBI can result in damage to several neurotransmitter networks, including serotonergic, dopaminergic, cholinergic, and noradrenergic systems [5], which are key in the pathophysiology of psychiatric disorders. Consequently, pharmacological treatment post-concussion may work by addressing these imbalances. Treatment of the psychiatric symptoms post-injury may also have the secondary benefit of improving other co-occurring symptoms. For instance, depression severity post-injury has been associated with burden of post-concussive symptoms and cognitive impairment [49], and successful depression treatment following mTBI has also been linked to improvement in cognitive function [50, 51]. Since sleep disturbances have been shown to impede functional recovery following mTBI [52], improvement of sleep by way of successful depression treatment may have a broader positive impact on recovery. Concurrent treatment of other post-concussive symptoms such as headache may also have the secondary impact of improving mood [53].

Unfortunately, whether due to the difficulty in identifying psychiatric comorbidities following concussion and disentangling them from post-concussive symptoms, or due to the stigma surrounding seeking help, treatment rates for psychiatric symptoms following concussion are low. One study reported that in the first year following traumatic brain injury (TBI), only 44% of individuals who developed MDD received pharmacological or psychological treatment [54]. Recognizing the early symptoms of psychiatric conditions post-injury is important for treatment, especially given that resolving psychiatric symptoms may be associated with resolution of other co-occurring symptoms. The literature on treatment recommendations for depression and PTSD is summarized below.

Depression

Evidence for effective antidepressant treatment post-concussion is limited, and published findings are mixed. The current literature base is heterogeneous, spanning a wide range of post-injury time periods from the acute post-concussion period to years post-injury, and often including TBIs of mixed severity [55]. In general, the majority of meta-analyses suggest that pharmacotherapy may be effective in reducing depressive symptoms following injury, particularly medications targeting the serotonin system [56, 57], though other reviews have found more limited benefit of antidepressants following TBI [58]. The Ontario Neurotrauma Foundation (ONF) recommends selective serotonin reuptake inhibitors (SSRIs) as first-line pharmacological treatments for depression following concussion [4, 59], with sertraline [51, 60, 61] and citalopram [62] specifically showing effectiveness in treating depression post-injury [14]. Tricyclic antidepressants (e.g., amitriptyline) have also been identified as a potentially suitable choice following concussion [4, 59], but their side effect profile, especially in combination with the physical and cognitive symptoms that often persist post-injury, makes them a less favorable option and a second-line treatment [4]. Preventative measures have also been investigated; a double-blind randomized trial of sertraline vs. placebo following TBI found sertraline to be effective in preventing onset of mood disorders at 24 weeks post-injury, with no deleterious neuropsychological effects [63].

In selecting an appropriate treatment, it is important to recognize that the clinical profile of MDD following concussion may differ from that of primary MDD with no injury-related comorbidity. A preliminary report found that, compared to primary MDD, post-TBI MDD was associated with more severe cognitive impairments, greater levels of hostility and social isolation, and less severe depressive symptoms [64]. This may be a potential mechanism behind the particular efficacy of SSRIs post-concussion, as low levels of serotonin have been associated with aggression, though this relationship is complex [65]. However, at this time there is no strong evidence that post-concussion MDD requires substantially different treatment from MDD in other contexts [4]. Indeed, some have argued that those with mTBI and comorbid MDD may be more similar to individuals in primary MDD trials than to those in placebo-controlled trials of TBI of mixed severity [66]. As with all antidepressant treatment, individual patient factors should be considered when selecting an appropriate intervention; these considerations may be especially relevant post-concussion, given that additional injury-related symptoms may also be present (e.g., cognitive, physical, sleep problems). In general, there is evidence that individuals with persisting post-concussive symptoms may be more sensitive to pharmacological side effects [67, 68]. Paroxetine has been suggested to exacerbate cognitive complaints, so it may be used with caution in individuals with cognitive impairments post-injury [14]. In addition, there are specific treatment interventions that may not be appropriate in this population, such as bupropion, because of a potential increase in seizure risk [69, 70]. Preliminary work has also demonstrated elevated seizure risk post-TBI among those taking SSRIs, though it is unclear whether these findings can be attributed to the medications themselves or the underlying psychiatric conditions for which they were prescribed [71].

Regarding non-pharmacological interventions, psychotherapy has been recommended as a standalone or adjunctive treatment post-concussion [4]. Cognitive behavioral therapy (CBT) has been the most frequently studied psychological treatment [57], and there is preliminary evidence to suggest that in-person [72], online

[73], and telephone-administered [74] forms of CBT may each be effective for depression following TBI. Some modifications to a CBT program may be needed to account for cognitive difficulties [4]. There is currently insufficient evidence to evaluate the effectiveness of other psychological treatments for depression post-concussion. Neurostimulation treatments have also been investigated for both depression and PTSD post-concussion, most notably repetitive transcranial magnetic stimulation (rTMS) to the dorsolateral prefrontal cortex (dlPFC) [75]. There is currently insufficient evidence to recommend rTMS as an effective treatment for depression or PTSD post-concussion, but early reports suggest no adverse safety outcomes related to the presence of concussion/TBI [75]. Further study is needed to determine whether existing rTMS treatment protocols can be used effectively or if tailoring is required for the complex symptom profiles of a post-concussion population [75].

PTSD

First-line psychological treatment for PTSD and other anxiety disorders includes CBT, trauma-focused therapy, and psychoeducation, which are also common recommendations for primary PTSD [4, 72, 76, 77]. Early investigation of mindfulness-based stress reduction in veterans with co-occurring PTSD and mTBI has also been promising, with symptom reduction shown both immediately and after 3 months and additional improvements shown to cognition [78]. Regarding pharmacological treatments, first-line options for PTSD following concussion are SSRIs, and the SNRI venlafaxine has been identified as a second-line treatment [4]. More recently, hyperbaric oxygen therapy has also been investigated as a potential treatment for post-TBI PTSD and post-concussive syndrome (PCS) [79–81]. Although eye movement desensitization and reprocessing (EMDR) has been evaluated extensively in PTSD [82, 83], its evaluation in a post-concussion population is limited [84]. Further, as with depression, preventative treatments may be effective. One report found that early CBT for individuals with acute stress disorder may be effective in preventing progression to PTSD following mTBI [85].

Toward Biomarkers of Psychiatric Disorders Post-Concussion

Beyond identifying changes that co-occur alongside psychiatric disorders following concussion, there is increasing interest in identifying biomarkers that may predict these psychiatric sequelae *before* their emergence. Such identification is of great clinical relevance, as it could lead to targeted follow-up and prevention measures among individuals identified at highest risk. Preventative treatment has been effective, as described above using sertraline for MDD [63] and CBT for PTSD [85], and taking a precision medicine approach to directing these prophylactic treatments toward those at highest risk could help to improve care. An emerging literature base has identified potential biomarkers across several levels of analysis, from genetics and molecular processes (e.g., inflammation) to neuroimaging. As with primary psychiatric disorders, there is unlikely to be one single marker with adequate sensitivity and specificity, and multimodal strategies incorporating indicators at multiple levels of analysis may be more useful [86].

Genetics

Several single nucleotide polymorphisms (SNPs) and other genetic variants have been associated with neuropsychiatric symptoms following concussion, including apolipoprotein E (APOE), catechol-O-methyltransferase (COMT), insulin-like growth factor 1 (IGF-1), and brain-derived neurotrophic factor (BDNF). The latter is implicated in neuroplasticity and in risk for neuropsychiatric disorders more broadly and has received particular interest as a potential genetic marker of psychiatric sequelae following concussion. One report found that individuals carrying the T allele of the BDNF Val66Met polymorphism had higher anxiety and depression scores in the first week following mTBI [87]. This is consistent with a subsequent study which identified a moderating effect of BDNF genotype on the associations between mTBI frequency and "brooding," and between mTBI frequency and cognitive flexibility [88]. Another BDNF SNP (rs1157659) was associated with greater hippocampal volume among those with mTBI [89]. Given that reduced hippocampal volume has been associated with both MDD [90] and PTSD [91], this may provide insight into one potential mechanism of psychopathology post-concussion.

Another promising genetic candidate is the APOE gene, involved in repair following brain injury, and specifically its APOE-ε4 allele which has been extensively associated with risk for age-related diseases including Alzheimer's disease [92, 93]. The APOE-ε4 allele has been associated with higher levels of depression, anxiety, and PTSD symptoms among veterans with a history of mild to moderate TBI [94], in addition to poorer neuropsychological performance [95] and increased post-concussive symptoms [96] among veterans with a remote history of mTBI. However, in a separate study, there was no negative psychiatric or neuropsychological impact of the APOE-ε4 allele 6 months post-mTBI [97].

Other emerging candidates include the COMT gene, involved in inactivating the catecholamines dopamine and norepinephrine [98], and IGF-1, a hormone involved in neuroprotection following TBI. A report from the TRACK-TBI group [99] found that the COMT Val158Met polymorphism, linked to greater bioavailability of catecholamines [98], was associated with lower PTSD incidence 6 months post-mTBI, while a separate report found that specific variants of IGF-1 were associated with the presence of multiple neuropsychiatric symptoms in acute mTBI [100].

Inflammatory Markers

Inflammation is a plausible pathogenic mechanism by which psychiatric symptoms could emerge following a concussion [101], as inflammation is a well-documented neural response following brain injury [102, 103] and has also been widely associated with psychiatric disorders [104, 105]. Indeed, there is an emerging literature base on inflammatory associations with neuropsychiatric symptoms post-concussion. Among military personnel with mTBI, greater PTSD symptoms have been associated with increased levels of exosomal interleukin-10 (IL-10) [106] and tumor necrosis factor-α (TNF-α) [107]. Another report found increased IL-6 and TNF-α, but not IL-10, among military personnel with high levels of PTSD post-TBI [108]. Baseline inflammatory markers may be of particular clinical relevance in early prediction of poor psychiatric outcomes: one report found that elevated C-reactive protein (CRP) at baseline was associated with increased incidence of persistent psychological problems 3 months post-mTBI [109]. However, this inflammation may also persist over the long term and correlate with ongoing psychiatric sequelae: a separate report found that increased plasma IL-10 levels at 6 months were associated with higher PTSD and depression scores at the same timepoint [110].

Neuroimaging

Significant overlap exists between the regions that are most vulnerable in concussion and those implicated in MDD and PTSD, suggesting a common neurobiology. While concussion may affect any area of the brain, certain areas are known to be more susceptible to damage, particularly frontotemporal regions such as the orbitofrontal cortex (OFC), dlPFC, and amygdala, in part due to their proximity to the bony protuberances of the skull base [14, 111]. These more vulnerable regions are also highly implicated in the neurobiology of MDD, PTSD, and other psychiatric disorders [112, 113]. For instance, the amygdala is involved in processing emotional valence [114], and the OFC is important for reward valuation and inhibitory control over stress reactivity [115]; both are key regions of the reward system, which is commonly dysregulated in MDD [116] and are also critical regions implicated in the hypervigilance and threat discrimination in PTSD [117].

Consistent with the idea of overlapping deficits, structural and functional neuroimaging studies of comorbid concussion and depression report worse outcomes in combined concussion-depression samples as compared to those with concussion alone. Assessment time post-injury is heterogeneous across these studies, ranging from 20 days to 10 years, and with variation both within and across studies. The majority of the studies relate to chronic depression effects, with participants assessed over 1-year post-injury. Structural neuroimaging studies of comorbid mTBI and depression using diffusion tensor imaging (DTI) indicate decreased fractional anisotropy (FA) along the superior longitudinal fasciculus and corpus callosum as compared to those with mTBI alone [118–120], and functional neuroimaging measures during cognitive and emotional face matching tasks showed reduced activity in key regions of depression pathophysiology, including the dlPFC and striatum, as compared to individuals with mTBI alone [121, 122]. The observed decreases in functional connectivity could reflect more severe axonal injury among those who develop MDD following injury, while the decreased task-related activity in the dlPFC and striatum is consistent with studies among individuals with MDD [123] and may suggest decreased cognitive control and reward function among those with MDD post-concussion. Similarly, in the PTSD literature, hyperconnectivity between the hippocampus and striatum has been identified as a potential neuroimaging biomarker for comorbid mTBI and PTSD, suggesting a potential brain mechanism by which it may be difficult for these individuals to disengage from traumatic memories [124]. A separate report found a negative correlation between resting-state functional connectivity of the default mode network (DMN) and severity of PTSD symptoms [125].

Considerations for Biomarker Research

Many of the markers associated with psychiatric disorders post-concussion have also been identified as risk markers for these conditions among those without brain injury; it is unclear whether there are unique factors associated with risk post-concussion that go beyond risk for these conditions in general. For example, the BDNF Val66Met polymorphism [126] and the APOE-ε4 allele [127, 128] have also been extensively studied in relation to psychiatric disorders in non-TBI populations, though some evidence has been mixed [129]. One way to address this limitation is through further studies comparing individuals with primary psychiatric diagnoses (e.g., MDD without concussion history) to those who have both MDD and concussion in order to investigate potential differences in presentation and underlying neurobiology; some investigators have already adopted this design [64, 119] and additional studies are underway [130].

Finally, it is important to consider the feasibility and clinical utility of these measures; for instance, blood markers may be more feasible and cost-effective to integrate into clinical practice than functional neuroimaging scans which present issues of both cost and access. Even for the more clinically feasible and cost-effective measures, however, it is critical to consider individuals' willingness and attitudes toward participation: a telephone survey of veterans found that approximately two-thirds of respondents were not interested in hypothetical genetic testing for PTSD (62%) or addiction (69%) [131].

Overlap Between PCS and Psychological Sequelae

In discussing psychological sequelae of concussion, it is important to consider the potential difficulty in teasing apart post-concussive symptoms from the symptoms of a psychiatric disorder occurring post-concussion. For instance, sleep disturbances, concentration difficulties, and loss of energy or fatigue are not only considered post-concussive symptoms but are also among the symptoms required to fulfill a diagnosis of MDD [132]. Similarly, in a study of post-concussion syndrome (PCS) and PTSD using correspondence analyses, the symptoms reported 3 months following mTBI which would traditionally be conceptualized as part of PCS may behave in a similar way to hyperarousal symptoms of PTSD [48]. Figure 7.2 illustrates the overlap between commonly used diagnostic criteria for MDD, PTSD, and PCS. This difficulty in distinguishing between PCS and psychiatric symptoms also raises the possibility of diagnostic misclassification in the literature. Misclassification may be more common with brain injury not involving direct contact with the head, which may not receive adequate diagnosis; for instance, individuals who have experienced undiagnosed blast-induced concussion may have their symptoms identified as PTSD, when they may in reality be experiencing persistent PCS.

Further uncertainty stems from whether these post-concussive sequelae are primarily biogenic, triggered by biological changes related to the impact (e.g.,

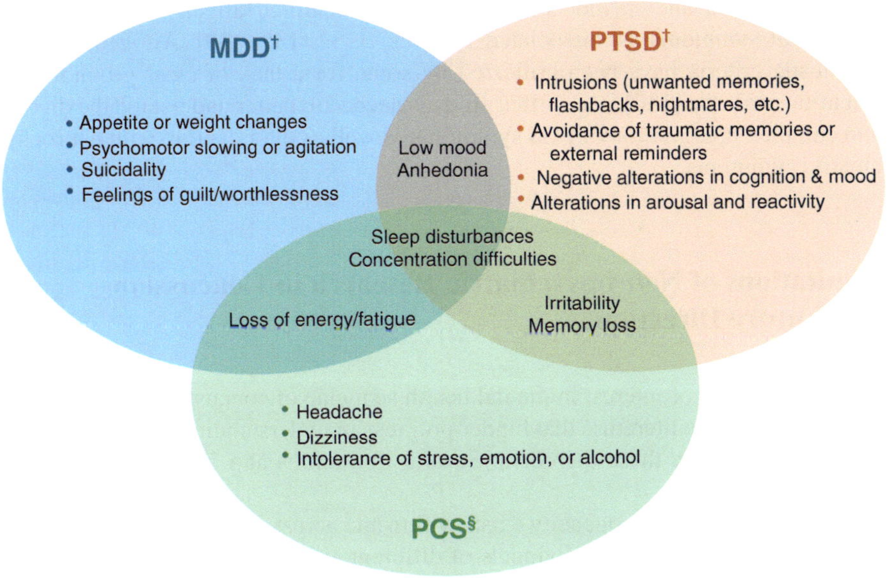

Fig. 7.2 Overlap in diagnostic criteria among MDD, PTSD, and PCS. † Based on DSM-5 criteria; § Based on ICD-10 criteria, as PCS was removed in the update from DSM-IV to DSM-5

inflammation, metabolic dysfunction), or whether they are primarily psychogenic, occurring due to psychological factors relating to the traumatic event but not specific to brain injury. Early models proposed that initial post-concussive symptoms are neurobiologically driven ("physiogenesis") and that their persistence over the long term is driven by psychological factors ("psychogenesis") [133], while later models support an early role of both neurobiological and psychological factors in the initial development of these symptoms [134].

This is complicated by the debate surrounding the definition of PCS itself, and whether it represents a distinct "syndrome" in its own right, or reflects a combination of cognitive, somatic, and psychological complaints that may present as part of other unrelated conditions [135]. This debate has been fueled by the lack of consistency in symptom presentation and clustering among individuals identified as having PCS [67, 135], and the low specificity of symptoms, given their high prevalence in non-TBI populations with other medical conditions, as well as in the general population [134, 135].

There is a lack of consensus about the diagnostic criteria for PCS between ICD-10 [136] and DSM-IV [137], where similar but distinct criteria may lead to significantly different prevalence rates: one study reported more than fivefold greater prevalence of PCS when using ICD-10 criteria as compared to DSM-IV [138]. The diagnosis of PCS has been deleted from DSM-5, and is replaced by Major or Mild "Neurocognitive Disorder due to Traumatic Brain Injury," consisting of substantially different criteria which center around objectively measured cognitive dysfunction and include other commonly identified affective, somatic, and behavioral symptoms as "associated features" [132, 135, 139]. Altogether, these diagnostic criteria have been criticized by some for a lack of clear grounding in scientific evidence [140], and further study is needed to better understand the diverse and complex presentation of PCS symptoms as well as their overlap with neuropsychiatric symptoms.

Limitations of Neuropsychiatric Research in Concussion and Future Directions

Despite the surge of interest in mental health sequelae of concussion, there are significant gaps in the literature that hinder progress in understanding the full extent of interaction between these disorders. Several limitations and future directions are highlighted below.

First, significant heterogeneity exists within and across studies. Inclusion criteria are often broad, including individuals of different age ranges, injury histories, and timeframes post-injury. Including individuals who have sustained their injuries both weeks ago and decades ago in the same group may limit comparability of the sample, though growing evidence supports the long-term impact of concussion and onset of mental health comorbidities years following injury [14, 16]. Furthermore, the majority of studies assessing psychiatric consequences of concussion are cross-sectional, limiting their ability to make causal inferences about the impact of concussion on psychiatric outcomes. For instance, the observations of increased inflammation or altered brain connectivity among those with psychiatric symptoms following concussion may reflect co-occurring processes rather than a biomarker that can predict the onset of these psychiatric symptoms. Future prospective studies assessing the same cohort longitudinally for long-term mental health outcomes following concussion would provide rich data allowing for the assessment of outcome trajectories.

Among neurobiological studies, there is also a question of whether the increased prevalence of psychiatric disorders observed following concussion reflects the underlying brain mechanisms of injury, other nonspecific physiological responses to injury, surrounding psychosocial factors relating to the injury, or a combination of multiple factors. An increasingly common practice to combat this limitation is to include comparison groups who have sustained a traumatic injury to another part of the body (e.g., orthopedic injury) to control for any secondary impact of the injury.

In order to better understand the interaction between concussion and psychiatric conditions, it would be valuable to examine how their combined impact differs from the presence of either condition alone. The majority of studies compare concussion with and without the presence of a given psychiatric disorder, whereas further studies may benefit from including groups with a psychiatric condition but no concussion history to better understand the additive impact of the two conditions.

Conclusion and Recommendations

Mental health consequences of head injuries are unfortunately common, and assessment for psychiatric disorders is a critical component of treating individuals with post-concussive complaints. Across all disorders, treatments should be selected based on individual patient needs, comorbid symptoms or diagnoses, symptom severity, and patient preference [4, 141]. In each of the disorders, there is no strong evidence to suggest that treatment post-concussion should substantially differ from current recommended treatments in the absence of concussion/TBI; however, caution should be used when prescribing medication post-concussion due to the possibility of side effects, so as not to cause or worsen physical or cognitive symptoms [4]. This means starting at the lowest effective dose and gradually increasing based on response and tolerability [4]. Regular follow-ups are also required to assess changes. In addition, since the circumstances surrounding the injury may be complex, it is important to consider other situational or psychosocial factors that may be contributing to the individual's symptom presentation (e.g., housing, finance, interpersonal stressors) and ensure that they are being addressed concurrently with psychiatric treatment [4, 142, 143]. Adopting a holistic view that encompasses biological, psychological, and social factors will help to optimize treatment outcomes in this often-complex population.

The quest for biomarkers of psychiatric conditions following concussion is an exciting but preliminary pursuit. Several candidate genetic, inflammatory, and neuroimaging markers of depression and PTSD have been identified, but future longitudinal studies in larger samples are needed to build on this promising initial work. The study of these overlapping conditions is complicated by the difficulty in distinguishing psychiatric symptoms from those thought to be occurring as part of PCS. A better understanding of PCS will in turn help to shed light on the complex interaction between psychiatric disorders and concussion.

Acknowledgments The authors thank the ASR Suicide & Depression Studies Program at St. Michael's Hospital and University of Toronto and the Canadian Biomarker Integration Network in Depression (CAN-BIND) for research support. CAN-BIND is an Integrated Discovery Program supported by the Ontario Brain Institute, which is an independent non-profit corporation, funded partially by the Ontario Government.

References

1. Kim E, Lauterbach EC, Reeve A, et al. Neuropsychiatric complications of traumatic brain injury: a critical review of the literature (A Report by the ANPA Committee on Research). J Neuropsychiatr. 2007;19:106–27. https://doi.org/10.1176/appi.neuropsych.19.2.106.
2. Der-Avakian A, Markou A. The neurobiology of anhedonia and other reward-related deficits. Trends Neurosci. 2012;35:68–77.
3. Hiser J, Koenigs M. The multifaceted role of the ventromedial prefrontal cortex in emotion, decision making, social cognition, and psychopathology. Biol Psychiatry. 2018;83:638–47.
4. Ontario Neurotrauma Foundation. Guidelines for concussion/mild traumatic brain injury & persistent symptoms. 2018. https://braininjuryguidelines.org/concussion/.
5. Polich G, Iaccarino MA, Zafonte R. Psychopharmacology of traumatic brain injury. In: Handbook of clinical neurology. Elsevier B.V; 2019. p. 253–67.
6. Emery CA, Barlow KM, Brooks BL, et al. A systematic review of psychiatric, psychological, and behavioural outcomes following mild traumatic brain injury in children and adolescents. Can J Psychiatry. 2016;61:259–69.
7. Gould KR, Ponsford JL, Johnston L, Schönberger M. The nature, frequency and course of psychiatric disorders in the first year after traumatic brain injury: a prospective study. Psychol Med. 2011;41:2099–109. https://doi.org/10.1017/S003329171100033X.
8. Whelan-Goodinson R, Ponsford J, Johnston L, Grant F. Psychiatric disorders following traumatic brain injury: their nature and frequency. J Head Trauma Rehabil. 2009;24:324–32. https://doi.org/10.1097/HTR.0b013e3181a712aa.
9. Jorge R, Robinson R, Arndt S, et al. Depression following traumatic brain injury: a 1 year longitudinal study. J Affect Disord. 1993;27:233–43.
10. Kreutzer J, Seel R, Gourley E. The prevalence and symptom rates of depression after traumatic brain injury: a comprehensive examination. Brain Inj. 2001;15:563–76.
11. Kaponen S, Taiminen T, Raija P, et al. Axis I and II psychiatric disorders after traumatic brain injury: a 30-year follow-up study. Am J Psychiatry. 2002;159:1315–21.
12. Rapoport M, Kiss A, Feinstein A. The impact of major depression on outcome following mild-to-moderate traumatic brain injury in older adults. J Affect Disord. 2006;92:273–6.
13. Pearson C, Janz T, Ali J. Mental and substance use disorders in Canada (Catalogue no.82-624-X). Stat Canada. 2013:1–8. https://doi.org/82-624-X.
14. Silver JM, McAllister TW, Arciniegas DB. Depression and cognitive complaints following mild traumatic brain injury. Am J Psychiatry. 2009;166:653–61. https://doi.org/10.1176/appi.ajp.2009.08111676.

15. Jorge RE, Robinson RG, Moser D, et al. Major depression following traumatic brain injury. Arch Gen Psychiatry. 2004;61:42–50. https://doi.org/10.1001/archpsyc.61.1.42.
16. Hellewell SC, Beaton CS, Welton T, Grieve SM. Characterizing the risk of depression following mild traumatic brain injury: a meta-analysis of the literature comparing chronic mTBI to non-mTBI populations. Front Neurol. 2020;11:1–14. https://doi.org/10.3389/fneur.2020.00350.
17. Jorge R, Robinson R, Arndt S, et al. Comparison between acute- and delayed-onset depression following traumatic brain injury. J Neuropsychiatry Clin Neurosci. 1993;5:43–9. https://doi.org/10.1176/jnp.5.1.43.
18. DeGutis J, Esterman M, McCulloch B, et al. Posttraumatic psychological symptoms are associated with reduced inhibitory control, not general executive dysfunction. J Int Neuropsychol Soc. 2015;21:342–52.
19. Rapoport MJ, McCullagh S, Shammi P, Feinstein A. Cognitive impairment associated with major depression following mild and moderate traumatic brain injury. J Neuropsychiatry Clin Neurosci. 2005;17:61–5. https://doi.org/10.1176/appi.neuropsych.17.1.61.
20. Theadom A, Barker-Collo S, Jones K, et al. Work limitations 4 years after mild traumatic brain injury: a cohort study. Arch Phys Med Rehabil. 2017;98:1560–6. https://doi.org/10.1016/j.apmr.2017.01.010.
21. Van Heeringen K, Mann JJ. The neurobiology of suicide. Lancet Psychiatry. 2014;1:63–72.
22. Mann JJ, Oquendo M, Underwood MD, Arango V. The neurobiology of suicide risk: a review for the clinician. J Clin Psychiatry. 1999;60 Suppl 2:7–11.
23. Fralick M, Thiruchelvam D, Tien HC, Redelmeier DA. Risk of suicide after a concussion. CMAJ. 2016;188(7):497–504.
24. Fralick M, Sy E, Hassan A, et al. Association of concussion with the risk of suicide: a systematic review and meta-analysis. JAMA Neurol. 2019;76:144–51. https://doi.org/10.1001/jamaneurol.2018.3487.
25. Bethune A, da Costa L, van Niftrik CHB, Feinstein A. Suicidal ideation after mild traumatic brain injury: a consecutive Canadian sample. Arch Suicide Res. 2017;21:392–402. https://doi.org/10.1080/13811118.2016.1199990.
26. Madsen T, Erlangsen A, Orlovska S, et al. Association between traumatic brain injury and risk of suicide. JAMA J Am Med Assoc. 2018;320:580–8. https://doi.org/10.1001/jama.2018.10211.
27. Brenner LA, Ignacio RV, Blow FC. Suicide and traumatic brain injury among individuals seeking veterans health administration services. J Head Trauma Rehabil. 2011;26:257–64. https://doi.org/10.1097/HTR.0b013e31821fdb6e.
28. Stanley IH, Joiner TE, Bryan CJ. Mild traumatic brain injury and suicide risk among a clinical sample of deployed military personnel: evidence for a serial mediation model of anger and depression. J Psychiatr Res. 2017;84:161–8. https://doi.org/10.1016/j.jpsychires.2016.10.004.
29. Bryan CJ, Clemans TA. Repetitive traumatic brain injury, psychological symptoms, and suicide risk in a clinical sample of deployed military personnel. JAMA Psychiat. 2013;70:686–91. https://doi.org/10.1001/jamapsychiatry.2013.1093.
30. World Health Organization. Preventing suicide: a global imperative; 2014.
31. Nock MK, Borges G, Bromet EJ, et al. Cross-national prevalence and risk factors for suicidal ideation, plans and attempts. Br J Psychiatry. 2008;192:98–105. https://doi.org/10.1055/s-0029-1237430.Imprinting.
32. Webner D, Iverson GL. Suicide in professional American football players in the past 95 years. Brain Inj. 2016;30:1718–21. https://doi.org/10.1080/02699052.2016.1202451.
33. Lehman EJ, Hein MJ, Gersic CM. Suicide mortality among retired national football league players who played 5 or more seasons. Am J Sports Med. 2016;44:2486–91. https://doi.org/10.1177/0363546516645093.
34. David Klonsky E, May AM. The three-step theory (3ST): a new theory of suicide rooted in the "ideation-to-action" framework. Int J Cogn Ther. 2015;8:114–29.
35. Van Orden KA, Witte TK, Cukrowicz KC, et al. The interpersonal theory of suicide. Psychol Rev. 2010;117:575–600. https://doi.org/10.1037/a0018697.

36. Martin RJ, Chaney BH. Exploration of the relationship between concussions and depression symptoms, anxiety symptoms, and hazardous drinking among a sample of college students. J Dual Diagn. 2018:1–8. https://doi.org/10.1080/15504263.2018.1473906. Online ahead of print.
37. Yang J, Peek-Asa C, Covassin T, Torner JC. Post-concussion symptoms of depression and anxiety in Division I collegiate athletes. Dev Neuropsychol. 2015;40:18–23. https://doi.org/10.1080/87565641.2014.973499.
38. Stein MB, Kessler RC, Heeringa SG, et al. Prospective longitudinal evaluation of the effect of deployment-acquired traumatic brain injury on posttraumatic stress and related disorders: results from the Army Study to Assess Risk and Resilience in Servicemembers (Army STARRS). Am J Psychiatry. 2015;172:1101–11. https://doi.org/10.1176/appi.ajp.2015.14121572.
39. Sofko CA, Currier JM, Hill BD, Drescher KD. History of loss of consciousness with mild traumatic brain injury affects PTSD symptom presentation in treatment-seeking Iraq/Afghanistan veterans. Brain Inj. 2016;30:1561–9. https://doi.org/10.1080/02699052.2016.1199897.
40. Roitman P, Gilad M, Ankri YLE, Shalev AY. Head injury and loss of consciousness raise the likelihood of developing and maintaining PTSD symptoms. J Trauma Stress. 2013;26:727–34. https://doi.org/10.1002/jts.21862.
41. Eskridge SL, MacEra CA, Galarneau MR, et al. Influence of combat blast-related mild traumatic brain injury acute symptoms on mental health and service discharge outcomes. J Neurotrauma. 2013;30:1391–7. https://doi.org/10.1089/neu.2012.2537.
42. Hardy M, Kennedy J, Reid M, Cooper D. Differences in posttraumatic stress disorder, depression, and attribution of symptoms in service members with combat versus noncombat mild traumatic brain injury. J Head Trauma Rehabil. 2020;35:37–45. https://doi.org/10.1097/HTR.0000000000000486.
43. Haarbauer-Krupa J, Taylor CA, Yue JK, et al. Screening for post-traumatic stress disorder in a civilian emergency department population with traumatic brain injury. J Neurotrauma. 2017;34:50–8. https://doi.org/10.1089/neu.2015.4158.
44. Brassil HE, Salvatore AP. The frequency of post-traumatic stress disorder symptoms in athletes with and without sports related concussion. Clin Transl Med. 2018 https://doi.org/10.1186/s40169-018-0200-y. [Online ahead of print].
45. Motzkin JC, Koenigs MR. Post-traumatic stress disorder and traumatic brain injury. In: Handbook of clinical neurology; 2015. p. 633–48.
46. Aase DM, Babione JM, Proescher E, et al. Impact of PTSD on post-concussive symptoms, neuropsychological functioning, and pain in post-9/11 veterans with mild traumatic brain injury. Psychiatry Res. 2018;268:460–6. https://doi.org/10.1016/j.psychres.2018.08.019.
47. Jackson CE, Green JD, Bovin MJ, et al. Mild traumatic brain injury, PTSD, and psychosocial functioning among male and female U.S. OEF/OIF veterans. J Trauma Stress. 2016;29:309–16. https://doi.org/10.1002/jts.22110.
48. Lagarde E, Salmi LR, Holm LW, et al. Association of symptoms following mild traumatic brain injury with posttraumatic stress disorder vs postconcussion syndrome. JAMA Psychiat. 2014;71:1032–40. https://doi.org/10.1001/jamapsychiatry.2014.666.
49. Terry DP, Brassil M, Iverson GL, et al. Effect of depression on cognition after mild traumatic brain injury in adults. Clin Neuropsychol. 2019;33(1):124–36. https://doi.org/10.1080/13854046.2018.1459853.
50. Fann JR, Uomoto JM, Katon WJ. Cognitive improvement with treatment of depression following mild traumatic brain injury. Psychosomatics. 2001;42:48–54. https://doi.org/10.1176/appi.psy.42.1.48.
51. Fann JR, Bombardier CH, Temkin N, et al. Sertraline for major depression during the year following traumatic brain injury: a randomized controlled trial. J Head Trauma Rehabil. 2017;32:332–42. https://doi.org/10.1097/HTR.0000000000000322.
52. Kalmbach DA, Conroy DA, Falk H, et al. Poor sleep is linked to impeded recovery from traumatic brain injury. Sleep. 2018;41(1):zsy147. https://doi.org/10.1093/sleep/zsy147.

53. Rabinowitz AR, Watanabe TK. Pharmacotherapy for treatment of cognitive and neuropsychiatric symptoms after mTBI. J Head Trauma Rehabil. 2020;35:76–83. https://doi.org/10.1097/HTR.0000000000000537.
54. Bombardier CH, Fann JR, Temkin NR, et al. Rates of major depressive disorder and clinical outcomes following traumatic brain injury. JAMA. 2010;03:1938–45. https://doi.org/10.1001/jama.2010.599.Rates.
55. Fann JR, Hart T, Schomer KG. Treatment for depression after traumatic brain injury. J Neurotrauma. 2009;26(2402):2383–402.
56. Salter KL, Andrew McClure J, Foley NC, et al. Pharmacotherapy for depression posttraumatic brain injury: a meta-analysis. J Head Trauma Rehabil. 2016;31:E21–32. https://doi.org/10.1097/HTR.0000000000000193.
57. Fann JR, Hart T, Schomer KG. Treatment for depression after traumatic brain injury: a systematic review. J Neurotrauma. 2009;26:2383–402. https://doi.org/10.1089/neu.2009.1091.
58. Kreitzer N, Ancona R, McCullumsmith C, et al. The effect of antidepressants on depression after traumatic brain injury: a meta-analysis. J Head Trauma Rehabil. 2019;34:E47–54. https://doi.org/10.1097/HTR.0000000000000439.
59. Warden DL, Gordon B, McAllister TW, et al. Guidelines for the pharmacologic treatment of neurobehavioral sequelae of traumatic brain injury. J Neurotrauma. 2006;23:1468–501. https://doi.org/10.1089/neu.2006.23.1468.
60. Fann JR, Uomoto JM, Katon WJ. Sertraline in the treatment of major depression following mild traumatic brain injury. J Neuropsychiatry Clin Neurosci. 2014;12:226–32. https://doi.org/10.1176/jnp.12.2.226.
61. Turner-Stokes L, Hassan N, Pierce K, Clegg F. Managing depression in brain injury rehabilitation: the use of an integrated care pathway and preliminary report of response to sertraline. Clin Rehabil. 2002;16:261–8. https://doi.org/10.1191/0269215502cr489oa.
62. Rapoport MJ, Chan F, Lanctot K, et al. An open-label study of citalopram for major depression following traumatic brain injury. J Psychopharmacol. 2008;22:860–4. https://doi.org/10.1177/0269881107083845.
63. Jorge RE, Acion L, Burin DI, Robinson RG. Sertraline for preventing mood disorders following traumatic brain injury: a randomized clinical trial. JAMA Psychiat. 2016;73:1041–7. https://doi.org/10.1001/jamapsychiatry.2016.2189.
64. Mauri MC, Paletta S, Colasanti A, et al. Clinical and neuropsychological correlates of major depression following post-traumatic brain injury, a prospective study. Asian J Psychiatr. 2014;12:118–24. https://doi.org/10.1016/j.ajp.2014.07.003.
65. Rosell DR, Siever LJ. The neurobiology of aggression and violence. CNS Spectr. 2015;20:254–79.
66. Silverberg ND, Panenka WJ. Antidepressants for depression after concussion and traumatic brain injury are still best practice. BMC Psychiatry. 2019;19:100. https://doi.org/10.1186/s12888-019-2076-9.
67. Arciniegas DB, Anderson CA, Topkoff J, McAllister TW. Mild traumatic brain injury: a neuropsychiatric approach to diagnosis, evaluation, and treatment. Neuropsychiatr Dis Treat. 2005;1:311–27.
68. Alderfer BS, Arciniegas DB, Silver JM. Treatment of depression following traumatic brain injury. J Head Trauma Rehabil. 2005;20:544–62.
69. Davidson J. Seizures and bupropion: a review. J Clin Psychiatry. 1989;50:256–61.
70. Jorge RE, Arciniegas DB. Neuropsychiatry of traumatic brain injury. Psychiatr Clin North Am. 2014;37:11–5.
71. Christensen J, Pedersen HS, Fenger-Grøn M, et al. Selective serotonin reuptake inhibitors and risk of epilepsy after traumatic brain injury – a population based cohort study. PLoS One. 2019;14. https://doi.org/10.1371/journal.pone.0219137.
72. Ponsford J, Lee NK, Wong D, et al. Efficacy of motivational interviewing and cognitive behavioral therapy for anxiety and depression symptoms following traumatic brain injury. Psychol Med. 2016;46:1079–90. https://doi.org/10.1017/S0033291715002640.

73. Topolovec-Vranic J, Cullen N, Michalak A, et al. Evaluation of an online cognitive behavioural therapy program by patients with traumatic brain injury and depression. Brain Inj. 2010;24:762–72. https://doi.org/10.3109/02699051003709599.
74. Fann JR, Bombardier CH, Vannoy S, et al. Telephone and in-person cognitive behavioral therapy for major depression after traumatic brain injury: a randomized controlled trial. J Neurotrauma. 2015;32:45–57. https://doi.org/10.1089/neu.2014.3423.
75. Oberman LM, Exley S, Philip NS, et al. Use of repetitive transcranial magnetic stimulation in the treatment of neuropsychiatric and neurocognitive symptoms associated with concussion in military populations. J Head Trauma Rehabil. 2020;35:388–400. https://doi.org/10.1097/HTR.0000000000000628.
76. Ragsdale KA, Voss Horrell SC. Effectiveness of prolonged exposure and cognitive processing therapy for U.S. veterans with a history of traumatic brain injury. J Trauma Stress. 2016;29:474–7. https://doi.org/10.1002/jts.22130.
77. Williams WH, Evans JJ, Wilson BA. Neurorehabilitation for two cases of post-traumatic stress disorder following traumatic brain injury. Cogn Neuropsychiatry. 2003;8:1–18. https://doi.org/10.1080/713752238.
78. Cole MA, Muir JJ, Gans JJ, et al. Simultaneous treatment of neurocognitive and psychiatric symptoms in veterans with post-traumatic stress disorder and history of mild traumatic brain injury: a pilot study of mindfulness-based stress reduction. Mil Med. 2015;180:956–63. https://doi.org/10.7205/milmed-d-14-00581.
79. Harch P, Andrews S, Fogarty E, et al. A phase I study of low-pressure hyperbaric oxygen therapy for blast-induced post-concussion syndrome and post-traumatic stress disorder. J Neurotrauma. 2012;29:168–85. https://doi.org/10.1089/neu.2012.2426.
80. Eve DJ, Steele MR, Sanberg PR, Borlongan CV. Hyperbaric oxygen therapy as a potential treatment for post-traumatic stress disorder associated with traumatic brain injury. Neuropsychiatr Dis Treat. 2016;12:2689–705. https://doi.org/10.2147/NDT.S110126.
81. Wang F, Wang Y, Sun T, Yu HL. Hyperbaric oxygen therapy for the treatment of traumatic brain injury: a meta-analysis. Neurol Sci. 2016;37:693–701. https://doi.org/10.1007/s10072-015-2460-2.
82. Novo Navarro P, Landin-Romero R, Guardiola-Wanden-Berghe R, et al. 25 years of Eye Movement Desensitization and Reprocessing (EMDR): the EMDR therapy protocol, hypotheses of its mechanism of action and a systematic review of its efficacy in the treatment of post-traumatic stress disorder. Rev Psiquiatr Salud Ment. 2018;11:101–14.
83. Wilson G, Farrell D, Barron I, et al. The use of Eye-Movement Desensitization Reprocessing (EMDR) therapy in treating post-traumatic stress disorder-a systematic narrative review. Front Psychol. 2018;9:923.
84. Gil-Jardiné C, Evrard G, Al Joboory S, et al. Emergency room intervention to prevent post concussion-like symptoms and post-traumatic stress disorder. A pilot randomized controlled study of a brief eye movement desensitization and reprocessing intervention versus reassurance or usual care. J Psychiatr Res. 2018;103:229–36. https://doi.org/10.1016/j.jpsychires.2018.05.024.
85. Bryant RA, Moulds M, Guthrie R, Nixon RDV. Treating acute stress disorder following mild traumatic brain injury. Am J Psychiatry. 2003;160:585–7. https://doi.org/10.1176/appi.ajp.160.3.585.
86. Scarr E, Millan MJ, Bahn S, et al. Biomarkers for psychiatry: the journey from fantasy to fact, a report of the 2013 CINP think tank. Int J Neuropsychopharmacol. 2015:1–9. https://doi.org/10.1093/ijnp/pyv042.
87. Wang YJ, Chen KY, Kuo LN, et al. The association between BDNF Val66Met polymorphism and emotional symptoms after mild traumatic brain injury. BMC Med Genet. 2018;19:13. https://doi.org/10.1186/s12881-017-0518-0.
88. Gabrys RL, Dixon K, Holahan MR, Anisman H. Self-reported mild traumatic brain injuries in relation to rumination and depressive symptoms: moderating role of sex differences and a brain-derived neurotrophic factor gene polymorphism. Clin J Sport Med. 2019;29:494–9. https://doi.org/10.1097/JSM.0000000000000550.

89. Hayes J, Reagan A, Logue M, et al. BDNF genotype is associated with hippocampal volume in mild traumatic brain injury. Genes Brain Behav. 2018;17:107–17.
90. Videbech P, Ravnkilde B. Hippocampal volume and depression: a meta-analysis of MRI studies. Am J Psychiatry. 2004;161:1957–66. https://doi.org/10.1176/appi.ajp.161.11.1957.
91. Logue MW, van Rooij SJH, Dennis EL, et al. Smaller hippocampal volume in posttraumatic stress disorder: a multisite ENIGMA-PGC study: subcortical volumetry results from posttraumatic stress disorder consortia. Biol Psychiatry. 2018;83:244–53. https://doi.org/10.1016/j.biopsych.2017.09.006.
92. Kim J, Basak JM, Holtzman DM. The role of apolipoprotein E in Alzheimer's disease. Neuron. 2009;63:287–303.
93. Verghese PB, Castellano JM, Holtzman DM. Apolipoprotein E in Alzheimer's disease and other neurological disorders. Lancet Neurol. 2011;10:241–52. https://doi.org/10.1016/S1474-4422(10)70325-2.
94. Merritt VC, Clark AL, Sorg SF, et al. Apolipoprotein e ϵ4 genotype is associated with elevated psychiatric distress in veterans with a history of mild to moderate traumatic brain injury. J Neurotrauma. 2018;35:2272–82. https://doi.org/10.1089/neu.2017.5372.
95. Merritt VC, Clark AL, Sorg SF, et al. Apolipoprotein E (APOE) ε4 genotype is associated with reduced neuropsychological performance in military veterans with a history of mild traumatic brain injury. J Clin Exp Neuropsychol. 2018;40:1050–61. https://doi.org/10.1080/13803395.2018.1508555.
96. Merritt VC, Lapira KM, Clark AL, et al. APOE-ϵ4 genotype is associated with elevated post-concussion symptoms in military veterans with a remote history of mild traumatic brain injury. Arch Clin Neuropsychol. 2019;34:706–12. https://doi.org/10.1093/arclin/acy082.
97. Chamelian L, Reis M, Feinstein A. Six-month recovery from mild to moderate traumatic brain injury: the role of APOE-ε4 allele. Brain. 2004;127:2621–8. https://doi.org/10.1093/brain/awh296.
98. Lachman HM, Papolos DF, Saito T, et al. Human catechol-O-methyltransferase pharmacogenetics: description of a functional polymorphism and its potential application to neuropsychiatric disorders. Pharmacogenetics. 1996;6:243–50. https://doi.org/10.1097/00008571-199606000-00007.
99. Winkler E, Yue J, Ferguson A, et al. COMT Val158Met polymorphism is associated with post-traumatic stress disorder and functional outcome following mild traumatic brain injury. J Clin Neurosci. 2017;35:109–16.
100. Wang YJ, Wong HSC, Wu CC, et al. The functional roles of IGF-1 variants in the susceptibility and clinical outcomes of mild traumatic brain injury. J Biomed Sci. 2019;26:1–12. https://doi.org/10.1186/s12929-019-0587-9.
101. Bodnar CN, Morganti JM, Bachstetter AD. Depression following a traumatic brain injury: uncovering cytokine dysregulation as a pathogenic mechanism. Neural Regen Res. 2018;13:1693–704.
102. Meier TB, Huber DL, Bohorquez-Montoya L, et al. A prospective study of acute blood-based biomarkers for sport-related concussion. Ann Neurol. 2020;87:907–20. https://doi.org/10.1002/ana.25725.
103. Shetty T, Cogsil T, Dalal A, et al. High-sensitivity C-reactive protein: retrospective study of potential blood biomarker of inflammation in acute mild traumatic brain injury. J Head Trauma Rehabil. 2019;34:E28–36. https://doi.org/10.1097/HTR.0000000000000450.
104. Dowlati Y, Herrmann N, Swardfager W, et al. A meta-analysis of cytokines in major depression. Biol Psychiatry. 2010;67:446–57. https://doi.org/10.1016/j.biopsych.2009.09.033.
105. Hori H, Kim Y. Inflammation and post-traumatic stress disorder. Psychiatry Clin Neurosci. 2019;73:143–53.
106. Gill J, Mustapic M, Diaz-Arrastia R, et al. Higher exosomal tau, amyloid-beta 42 and IL-10 are associated with mild TBIs and chronic symptoms in military personnel. Brain Inj. 2018;32:1277–84. https://doi.org/10.1080/02699052.2018.1471738.
107. Kanefsky R, Motamedi V, Mithani S, et al. Mild traumatic brain injuries with loss of consciousness are associated with increased inflammation and pain in military personnel. Psychiatry Res. 2019;279:34–9. https://doi.org/10.1016/j.psychres.2019.07.001.

108. Devoto C, Arcurio L, Fetta J, et al. Inflammation relates to chronic behavioral and neurological symptoms in military personnel with traumatic brain injuries. Cell Transplant. 2017;26:1169–77. https://doi.org/10.1177/0963689717714098.
109. Su SH, Xu W, Li M, et al. Elevated C-reactive protein levels may be a predictor of persistent unfavourable symptoms in patients with mild traumatic brain injury: a preliminary study. Brain Behav Immun. 2014;38:111–7. https://doi.org/10.1016/j.bbi.2014.01.009.
110. Vedantam A, Brennan J, Levin HS, et al. Early versus late profiles of inflammatory cytokines after mild traumatic brain injury and their association with neuropsychological outcomes. J Neurotrauma. 2021;38:53–62. https://doi.org/10.1089/neu.2019.6979.
111. Bigler ED. Anterior and middle cranial fossa in traumatic brain injury: relevant neuroanatomy and neuropathology in the study of neuropsychological outcome. Neuropsychology. 2007;21:515–31. https://doi.org/10.1037/0894-4105.21.5.515.
112. Mayberg HS. Limbic-cortical dysregulation: a proposed model of depression. J Neuropsychiatry Clin Neurosci. 1997;9:471–81.
113. Drevets WC, Price JL, Furey ML. Brain structural and functional abnormalities in mood disorders: implications for neurocircuitry models of depression. Brain Struct Funct. 2008;213:93–118.
114. McClure SM, York MK, Montague PR. The neural substrates of reward processing in humans: the modern role of fMRI. Neuroscientist. 2004;10:260–8. https://doi.org/10.1177/1073858404263526.
115. Berridge K, Kringelbach M. Affective neuroscience of pleasure: reward in humans and animals. Psychopharmacology (Berl). 2008;199:457–80.
116. Zhang W, Chang S, Guo L, et al. The neural correlates of reward-related processing in major depressive disorder: a meta-analysis of functional magnetic resonance imaging studies. J Affect Disord. 2013;151:531–9.
117. Sherin JE, Nemeroff CB. Post-traumatic stress disorder: the neurobiological impact of psychological trauma. Dialogues Clin Neurosci. 2011;13:263–78. https://doi.org/10.31887/dcns.2011.13.2/jsherin.
118. Matthews SC, Spadoni AD, Lohr JB, et al. Diffusion tensor imaging evidence of white matter disruption associated with loss versus alteration of consciousness in warfighters exposed to combat in operations enduring and Iraqi freedom. Psychiatry Res Neuroimaging. 2012;204:149–54. https://doi.org/10.1016/j.pscychresns.2012.04.018.
119. Maller JJ, Thomson RHS, Pannek K, et al. The (Eigen)value of diffusion tensor imaging to investigate depression after traumatic brain injury. Hum Brain Mapp. 2014;35:227–37. https://doi.org/10.1002/hbm.22171.
120. Alhilali LM, Delic JA, Gumus S, Fakhran S. Evaluation of white matter injury patterns underlying neuropsychiatric symptoms after mild traumatic brain injury. Radiology. 2015;277:793–800. https://doi.org/10.1148/radiol.2015142974.
121. Matthews SC, Strigo IA, Simmons AN, et al. A multimodal imaging study in U.S. veterans of Operations Iraqi and Enduring Freedom with and without major depression after blast-related concussion. Neuroimage. 2011;54 Suppl 1:S69–75. https://doi.org/10.1016/j.neuroimage.2010.04.269.
122. Chen JK, Johnston KM, Petrides M, Ptito A. Neural substrates of symptoms of depression following concussion in male athletes with persisting postconcussion symptoms. Arch Gen Psychiatry. 2008;65:81–9. https://doi.org/10.1001/archgenpsychiatry.2007.8.
123. Hamilton JP, Etkin A, Furman DJ, et al. Functional neuroimaging of major depressive disorder: a meta-analysis and new integration of base line activation and neural response data. Am J Psychiatry. 2012;169:693–703. https://doi.org/10.1176/appi.ajp.2012.11071105.
124. Rangaprakash D, Deshpande G, Daniel TA, et al. Compromised hippocampus-striatum pathway as a potential imaging biomarker of mild-traumatic brain injury and posttraumatic stress disorder. Hum Brain Mapp. 2017;38:2843–64. https://doi.org/10.1002/hbm.23551.
125. Santhanam P, Wilson SH, Oakes TR, Weaver LK. Effects of mild traumatic brain injury and post-traumatic stress disorder on resting-state default mode network connectivity. Brain Res. 1711;2019:77–82. https://doi.org/10.1016/j.brainres.2019.01.015.

126. Youssef MM, Underwood MD, Huang YY, et al. Association of BDNF Val66MET polymorphism and brain BDNF levels with major depression and suicide. Int J Neuropsychopharmacol. 2018;21:528–38. https://doi.org/10.1093/ijnp/pyy008.
127. Feng F, Lu SS, Hu CY, et al. Association between apolipoprotein e gene polymorphism and depression. J Clin Neurosci. 2015;22:1232–8. https://doi.org/10.1016/j.jocn.2015.02.012.
128. Wang W, Liu X, Ruan Y, et al. Depression was associated with apolipoprotein E ε4 allele polymorphism: a meta-analysis. Iran J Basic Med Sci. 2019;22:112–7. https://doi.org/10.22038/ijbms.2018.30825.7436.
129. Burns RA, Andrews S, Cherbuin N, Anstey KJ. Role of apolipoprotein E epsilon 4 (APOE *ε4) as an independent risk factor for incident depression over a 12-year period in cognitively intact adults across the lifespan. BJPsych Open. 2020;6:e47. https://doi.org/10.1192/bjo.2020.29.
130. MacQueen GM, Hassel S, Arnott SR, et al. The Canadian Biomarker Integration Network in Depression (CAN-BIND): magnetic resonance imaging protocols. J Psychiatry Neurosci. 2019;44:223–36. https://doi.org/10.1503/jpn.180036.
131. Lent MR, Hoffman SN, Lester Kirchner H, et al. Attitudes about future genetic testing for posttraumatic stress disorder and addiction among community-based veterans. Front Psych. 2017;8:1–7. https://doi.org/10.3389/fpsyt.2017.00076.
132. American Psychiatric Association. Diagnostic and statistical manual of mental disorders: DSM-5. Arlington, VA; 2013.
133. Lishman WA. Physiogenesis and psychogenesis in the "post-concussional syndrome". Br J Psychiatry. 1988;153:460–9. https://doi.org/10.1192/bjp.153.4.460.
134. Silverberg ND, Iverson GL. Etiology of the post-concussion syndrome: Physiogenesis and psychogenesis revisited. NeuroRehabilitation. 2011;29:317–29. https://doi.org/10.3233/NRE-2011-0708.
135. Polinder S, Cnossen MC, Real RGL, et al. A multidimensional approach to post-concussion symptoms in mild traumatic brain injury. Front Neurol. 2018;9:1–14. https://doi.org/10.3389/fneur.2018.01113.
136. World Health Organization. The ICD-10 classification of mental and behavioural disorders: diagnostic criteria for research. Geneva: WHO; 1993.
137. American Psychiatric Association. Diagnostic and statistical manual of mental disorders: DSM-IV. Washington, DC; 1994.
138. Boake C, McCauley SR, Levin HS, et al. Diagnostic criteria for postconcussional syndrome after mild to moderate traumatic brain injury. J Neuropsychiatry Clin Neurosci. 2005;17:350–6. https://doi.org/10.1176/jnp.17.3.350.
139. Wortzel HS, Arciniegas DB. The DSM-5 approach to the evaluation of traumatic brain injury and its neuropsychiatric sequelae. NeuroRehabilitation. 2014;34:613–23. https://doi.org/10.3233/NRE-141086.
140. Tator CH, Davis HS, Dufort PA, et al. Postconcussion syndrome: demographics and predictors in 221 patients. J Neurosurg. 2016;125:1206–16. https://doi.org/10.3171/2015.6.JNS15664.
141. Kennedy SH, Lam RW, McIntyre RS, et al. Canadian Network for Mood and Anxiety Treatments (CANMAT) 2016 clinical guidelines for the management of adults with major depressive disorder: section 3. Pharmacological treatments. Can J Psychiatry. 2016;61:540–60. https://doi.org/10.1177/0706743716659417.
142. The Management of Concussion-Mild Traumatic Brain Injury Working Group. VA / DoD clinical practice guideline for the management of concussion-mild traumatic brain injury; 2016.
143. Silver J, Arciniegas D, Yudosky S. Psychopharmacology. In: Silver J, Arciniegas D, Yudovsky S, editors. Textbook of traumatic brain injury. Arlington: American Psychiatric Publishing Inc; 2005. p. 609–40.

Chapter 8
Management of Concussion and Persistent Post-concussion Symptoms

Shawn Marshall and Jacqueline van Ierssel

Introduction

This chapter focuses on approaches to diagnosis, assessment, and treatment for persons who have sustained a concussion. While most persons sustaining a concussion will have early and complete recovery, a minority of patients will have symptoms beyond the expected four-week period of recovery. Efforts to identify and manage those persons at risk of prolonged recovery are warranted to mitigate the significant disability that can be associated with persistent symptoms such as headache, fatigue, vision impairments, balance impairments, and the potential for onset of mood disorders. Several evidence-based clinical guidelines, professional position statements, and consensus statements have been developed to review the definition, pathophysiology, criteria for recovery, risks of recurrent concussion, prevention strategies, to provide strategies for managing acute and persistent post-concussion symptoms, and to guide return to sport, return to school, and return to work; these are typically updated on a regular basis [1–6].

S. Marshall (✉)
Division of Physical Medicine and Rehabilitation, Department of Medicine, Ottawa Hospital Research Institute, Ottawa, ON, Canada

Bruyere Research Institute, Ottawa, ON, Canada

University of Ottawa, Ottawa, ON, Canada
e-mail: smarshall@toh.ca; smarshall@Ottawahospital.on.ca

J. van Ierssel
Research Institute, Children's Hospital of Eastern Ontario, Ottawa, ON, USA

T. A. Schweizer, A. J. Baker (eds.), *Tackling the Concussion Epidemic*,
https://doi.org/10.1007/978-3-030-93813-0_8

Guidelines

Consensus-based clinical practice guidelines exist specifically for the diagnosis, assessment, and management of concussion and mild traumatic brain injury with persistent post-concussion symptoms [1, 7]. The strength of clinical practice guidelines is that they are updated on a regular basis by expert consensus in the field, and in some instances, there is a move toward "Living Guidelines" where recommendations are updated on a continuous basis as the evidence evolves. These guidelines also often serve as a resource for the most up-to-date patient information resources, assessment tools, and outcome measures. A review by Silverberg et al. [1], sponsored by the American Congress of Rehabilitation Medicine (ACRM), highlighted and compared recommendations proposed by the following five guidelines:

1. Ontario Neurotrauma Foundation Guidelines for Concussion/Mild Traumatic Brain Injury & Prolonged Symptoms 3rd Edition, for Adults Over 18 Years of Age [5]
2. Ontario Neurotrauma Foundation Living Guideline for Diagnosing and Managing Pediatric Concussion [8]
3. VA/DoD Clinical Practice Guidelines: Management of Concussion-Mild Traumatic Brain Injury (mTBI) (2016) [3]
4. Consensus Statement on Concussion in Sport – The 5th International Conference on Concussion in Sport Held in Berlin, October 2016 [2]
5. Centers for Disease Control and Prevention Guideline on the Diagnosis and Management of Mild Traumatic Brain Injury Among Children [9]

Diagnosis and Assessment of Concussion

As described in the previous chapters, concussion can be defined as "an injury to the brain that results in temporary loss of normal brain function. Medically, it is defined as a clinical syndrome characterized by immediate and transient alteration in brain function, including alteration of mental status or level of consciousness, that results from mechanical force or trauma" [10]. Much emphasis has been placed on the recognition of a potential concussion particularly in sport where there is high risk of injury. There is a risk of concussion under-reporting in sport-related concussions, as athletes may be hesitant to report symptoms out of concern for being removed from play, not wanting to let their team down, perceived pressure, or not thinking the injury was serious enough, or being unable to recognize the injury in the moment due to confusion or alteration in consciousness [11].

Prompt diagnostic evaluation is a key element in managing concussion [3, 5]. While concussion is typically a self-limited condition from which most injured persons will fully recover, medical evaluation by a physician or nurse practitioner is necessary to the injured person to rule out more significant brain injury or associated injuries such as cervical injury where further investigation or evaluation may

Step 1. Establish Plausible Injury Mechanism.

- Ask the patient to describe the sequence of events surrounding the injury.
- Listen carefully and query as necessary for a concussive force (eg, Did your head jolt back and forth?) and its intensity (eg, From what height did you fall?).
- Distinguish the patient's personal memories from facts he or she inferred or learned from other people afterward.

Step 2. Query Signs and Symptoms.

Determine whether the patient's mental status was altered immediately after the impact. Example questions:

- Do you remember the impact and moments just after?
- Did anyone see you lay still and unresponsive right after the accident?
- Were you confused or unsure about where you were and what was happening?
- Were you able to think clearly about what to do after the accident?
- Were you able to answer questions appropriately and follow instructions from people at the scene?
- Did anyone tell you that your speech was incoherent or not making sense?

Step 3. Rule Out Confounding Factors

Check whether factors other than brain injury can account for the acute alteration in mental status. Example questions:

- Were you drinking alcohol or using drugs just before the accident?
- Did you see the impact coming? Did you think that you or others would be seriously injured or killed? Did you feel panicked or scared?
- Did you injure other parts of your body? Were you in severe pain?

Fig. 8.1 Diagnostic interview for mild traumatic brain injury. (Reprinted from Silverberg et al. [1]. With permission from Elsevier.)

be necessary. The key elements of the assessment include a detailed history of the event, neurological examination, and formal neck examination due to the high rate of associated cervical injury with concussion. Questions asked in the diagnostic interview (Fig. 8.1) [1] typically focus on the possible mechanism of injury, the patient's symptoms post-injury, and confounding factors such as concurrent injury or intoxication at the time of the injury. Formal cognitive screening as well as a standardized, age-appropriate symptom questionnaire to document current symptoms is also important to assist in monitoring symptoms and overall recovery. Commonly used assessment tools include the Sport Concussion Assessment Tool, 5th edition (SCAT5) [12], the Rivermead Post-Concussion Symptom Questionnaire [13], or the Post-Concussion Symptom Scale [14]. The need for early neuroimaging with CT or MRI should be determined using formal criteria such as the Canadian CT Head rule [15] (ages 16–64), the PECARN (Pediatric Emergency Care Applied Research Network) algorithm (up to 18 years) [16], or the CATCH2 (Canadian Assessment of Childhood Head Injury) rule [17]. Routine neuroimaging for concussion is not recommended, given the associated costs and limited diagnostic or prognostic utility [3, 5].

Following completion of the initial assessment, patient with a confirmed concussion and support person should be provided with both verbal and written information (e.g., the Concussion Information Package for Parents/Caregivers available from www.cattonline.com [18]) regarding their diagnosis of concussion, expected symptoms, prognosis, and timeline for recovery. The patient should be reassured that the majority of adults recover within the first two weeks post-concussion and youth within the first four weeks, although female adolescents may take longer. Formal information provided should highlight the initial need for rest during the first 24 to 48 hours post-injury [19] with graduated return to activities as tolerated thereafter. Specifically, the patient should not return to work, school, or driving for at least 24 hours post-injury. Driving is a cognitively demanding activity, and evidence has shown that driving performance is negatively impacted for at least 24 hours post-injury [20–22]. According to expert consensus recommendations, driving should be avoided for the first 24 to 48 hours and may be resumed once the patient is able to concentrate and safely operate a vehicle without symptom exacerbation [8]. For athletes, return to sport should be guided by the International Consensus Statement on Concussion in Sport [2], which provides specific guidance on how to safely progress through a six-step graded exertional strategy.

A follow-up assessment is recommended for patients who continue to experience symptoms or have difficulty progressing through return-to-school, return-to-work, or return-to-sports stages. While most patients are expected to recover in the first few weeks, 20–30% of patients will continue to have symptoms beyond the one-month period [23–25]. Identification of both medical and psychosocial risk factors including previous history of concussion, psychiatric history, substance use, high number of symptoms, and lower economic status can help determine whether an individual is at risk for prolonged recovery [5]. A review by Silverberg et al. [26] identified that female sex, pre-injury mental health problems, early post-injury cognition impairment, and acute psychological distress put the individual at risk for a worse prognosis. Prospective research by Cnossen et al. [27] demonstrated that female sex, neck pain, three or more post-concussion symptoms worse than pre-injury and post-traumatic stress at two weeks post-injury were the most predictive factors for persistent symptoms at 6 months post-injury. Separate clinical prediction rules have been developed for children and adults presenting to the emergency department or primary care physician for an acute concussion to determine the risk of developing persistent post-concussion symptoms. The 5P clinical risk score developed by Zemek et al. [24] has shown moderate discrimination to categorize acutely concussed children and adolescents into low-, medium-, or high-risk groups based on age, sex, history of prior concussion symptom duration, migraine history, answering questions slowly, balance errors, headache, sensitivity to noise, and fatigue. The Toronto Rehabilitation Institute Concussion Outcome Risk Determination & Rehab Recommendations (TRICORDRR) clinical risk score developed by Langer et al. [28] calculates an adult's risk of requiring treatment for post-concussion symptoms at six months based on premorbid psychiatric conditions, pre-injury health system usage, and older age.

Key Recommendations for Management

Education is a key component for managing concussion following diagnosis [5, 29] and should include instructions for the short-term following injury, anticipated symptoms, self-management strategies for symptoms, and information regarding expectations for recovery. At the initial assessment, information should be provided in both verbal and written format that includes links to online educational resources for the patient and family. Although it is very uncommon to have complications post-concussion, patients and families should be advised of potential warning signs such as seizure, progression of symptoms, or deterioration in cognition or physical abilities that may warrant repeat medical evaluation. The education materials should also provide information for the initial 24- to 48-hours post-injury where rest should be emphasized, including avoidance of alcohol and driving. Efforts should be made to normalize the symptoms experienced following concussion by providing a description of common, anticipated symptoms. This includes somatic symptoms such as headache, fatigue, sleep impairment, tinnitus, noise and light sensitivity, and dizziness/balance difficulties; affective symptoms such as irritability, emotional lability, or feeling anxious or worried; and cognitive symptoms such as fogginess, decreased concentration, and memory. The most important element of the education is to validate the symptoms that the patient is experiencing acutely and to emphasize that recovery will likely occur over the ensuing days to weeks, which is typical for most persons who experience concussion.

Follow-up provides the opportunity to monitor progression and once again intervene with education for the patient. Upon discharge from the emergency department or following the initial clinic assessment, arrangements should be made for follow-up either in person or by telephone/virtual [30] and should be provided over the initial 12 weeks post-injury as required [5]. A standardized assessment tool should continue to be used at each follow-up appointment to monitor recovery progression and guide care. Routine screening for depression and anxiety should also be completed as part of these assessments, since mood can impact overall recovery and be an indication for referral to either a specialist or specialized concussion program.

Return to Activity

The goal for patients is to recover and return to their normal functional activities including sport, recreational and social activities, school and work. Following assessment and confirmation of a concussion diagnosis, rest is recommended for the first 24 to 48 hours post-injury, after which graduated return to regular routines based on "activity as tolerated" is recommended. Explaining this term to patients can facilitate a shared understanding of the goals or expectations for recovery. Activity as tolerated is sometimes described as the "Goldilocks Approach" of just the right amount of activity that allows a patient to resume normal daily activities with a mild amount of discomfort, as long as the symptoms are not functionally limiting and resolve quickly with rest. Too little activity may increase the risk for persistent post-concussion symptoms, while too much activity may result in a significant or persistent exacerbation of symptoms that is functionally limiting. Existing evidence demonstrates that complete rest beyond the 24 to 48-hour period post-injury is not beneficial and may in fact be harmful [31]. Leddy et al. [32] have advanced this concept by demonstrating the added value of a formal exercise strategy post-concussion to accelerate recovery and potentially reduce the incidence of persistent symptoms. Sub-symptom threshold exercises are prescribed at 80–90% of the heart rate at which symptoms start or worsen using the Buffalo Concussion Treadmill test protocol with recent systematic reviews showing promising results for patient recovery [33, 34]. Resuming normal activity, whether return to sport, hobby, school, or work, is critical to recovery as with any injury. Patients may find it useful to be provided an analogy such as a comparable and more comprehendible musculoskeletal injury.

Analogy Example: An Acute Ankle Sprain

While a sprained ankle may vary between patients in terms of symptoms and severity, the goals of recovery and return to normal function are consistent. Rest is recommended early on when symptoms are most severe, but full recovery is expected. If a patient rests their ankle completely until all the pain is gone ("too little"), there is risk for muscle deconditioning and prolongation of symptoms. If the patient returns to activity too soon or too aggressively, the ankle may become inflamed, swollen, and patient may not be able to walk the next day ("too much"). Resuming activity as tolerated refers to being able to walk on the injured ankle with minor discomfort or limping and continued functional improvement each day. If the patient walked too much one day, then no specific harm would come, but would need to reduce their amount of walking and intersperse rest to allow for healing.

Similarly, cognitive and physical rest is recommended early on following a concussion to minimize symptoms, allow recovery, and prevent reinjury. If a patient returns to activity too soon, symptoms may worsen and limit their ability to participate in daily activities. Following the initial period of rest, patients are encouraged to resume activities they can tolerate with minimal worsening of symptoms, so long as the symptoms settle easily afterward. Planning frequent breaks between short periods of activity can help minimize symptoms and promote recovery.

Referral for More Specialized Assessment

Patients who continue to experience sustained or worsening symptoms beyond the normal period of expected recovery should be referred to an interdisciplinary concussion clinic for specialized treatment to advance their recovery. Early referral to a specialized concussion clinic may also be considered for patients with previously identified risk factors such as history of previous concussions with prolonged recovery, pre-existing mental health disorders, addictions, or high initial symptom severity, as both medical and therapy expertise are available to assess and manage the patient [35]. For patients with deteriorating mental health, referral for psychology or psychiatry assessment may also be considered. Currently, there is no standardized treatment pathway for patients with persistent post-concussion symptoms; therefore, referral decisions should be individualized and based on the underlying etiology [36]. For example, dizziness may be migraine-related, vestibular, or psychogenic in origin, necessitating a referral to the most appropriate health professional(s). Since there is often considerable overlap in symptom domains, patients with persistent post-concussion symptoms are best managed by an interdisciplinary team, and healthcare professionals should adhere to their scope of practice and clinical expertise [36]. In order ensure appropriate referrals for specialized concussion care, referrals should include information on patient demographics, specific reasons for the referral, results of previous investigations and interventions, risk factors for prolonged recovery, pre-existing medical conditions, symptoms, functional limitations, and patient goals [37].

The Ontario Neurotrauma Foundation has outlined the following core functions a concussion clinic should be able to provide in their Standards of Post-concussion Care: (1) diagnosis and medical treatment decisions, (2) physical treatment, (3) cognitive, functional, emotional support, (4) coordination of care function, and (5) education [36].

Monitoring of Symptoms

Unlike classic musculoskeletal injuries, where the anticipated symptoms and functional limitations tend to be localized to the point of injury, concussion affects multiple symptom domains, including somatic, affective, cognition, and sleep. Monitoring these symptoms post-concussion is important in order to (1) monitor recovery progression and (2) identify specific symptoms relevant to the patient for which intervention, starting with education, may need to be initiated. Current guidelines recommend the use of age-appropriate validated symptom assessments [1] such as the Rivermead Post-concussion Symptom Questionnaire (RPQ) [13], the Post-concussion Symptom Scale [14], SCAT5 Symptom Scale [12], or the Post-concussion Symptom Inventory [38]. These symptom scales are typically self-administered at the initial assessment and follow-up assessments to provide a summative score for the number and severity of symptoms as a measure of recovery. While these scores represent the degree of symptoms or impairments experienced by the patients, they do not necessarily demonstrate the impact on abilities or function for patients. The Concussion Recovery Questionnaire (CORE-Q) [39] is a more recently developed assessment tool that measures functional status post-concussion by assessing how long a patient can perform patient-relevant tasks such as use of screen time, reading, exercising, and socializing before the onset or worsening of symptoms.

Persistent Post-concussion Symptoms

Historically, there remains debate over the duration of symptoms following concussion and the specific diagnosis of post-concussion syndrome itself. Figure 8.2 [40] highlights the comparisons between different diagnostic criteria, including Post-concussion Syndrome (International Classification of Diseases (ICD) 10), which relies heavily on symptoms with a psychological underpinning, and Post-concussional Disorder (Diagnostic and Statistical Manual of Mental Disorders (DSM)-IV), which depends on cognitive impairments and has subsequently been replaced by Major or Mild Neurocognitive Disorder due to traumatic brain injury (DSM-V). More recently, Lagacé-Legendre et al. [41] used a Delphi method to create an expert consensus-based definition for persistent post-concussion symptoms that emphasizes the presence of at least one key symptom at three months post-injury, where there is an impact on the patient's quality of life and functioning (Fig. 8.3). There have been significant efforts to highlight that the symptoms following concussion are non-specific, and comparisons have been made to chronic pain populations [42] and even student populations where a high prevalence of similar symptoms are found [43, 44]. Voormolen et al. [45] have also demonstrated a high prevalence of concussion-related symptoms in the general population in Italy, the Netherlands, and the United Kingdom. Despite debate as to the association between persistent post-concussion symptoms and factors such as litigation, solely mental

	ICD-10	DSM-IV	DSM-V	5th International Consensus Conference on Concussion in Sport
Terminology	Postconcussion syndrome	Postconcussional disorder	Major or mild neurocognitive disorder: traumatic brain injury	Sports-related concussion: symptoms and signs
Trauma	History of head trauma	History of head injury	Impact to head or rapid movement/displacement of brain	Impulsive force trasmitted to the head
Loss of consciousne (LOC)	"Usually sufficiently severe to result in loss of consciousness"	Suggested criterion: > 5 minutes	Not required	Not required
Altered consciousness / cognitive impairment	Yes	Relative attention or memory impairment on neuropsychologie testing	Yes, or (+) imaging/ neurologic exam	"Impairment of neurologic functioning"
Maximum symptom delay for attribution to trauma	4 weeks	N/A	Immediate or when conscious	Minutes to hours
Minimum duration	N/A	3 months	"Past the acute injury phase"	Adults: 10–14 days Children: 4 weeks
Objective evidence	Not required	Required	Not required	Not required

DSM-IV, *Diagnostic and Statistical Manual of Mental Disorders,* fourth edition; DSM-V, *Diagnostic and Statistical Mannual of Mental Disorders* fifth edition: ICD-10, *International Statistical Classification of Diseases and Related Heath Problems,* 10th revision.

Fig. 8.2 Definitions for con cussion and persistent post-concussion-related symptoms. ([Reprinted from Dwyer and Katz [139]. With permission from Elsevier.)

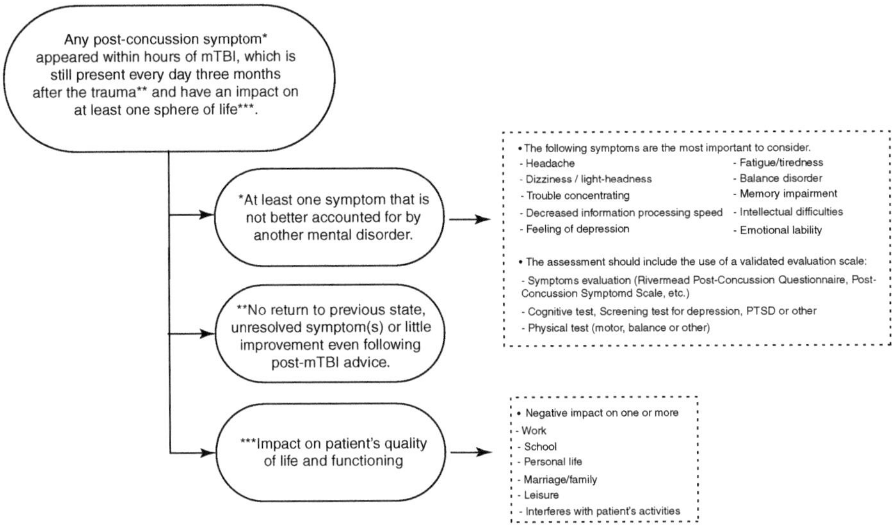

Fig. 8.3 Consensual definition of PPCS. mTBI, mild traumatic brain injury; PTSD, post-traumatic stress disorder. (Reprinted from Lagacé-Legendre et al. [41]. With permission from Wolters Kluwer Health, Inc.)

health or psychological factors, or even malingering [46, 47], evidence indicates that 10–30% of patients continue to have persistent symptoms beyond three months post-injury [48–52]. As in other medical conditions and injuries, persistent post-concussion symptoms can significantly affect a patient's level of functioning and quality of life. Even when on the surface it appears that full recovery has been made, with return to full academic load or work duties, patients may continue to have struggles with decreased efficiency and productivity. This was highlighted by Silverberg et al. [53] who demonstrated that history of concussion was often associated with underemployment (i.e., not returning to previous level of employment) and decreased work productivity even in those individuals who had returned to work. Efforts to prevent persistent post-concussion symptoms or to mitigate the degree of symptomatology, thus, have implications for improving functioning, disability, and health.

Analogy for Patients with Persistent Post-concussion Symptoms Affecting Function: A Chronic Ankle Sprain

Healthy, uninjured individuals would typically be able to run once around a track without a challenge. Patients with a chronic ankle sprain would hobble around the track successfully- job completed. However, the patient would likely require a much longer time to complete the run (decreased efficiency), be more fatigued (more energy expended for the task), require increased rest in order to recover, and experience too much discomfort to complete other functional tasks afterward. Similarly, patients with persistent post-concussion symptoms may be able to complete the task, but with less efficiency or accuracy, need increased effort than prior to injury, experience an increase in symptoms, and require longer time to recover post-exertion.

Management of Persistent Post-concussion Symptoms

Initial and follow-up management of acute concussion should normalize concussion symptoms and emphasize that symptoms are typically self-limited and resolve within days to weeks. The focus of approach remains on proper assessment by a medical professional to confirm diagnosis, rule out more serious injury, provide appropriate education to the patient and family, and follow-up to allow monitoring of recovery. Primary messages include managing expectations of recovery and progressively increasing activity as tolerated throughout the recovery period. However, a high initial severity of symptoms has been shown to be the strongest predictor of prolonged recovery, and these patients should be considered for early referral to a specialized concussion clinic for more specific interventions. Clinical experience suggests that the three most important symptoms to be addressed early on are sleep impairment, post-traumatic headache, and mood. Often these symptoms present together and are highly interconnected, where the worsening of one symptom directly negatively impacts another. This has been described as a vicious cycle (Fig. 8.4) [54], which can lead to a worsening of symptoms and decreased functioning if not recognized or addressed. One challenge for managing symptoms post-concussion is the limited evidence for symptom-specific treatment in a concussion population. While scientific literature on concussion diagnosis and treatment is rapidly increasing, the paucity of research examining concussion-specific treatment

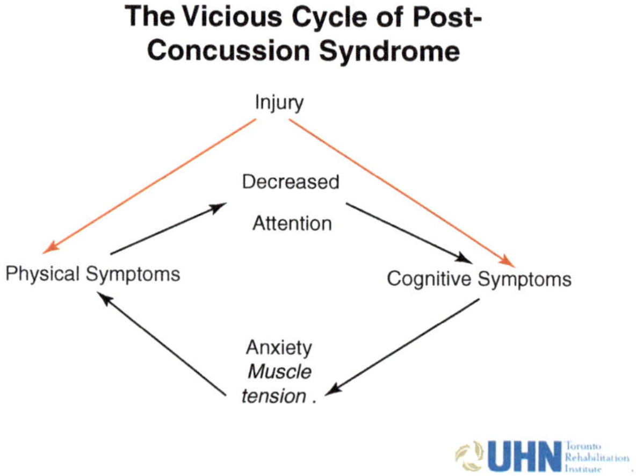

Fig. 8.4 The vicious cycle of postconcussion syndrome. (Courtesy of Dr. Mark Bayley)

results in many guideline recommendations that are of lower grade, despite strong evidence of treatment effectiveness in other diagnostic populations.

Post-traumatic Headache

Headache is the most common symptom following concussion with 30 to 90% of patients reporting post-concussion headache [23, 55–58]. Post-concussion headache can be highly disabling, since it is frequently the limiting symptom for patients when increasing their activity level. As with all post-concussion symptoms, it is anticipated that with a limited initial period of rest followed by a gradual resumption of activities, headache symptoms will resolve. Post-traumatic headaches are a secondary type of headache that the International Classification of Headache Disorders (ICHD) [59] defines as a new or worsening headache that occurs in close temporal relation to a head injury. Post-traumatic headaches are classified as acute when occurring within the first 3 months of injury and persistent if headaches persist beyond 3 months post-injury. Evaluation should focus on determining the headache pattern, including migraine, tension-type, cervicogenic, medication overuse, or exertional headache.

All patients who present with headaches at their follow-up assessment should undergo a comprehensive medical assessment that includes a focused clinical history, physical examination, and diagnostic imaging of the brain or cervical spine in individuals with focal neurological deficits or worrisome findings. A focused headache history should probe headache severity, frequency, location, duration, and quality of pain. Personal and family history of headache, as well pre-existing medical conditions should be established in order to identify co-morbid medical conditions and for mitigation of factors that could contribute to post-traumatic headaches. Information regarding relieving and aggravating factors can also add to determining the headache phenotype, which can aid in management. Physical examination should include a complete neurological examination, along with examination of the cervical spine and vestibular system.

An important first, and often overlooked, step in managing headache is lifestyle modification. Information regarding getting adequate sleep, eating regular meals, maintaining hydration, reducing stress, and achieving exercise are important self-management strategies to reduce headache symptoms. Printable concussion information packages for athletes, parents, and caregivers have been developed by the Concussion Awareness Training Tool [18] and Parachute Organization [60] to provide athletes, parents, and caregivers with easily accessible and understandable information on what a concussion is, when to seek urgent medical help, expected symptoms, and when to resume normal activities. Pharmacologic management may be necessary if the symptoms are pronounced and limiting function. Simple over-the-counter analgesics may be indicated such as ibuprofen or acetaminophen; however, in order to avoid the undesired phenomenon of rebound (medication overuse) headaches, these medications should not be used more than 15 days in a month [61].

If there are features similar to migraine-type headaches, then migraine abortive triptans may be effective, with the similar caveat that these should not be used more than 10 times per month, to mitigate the risk of developing medication-induced headaches. The use of narcotic medications should be avoided as a headache treatment. Prophylactic medications including low-dose tricyclic antidepressant medications to start, such as amitriptyline or nortriptyline [62], may be indicated, although propranolol may also be considered; second-line therapy medications include topiramate or gabapentin [62]. When deciding upon medications, it is often worthwhile considering other symptoms that the patient may have; for example, amitriptyline can be of assistance when sleep initiation is a problem and topiramate may be more useful where oversedation or weight gain may be concerns. Ultimately, post-traumatic headache can significantly impact the course of recovery and referral to a concussion specialty clinic, neurologist, or headache clinic may be indicated if symptoms do not improve.

Sleep Disturbances

Sleep disturbances are common following a concussion, with upward of 40–50% of patients reporting sleep issues [23, 63, 64]. Sleep disturbances include insomnia, hypersomnia, obstructive sleep apnea, poor sleep maintenance, poor sleep efficiency, early awakening, delayed sleep onset, or alterations in circadian cycle. Initially, post-injury hypersomnolence tends to be more problematic, with insomnia becoming a more prevalent issue later in recovery. Achieving a successful, restorative sleep pattern is critical when recovering from concussion, since it has the potential to impact mood, fatigue, headache [65–67] and has been demonstrated to be a prognostic indicator for negative post-concussion outcomes beyond one-year post-injury [68]. Patients should be provided with information regarding post-concussion expectations to make them aware that sleep disturbances are common and expected.

When sleep disturbances present as a prominent symptom for the patient, a detailed history focused on identifying pre-existing sleep issues is recommended, since previously benign sleep problems may be exacerbated during recovery. Relevant sleep history details such as sleep schedule, onset, sleep maintenance, and awakening will help clarify patient diagnosis and management. Patients should be screened for medication use, both prescription and over the counter, as well as potential mood disorders such as post-traumatic stress disorder, depression, and anxiety as they may impact sleep patterns. When issues such as sleep-disordered breathing, nocturnal seizures, or narcolepsy are suspected, referral for specialized testing should be considered, including polysomnography [69].

Management of post-concussion sleep-wake disturbances is recommended and should start with education regarding establishing good sleep hygiene, ideally provided to the patient in a handout or web link format. Education should focus on setting a routine around sleep, so that sleep and wake times are consistent daily, including weekends. A bedtime routine for preparing to go to sleep should also be in place that could involve steps such as taking a warm bath or massage. Daytime naps should be limited in order to promote nighttime sleepiness, but, if necessary, then should not exceed 30 minutes once per day. Patients should be encouraged to avoid caffeinated beverages and alcohol before sleep, engage in regular moderate to vigorous physical activity as tolerated, and spend time outdoors during daylight hours to help regulate their internal clock. Sleep environment is a key component of sleep hygiene and patients should be encouraged to avoid electronics one to two hours prior to bedtime, maintain the bedroom exclusively for sleep, keep the room a cool, dark, and quiet place.

Active interventions for sleep may sometimes be necessary. Studies have demonstrated the effectiveness of CBT for sleep, and while in-person therapy has been the standard [70–72], online resources are becoming increasingly available. Medication intervention may be necessary to establish and maintain sleep when non-pharmacological interventions have not been completely successful. The aim is to use medications at a low dose to minimize the risk of dependency. Trazodone or doxepin at low dose can be considered and while benzodiazepines should be avoided, non-benzodiazepine medications such as zopiclone, eszopiclone, or zolpidem may be appropriate in the short term.

Mental Health Disorders

Mental health status is an important consideration in managing patients with concussion, in order to promote timely and maximal recovery. Chapter 7 provides a detailed focus on how mental health relates to concussion and recovery. From a practical perspective, mental health and mood disorders are particularly important to recognize and address post-concussion, since pre-existing mental health disorders may put the patient at higher risk for persistent symptoms [26, 73, 74]. Associated symptoms and psychosocial consequences of concussion, such as isolation and removal from sport or work, may further lead to the onset of new mental health problems such as anxiety, depression, and post-traumatic stress. Mental health disorders are often interrelated with headache and sleep disturbances and can contribute to the "vicious cycle" of post-concussion where mood impacts sleep, which may impact headache, which may further contribute to deterioration in mood and function. A standardized assessment for mood is recommended for all patients where there may be concerns of anxiety or depression [75]. If present, establishing the degree of mental health involvement will guide whether non-pharmacological or pharmacological approaches for management should be considered. Currently, there is no evidence to suggest that mental health disorders following concussion should be treated any differently than for patients where the etiology is not due to concussion.

For nonpharmacological approaches, referral to a mental health professional may be necessary, where approaches such as CBT have been shown to be effective

[72, 75]. In some instances where symptoms are in the moderate to severe range, pharmaceutical approaches may be necessary, such as the use of antianxiety or antidepressant medications.

Fatigue

Fatigue is a symptom in which a person experiences tiredness or exhaustion either physically or mentally that subsequently impacts their ability to function. It is one of the most common symptoms experienced following concussion, with upward of 40% of patients with concussion continuing to experience this symptom six months post-injury [23, 27, 76]. While fatigue is often intertwined with other symptoms including headache, sleep, mood, dizziness, vision symptoms, and cognition symptoms, it can also be a primary symptom following concussion, independent of these other symptoms [77, 78]. Commonly, fatigue is exacerbated by cognitive effort and environmental stimuli such as noise and light.

When patients present with fatigue, they should be advised that fatigue is one of the most common symptoms experienced post-concussion. Following the recommended physical and cognitive rest in the first 24–48 hours post-injury, patients should be encouraged to resume low-risk daily activities as tolerated. If fatigue remains a prominent symptom during recovery, the clinician should consider additional potential causes of fatigue such as medication side effects, mood disorder, or possible endocrine abnormalities; however, more recent research has been less supportive of endocrine abnormalities likely contributing to fatigue post mTBI/concussion [79–81].

Often, fatigue can be a particularly challenging symptom to manage. A recent systematic review by Sullivan et al. [82] concluded that psychological approaches (CBT, counseling or education) for treating sleep and fatigue post-concussion showed small improvements in outcome, compared to standard care or giving generic advice. Management strategies should focus on environmental modifications, lifestyle factors, and energy conservation in order to maximize function. Environmental modifications that include a non-distracting, quiet environment with tolerable lighting such as incandescent versus fluorescent tend to reduce symptoms associated with light sensitivity. Similarly, an ergonomic workstation assessment should be considered, and adjustments made where appropriate to minimize musculoskeletal pain due to neck injury, which frequently occurs during concussion injuries. Healthy lifestyle strategies, such as adequate hydration, healthy nutrition, and symptom-free aerobic exercise as tolerated, are additional key elements for managing fatigue. Energy conservation, or the “Four P’s” is one of the most important factors in managing fatigue: prioritizing, planning, pacing, and positioning. Patients should be encouraged to prioritize tasks based on their importance and symptom provocation, asking for help or “dumping” those tasks that are unimportant or could be done by others. Planning out necessary tasks at times of the day when the patient has the most energy with frequent scheduled breaks can save mental energy. Pacing can be described as "spreading out activity and alternating it with rest periods so that you are able to continue for longer" [83]. While it is a simple concept that stretches back in time to the fable of the Tortoise and the Hare, in today's age, it is an important concept to emphasize for patients suffering from fatigue. Positioning refers to choosing quiet, non-distracting environments that allow for easier concentration. Formal information and education around energy conservation is critical for patients to avoid pushing through symptoms and creating a common pattern of significant symptom spikes and subsequent crashes that require a prolonged recovery time. The goal is for patients to increase their activity tolerance by remaining below their symptom threshold. As symptoms become better controlled, patients should be encouraged to gradually increase their level of activity while continuing to self-pace and self-monitor their symptoms.

In terms of pharmacologic management, stimulant medications are a potential approach to aid common symptoms experienced post-concussion such as fatigue, attention deficits, and slowed information processing. Recent trials for pharmacologic management using stimulant medications [84, 85] have shown improvement in both fatigue symptoms and cognitive symptoms. However, a recent systematic review by Iaccarino et al. [86] concluded that there was not sufficient evidence for the routine use of stimulant medications, specifically methylphenidate, for mTBI as most of the existing studies had small sample sizes and heterogeneous populations. They further emphasized that future studies would specifically need to address pre-injury history of ADHD which is overrepresented in the concussion population.

Cognition

Cognitive symptoms, along with headache and fatigue, are commonly reported immediately following concussion, where specific symptoms have been classically identified as taking longer to think (60%), forgetfulness or poor memory (56%), and poor concentration (52%) [23]. Impaired cognition aligns with the pathophysiologic understanding of concussion where there is disruption of brain metabolic activity and subsequent function due to trauma, with effects that appear to last from one week up to six months post-injury [87–89]. While most patients with concussion will completely recover, between 15 and 33% of patients will have ongoing cognitive symptoms beyond the six-month time frame [90, 91]. Controversy remains as to whether concussion can lead to permanent versus transient impairments in cognitive function, with several systematic reviews failing to identify significant neuropsychological deficits post-concussion [92–96]. However, Ruff and Weyer Jamora [46], in their review focusing on myths and mild TBI, argue that systematic reviews and meta-analyses in fact are problematic in teasing out or confirming the deficits related to concussion based on issues with inconsistent testing protocols, challenges with the patients studied where in many instances litigants versus non-litigants were not identified and pooling of data leads to loss of demonstration of effect recognizing the majority of persons following concussion do experience full recovery .

Bigler [97] further provides a critical review of neuropsychology testing and the field's approach to concussion that supports the position that post-concussion neurocognitive deficits may persist post-injury.

Screening through formal instruments assists in identifying cognitive symptoms that may significantly impact the ability to return to school and work. It is expected that cognition post-injury will improve over time and evidence suggests that early education and management of expectations regarding full recovery contribute to improved outcomes and to avoiding the misattribution of symptoms [98–100]. As with other symptoms following concussion, cognitive issues may stem from other symptoms, including headache pain, fatigue, mood alteration, and sensitivity to environmental stimuli such as light and noise. The impact of these factors on cognition needs to be taken into consideration when there is evidence of declining cognitive function over time.

More formal cognitive assessments should be considered for those patients with ongoing cognitive symptoms beyond the three-month time frame, when the ability to return to work or school is directly impacted [101]. Neuropsychological testing may assist in the identification of cognitive strengths and challenges and the subsequent impact on independent functioning. Identification of areas of relative weakness can help to guide rehabilitation therapy interventions with a goal of improving function.

Vestibular and Vestibular Ocular Reflex Dysfunction

Dizziness, including vertigo, is the second most common symptom following concussion, with 52% of patients reporting dizziness acutely, and up to 28% continue to report dizziness 1-year post-injury [23]. The vestibular system plays a key role in spatial orientation and the maintenance of balance by sensing motion of the head in space [102]. If the vestibular system is injured following a concussion, altered signals of head motion and position can result in symptoms of vertigo or dizziness, and functional balance impairment that can significantly interfere with work, school, sport, and family life [23, 27].

Dizziness and vertigo may be broadly classified as either peripheral vestibular, central nervous system, musculoskeletal (cervicogenic), or psychogenic in origin [3]. Peripheral vestibular disorders involve the vestibular organs of the inner ear or the vestibular branch of the 8th cranial nerve, and include benign paroxysmal positional vertigo (BPPV), labyrinthine concussion, perilymphatic fistula, or post-traumatic endolymphatic hydrops [3, 103]. The most common cause of peripheral vestibular impairment following concussion is benign paroxysmal positional vertigo (BPPV) [104], which has been reported in 28% of military personnel and 29% of children and adolescents with post-concussion dizziness [105, 106]. Post-concussion BPPV occurs when an impact to the head dislodges calcium crystals contained within the inner ear allowing them to float freely within one of three fluid-filled semi-circular canals, most commonly the posterior semi-circular canal

[106]. Changes in head position cause these crystals to shift, sending false signals of movement to the brain that induce a sensation of spinning (vertigo) along with nystagmus (involuntary, rapid movement of the eyes). Clinically, these episodes are characterized as brief (less than one minute), intense, triggered by head movement in relation to gravity, such as rolling in bed or bending over, and separated by periods of remission. Diagnosis is made based on patient history and physical examination.

Occasionally, a previous episode of vertigo following a vestibular disorder, anxiety, or concussion will trigger Persistent Postural-Perceptual Dizziness (PPPD or 3-P-D). Previously referred to as Chronic Subjective Dizziness, PPPD was defined as a disorder in 2015 to describe symptoms of dizziness, non-spinning vertigo, unsteadiness (feeling like you are going to fall), light headedness (feeling woozy), and mild dissociation (feeling spaced out) on most days for at least three months [107]. These symptoms are often made worse when a person stands or sits upright, is in a busy visual environment, sees complex visual patterns, or is traveling in a car or bus. There is no specific test for PPPD, and a diagnosis is made based on clinical presentation. Limited evidence suggests that PPPD is treatable using a combination of vestibular rehabilitation, antidepressant medication, CBT, and anxiety management strategies [108]. Further study is needed to evaluate the benefits of these treatments.

The vestibular system also acts to maintain gaze stabilization during head movement via the vestibular ocular reflex (VOR) [102]. Each pair of semi-circular canals is connected via motor neurons to a pair of extraocular muscles. When the VOR is intact, angular movement is sensed by semi-circular canals, which drives the eyes to move in an equal and opposite direction to the plane of movement [109]. Evidence suggests that VOR dysfunction occurs in 69% of adolescents following concussion and is associated with prolonged recovery and neurocognitive function [110]. An abnormal VOR following concussion can result in symptoms of dizziness, visual instability, and motion sensitivity.

Assessments of dizziness and balance impairment should include a thorough history and neurological examination focusing on gait, balance, vision, and vestibular disorders, including the Dix-Hallpike maneuver to examine for posterior canal BPPV. Use of a formal questionnaire such as the Dizziness Handicap Inventory can assist in evaluating the functional impact of dizziness [111] and assess the effectiveness of treatment interventions. Standardized measures of static balance such as the Balance Error Scoring System (BESS) [112] and dynamic balance using the Community Balance and Mobility Measure [113] can provide comparisons against baseline or normative data and assist in monitoring recovery. The head thrust test (HTT) is a simple clinical test of VOR function that assesses the ability of the patient to keep their eyes focused on a target during a rapid, unpredictable passive rotation of the head to the right or left. A rapid, corrective saccade to regain focus on the target indicates an abnormal VOR [114]. Dynamic visual acuity (DVA) is a functional measure of VOR that assesses the difference in visual stability while reading with the head still versus during rapid head turning [114].

Current international consensus statements and national concussion guidelines recommend vestibular rehabilitation be considered as part of a multimodal treatment program for post-concussion dizziness, vertigo, VOR dysfunction, and problems with balance and gait [2, 5]. Although the current literature base for the use of vestibular rehabilitation in patients with dizziness and balance problems following concussion is limited and of low quality, emerging evidence suggests that vestibular rehabilitation is an effective treatment strategy when symptoms persist [115]. When BPPV is present, a series of head movements is used to mechanically reposition the misplaced calcium crystals using canalith repositioning maneuvers, such as the Epley maneuver for posterior canal BPPV or the Barbeque maneuver for horizontal canal BPPV [116]. Canalith repositioning maneuvers are considered safe and effective, and overall prognosis is excellent, with the majority of patients experiencing symptom resolution following the initial treatment [116, 117]. A Cochrane review by McDonnell et al. [118] found that vestibular rehabilitation improved patient outcomes for unilateral peripheral vestibular dysfunction when combined with repositioning maneuvers for BPPV.

The goal of active vestibular rehabilitation is to compensate for functional vestibular deficits through the principles of adaptation, habituation, and substitution. Adaptation exercises are the most common home exercise program prescribed to retrain the vestibular ocular reflex following concussion [119]. Exercises involve coordinated eye-head movements where the patient turns their head from side to side while focusing on a target to promote gaze stabilization (e.g., VORx1). Exercises are progressed by increasing the speed of head movement, adding a visually distracting background, and starting with a stable base of support (e.g., sitting) to performing the exercise while walking. Habituation exercises involve small doses of repeated exposure to provocative movements to induce mild and tolerable symptoms of dizziness or unsteadiness. Symptom provocation should be brief, and the patient should recover quickly with rest. Exercises are designed to desensitize the central processing system in order to reduce the pathological response to stimuli [120]. An exercise program may be developed based on either the functional limitations of the patient or their response to the standardized Motion Sensitivity Quotient/Test [121]. Substitution exercises are designed to compensate for vestibular dysfunction by strengthening the proprioceptive system and providing visual strategies to improve postural stability [120]. Most commonly, the balance system is initially challenged with standing static balance exercises on level or foam surfaces, progressing from a stable base of support (e.g., double leg stance) to a more unstable base of support (e.g., single leg stance) [119]. Visual input can be removed by asking the patient to close their eyes to increase reliance on the proprioceptive system and mimic more challenging functional tasks like maintaining balance in a dark environment. The functional needs of the patient should be considered when prescribing dynamic balance exercises. These exercises should challenge the balance system in a way that mimics the activities to which the patient needs or wants to return. Common examples include walking with head turning, tandem gait, and obstacle avoidance [122]. The general framework for prescribing vestibular rehabilitation is based on a problem-oriented approach, whereby (1) vestibular impairments are identified on initial assessment,(2) an individualized exercise program is prescribed to address patient-specific symptoms and functional limitations, and (3) exercises are progressed by increasing exercise speed, adding functional activities, decreasing compensatory strategies, and incorporating environmental contexts [109]. Further research is required to determine the optimal exercise parameters (frequency, intensity, type, and time) and progression patterns to effectively treat vestibular dysfunction following concussion [115].

Vision and Ocular Motor Deficits

Vision problems are common following concussion and have been shown to predict prolonged recovery following concussion [123]. The incidence of vision problems has been reported in up to 75% of military individuals following blast injury [124] and as high as 90% in adults following concussion, compared with 22% of patients in a general optometric clinic [125]. A similarly high incidence of vision problems has been reported in up to 88% of adolescents following concussion [123, 126]. Vision problems may result from neurological damage to the cranial nerves or white matter tracts along either afferent or efferent cortical pathways [127]. This can contribute to ocular motor problems with accommodation, version deficits (pursuits, saccades, nystagmus), or vergence dysfunction (convergence, insufficiency) [128]. Problems with accommodation may result in difficulty focusing on near visual work or shifting focus from near to far. Version deficits can contribute to problems with reading speed, accuracy, or efficiency, tracking a moving object such as a puck during a hockey game, or scanning the environment when driving or looking for an object on a shelf. Vergence dysfunction can lead to difficulty keeping an object in singular vision as it moves closer or farther away. These functional deficits can contribute to symptoms such as blurred or double vision, eye strain, visual fatigue, motion sensitivity, headaches, dizziness, problems with balance, or poor concentration [123, 129], which can significantly impact a person's ability to return to visually intensive activities such as driving, work or school performance, or screen-based tasks [123, 129]. Non-ocular motor deficits include photosensitivity, visual field deficits, and visual-midline shift, a perceived shifting of the midline for an individual [130].

A comprehensive assessment for patients with persistent visual problems following concussion includes a focused clinical history, a functional-based symptom questionnaire, and basic eye/vision examination [131]. Clinical practice guidelines from the US Department of Veteran Affairs and Department of Defense (VA/DoD) recommend that the clinical history should obtain information regarding the mechanism of injury, associated injuries and symptoms, comorbidities, and medication use [3]. Clinicians should inquire as to visual problems or symptoms associated with reading, computer use, or driving. Standardized questionnaires such as the Convergence Insufficiency Symptom Survey [132, 133] or Pediatric Visually Induced Dizziness Questionnaire [134] can assist in identifying functional vision problems in adults and children with persistent post-concussion symptoms. A basic eye/vision assessment includes an examination of visual acuity, visual fields, pupillary response, eye movement, nystagmus, an external exam, ocular motor function, and fundoscopy [131]. The Vestibular Ocular Motor Screening (VOMS) is a brief clinical tool for vestibular and ocular motor impairment that assesses symptom provocation with the following tests: (1) smooth pursuits, (2) horizontal and vertical saccades, (3) convergence, (4) horizontal VOR, and (5) visual motion sensitivity [135]. The VOMS has been shown to differentiate between concussed athletes and healthy controls and is associated with increased recovery time when symptom provocation or clinical abnormalities are present [135, 136].

A systematic review of ocular motor-based vision assessments by Hunt et al. [127] demonstrated that measurements of saccades, smooth pursuits, and vergence have the potential to detect vision changes following concussion; however, the authors concluded that current evidence is insufficient to warrant clinical recommendations regarding diagnosis or clinical recovery. A systematic review and meta-analysis that specifically focused on eye-tracking technology similarly by Snegireva et al. [137] concluded that while this technology may be promising for distinguishing those who have sustained a concussion from those who have not, there is insufficient evidence to support its clinical practice currently and further research is required.

Rehabilitation for vision problems has garnered significant interest for both patients and clinicians. Recent reviews have examined the evidence to support the assumption that improving ocular motor dysfunction post-concussion will reduce vision-related symptoms [129, 138]. Primary types of intervention include vision therapy and optical devices [129]. Vision therapy involves active exercises designed to train versional (smooth pursuits, saccades) and vergence (convergence insufficiency) ocular motor deficits, accommodation, and reading ability using home-based, office-based, or computer-based programs [129, 138]. Proposed optical devices include the use of spectacles, base-in or yolked prisms, binasal occlusion, and photochromatic lenses [129]. These reviews concluded that evidence supports the use of vision therapy to treat primary convergence insufficiency and accommodative insufficiency; however, current evidence on the efficacy of vision therapy and optical devices following concussion is limited, of poor quality, and insufficient to support clinical recommendations [138]. These findings are consistent with clinical practice guidelines by the Veteran Affairs/Department of Defense (VA/DoD) [3] that "limited evidence exists to demonstrate that vision rehabilitation reduces or eliminates functionally-limiting vision symptoms following mTBI." Given the limited evidence, the guidelines recommend that, for patients with prominent vision-related symptoms, a time-limited intervention of goal-oriented vision therapy may be considered and continued only if objective improvements are achieved.

Vision symptoms post-concussion remain a common symptom that requires a medical assessment to rule out significant issues such as ocular injury, cranial nerve injury, or visual field deficit. However, for the more common symptoms of light sensitivity, blurred vision, or tracking difficulties, no specific therapy has been objectively demonstrated to improve natural recovery.

Conclusion

Most individuals completely recover following concussion; however, up to one-third will experience ongoing physical, emotional, and cognitive symptoms beyond 3-months. Persistent post-concussion symptoms can negatively affect an individual's ability to work and participate in social activities, level of function, and quality of life. Objective biomarkers have an important role in understanding the pathophysiology of recovery but are not yet suitable for clinical application in the assessment or management of persistent post-concussion symptoms.

Identifying the cause of persistent post-concussion symptoms can therefore be challenging, as a combination of physiological, psychological, and comorbid causes may contribute to symptom profiles. Current management strategies highlight the importance of early education and an individualized approach to symptom-based treatment by an interdisciplinary team. While the recommendations in this chapter are based on current evidence, professional position statements, and international and national consensus-based guidelines, the current evidence base is limited and often of low quality. As concussion research is expanding at a rapid pace, it is expected that these recommendations will continue to evolve. Further research is needed to identify the pathophysiological cause of persistent post-concussion symptoms and the effectiveness of concussion-specific interventions.

References

1. Silverberg ND, Iaccarino MA, Panenka WJ, Iverson GL, McCulloch KL, Dams-O'Connor K, et al. Management of concussion and mild traumatic brain injury: a synthesis of practice guidelines. Arch Phys Med Rehabil. 2020;101(2):382–93.
2. McCrory P, Meeuwisse W, Dvořák J, Aubry M, Bailes J, Broglio S, et al. Consensus statement on concussion in sport-the 5th international conference on concussion in sport held in Berlin, October 2016. Br J Sports Med. 2017;51(11):838–47.
3. Veterans Affairs/Department of Defense. Management of Concussion-Mild Traumatic Brain Injury (MTBI) clinical practice guidelines [Internet]. VA/DoD Clinical Practice Guidelines. 2016 [cited 2018 Jun 11]. Available from: https://www.healthquality.va.gov/guidelines/rehab/mtbi/.
4. Ontario Neurotrauma Foundation. Guidelines on diagnosing and managing pediatric concussion [Internet]. 2014 [cited 2018 Jun 11]. Available from: https://web.archive.org/web/20180416105612/http://onf.org/documents/guidelines-diagnosing-and-managing-pediatric-concussion.
5. Ontario Neurotrauma Foundation. Guidelines for concussion/mild traumatic brain injury & persistent symptoms. 2018 [cited 2021 Aug 17]. Available from: https://braininjuryguidelines.org/concussion/.
6. Harmon KG, Clugston JR, Dec K, Hainline B, Herring S, Kane SF, et al. American Medical Society for Sports Medicine position statement on concussion in sport. Br J Sports Med. 2019;53(4):213–25.
7. The New England Journal of Medicine. Concussion: guidelines and best practices [Internet]. Massachusetts Medical Society; [cited 2021 Aug 17]. Available from: https://web.archive.org/web/2015*/collections.nejm.org.

8. Reed N, Zemek R, Dawson J, Ledoux AA, Provvidenza C, Paniccia M, et al. Living guideline for diagnosing and managing pediatric concussion. Toronto [Internet]. Ontario Neurotrauma Foundation; [cited 2021 Aug 17]. Available from: https://braininjuryguidelines.org/pediatricconcussion/.
9. Lumba-Brown A. Centers for Disease Control and Prevention guideline on the diagnosis and management of mild traumatic brain injury among children. JAMA Pediatr. 2018;172:e182853.
10. American Association of Neurological Surgeons. Concussion [Internet]. [cited 2021 Aug 17]. Available from: https://www.aans.org/en/Patients/Neurosurgical-Conditions%20and%20Treatments/Concussion.
11. Ferdinand Pennock K, McKenzie B, McClemont Steacy L, Mainwaring L. Under-reporting of sport-related concussions by adolescent athletes: a systematic review. Int Rev Sport Exerc Psychol. 2020;6:1–27.
12. The Concussion in Sports Group. Sport concussion assessment tool - 5th edition. Br J Sports Med. 2017 Apr 26;bjsports-2017-097506SCAT5.
13. King NS, Crawford S, Wenden FJ, Moss NE, Wade DT. The Rivermead Post Concussion Symptoms Questionnaire: a measure of symptoms commonly experienced after head injury and its reliability. J Neurol. 1995;242(9):587–92.
14. Lovell MR, Collins MW. Neuropsychological assessment of the college football player. J Head Trauma Rehabil. 1998;13(2):9–26.
15. Stiell IG, Wells GA, Vandemheen K, Clement C, Lesiuk H, Laupacis A, et al. The Canadian CT Head Rule for patients with minor head injury. Lancet Lond Engl. 2001;357(9266):1391–6.
16. Kuppermann N, Holmes JF, Dayan PS, Hoyle JD, Atabaki SM, Holubkov R, et al. Identification of children at very low risk of clinically-important brain injuries after head trauma: a prospective cohort study. Lancet Lond Engl. 2009;374(9696):1160–70.
17. Osmond MH, Klassen TP, Wells GA, Correll R, Jarvis A, Joubert G, et al. CATCH: a clinical decision rule for the use of computed tomography in children with minor head injury. CMAJ Can Med Assoc J J Assoc Medicale Can. 2010;182(4):341–8.
18. Concussion Awareness Training Tool. Concussion resources for parents & caregivers [Internet]. [cited 2021 Aug 17]. Available from: https://webcache.googleusercontent.com/search?q=cache:YIgqBkX2wN0J:https://cattonline.com/wp-content/uploads/2021/05/Concussion-Resources-for-Parents-and-Caregivers-CATT-V3-April-2021-.pdf+&cd=1&hl=en&ct=clnk&gl=ca.
19. Silverberg ND, Iverson GL. Etiology of the post-concussion syndrome: physiogenesis and psychogenesis revisited. NeuroRehabilitation. 2011;29(4):317–29.
20. Preece MHW, Geffen GM, Horswill MS. Return-to-driving expectations following mild traumatic brain injury. Brain Inj. 2013;27(1):83–91.
21. Preece MHW, Horswill MS, Geffen GM. Driving after concussion: the acute effect of mild traumatic brain injury on drivers' hazard perception. Neuropsychology. 2010;24(4):493–503.
22. Baker A, Unsworth CA, Lannin NA. Fitness-to-drive after mild traumatic brain injury: mapping the time trajectory of recovery in the acute stages post injury. Accid Anal Prev. 2015;79:50–5.
23. Theadom A, Parag V, Dowell T, McPherson K, Starkey N, Barker-Collo S, et al. Persistent problems 1 year after mild traumatic brain injury: a longitudinal population study in New Zealand. Br J Gen Pract J R Coll Gen Pract. 2016;66(642):e16–23.
24. Zemek R, Barrowman N, Freedman SB, Gravel J, Gagnon I, McGahern C, et al. Clinical risk score for persistent postconcussion symptoms among children with acute concussion in the ED. JAMA. 2016;315(10):1014–25.
25. Varner C, Thompson C, de Wit K, Borgundvaag B, Houston R, McLeod S. Predictors of persistent concussion symptoms in adults with acute mild traumatic brain injury presenting to the emergency department. CJEM. 2021;23(3):365–73.
26. Silverberg ND, Gardner AJ, Brubacher JR, Panenka WJ, Li JJ, Iverson GL. Systematic review of multivariable prognostic models for mild traumatic brain injury. J Neurotrauma. 2015;32(8):517–26.

27. Cnossen MC, van der Naalt J, Spikman JM, Nieboer D, Yue JK, Winkler EA, et al. Prediction of persistent post-concussion symptoms after mild traumatic brain injury. J Neurotrauma. 2018;35(22):2691–8.
28. Langer LK, Alavinia SM, Lawrence DW, Munce SEP, Kam A, Tam A, et al. Prediction of risk of prolonged post-concussion symptoms: derivation and validation of the TRICORDRR (Toronto Rehabilitation Institute Concussion Outcome Determination and Rehab Recommendations) score. PLoS Med. 2021;18(7):e1003652.
29. Ponsford J, Willmott C, Rothwell A, Cameron P, Kelly A-M, Nelms R, et al. Impact of early intervention on outcome following mild head injury in adults. J Neurol Neurosurg Psychiatry. 2002;73(3):330–2.
30. Thibault-Halman G, Fenerty L, Taylor P, Kureshi N, Walling S, Clarke D. P.016 early telephone follow-up for traumatic brain injury patients using the Rivermead post-concussion symptoms questionnaire. Can J Neurol Sci J Can Sci Neurol. 2017;44(S2):S17–8.
31. Silverberg ND, Iverson GL. Is rest after concussion "the best medicine?": recommendations for activity resumption following concussion in athletes, civilians, and military service members. J Head Trauma Rehabil. 2013;28(4):250–9.
32. Leddy JJ, Haider MN, Ellis M, Willer BS. Exercise is medicine for concussion. Curr Sports Med Rep. 2018;17(8):262–70.
33. Leddy JJ, Willer B. Use of graded exercise testing in concussion and return-to-activity management. Curr Sports Med Rep. 2013;12(6):370–6.
34. McIntyre M, Kempenaar A, Amiri M, Alavinia SM, Kumbhare D. The role of subsymptom threshold aerobic exercise for persistent concussion symptoms in patients with postconcussion syndrome: a systematic review. Am J Phys Med Rehabil. 2020;99(3):257–64.
35. Ontario Neurotrauma Foundation. Standards for post-concussion care [Internet]. 2017 [cited 2021 Aug 17]. Available from: https://concussionsontario.org/wp-content/uploads/2017/06/ONF-Standards-for-Post-Concussion-Care-June-8-2017.pdf.
36. Concussions Ontario. Standards for high quality post-concussion services and concussion clinics [Internet]. [cited 2021 Aug 17]. Available from: http://concussionsontario.org/healthcareprofessionals/standards/standards-for-high-quality-post-concussion-services-and-concussion-clinics/.
37. Concussions Ontario. Referral indicators [Internet]. [cited 2021 Aug 17]. Available from: https://concussionsontario.org/healthcareprofessionals/standards/tools-resources/referral-indicators/.
38. Sady MD, Vaughan CG, Gioia GA. Psychometric characteristics of the postconcussion symptom inventory in children and adolescents. Arch Clin Neuropsychol Off J Natl Acad Neuropsychol. 2014;29(4):348–63.
39. Van Ierssel J, Sveistrup H, Marshall S, Graham I. The concussion recovery questionnaire (CORE-Q): conceptual model development and item generation of a concussion-specific measure of functional status. Brain Inj. 2020;34(5):619–29.
40. Dwyer B, Katz DI. Postconcussion syndrome. In: Handbook of clinical neurology [Internet]. Elsevier; 2018. p. 163–78. Available from: https://linkinghub.elsevier.com/retrieve/pii/B9780444639547000173.
41. Lagacé-Legendre C, Boucher V, Robert S, Tardif P-A, Ouellet M-C, de Guise E, et al. Persistent postconcussion symptoms: an expert consensus-based definition using the Delphi method. J Head Trauma Rehabil. 2021;36(2):96–102.
42. Iverson GL, McCracken LM. "Postconcussive" symptoms in persons with chronic pain. Brain Inj. 1997;11(11):783–90.
43. Iverson GL, Lange RT. Examination of "postconcussion-like" symptoms in a healthy sample. Appl Neuropsychol. 2003;10(3):137–44.
44. McCrea M, Iverson GL, McAllister TW, Hammeke TA, Powell MR, Barr WB, et al. An integrated review of recovery after mild traumatic brain injury (MTBI): implications for clinical management. Clin Neuropsychol. 2009;23(8):1368–90.
45. Voormolen DC, Cnossen MC, Polinder S, Gravesteijn BY, Von Steinbuechel N, Real RGL, et al. Prevalence of post-concussion-like symptoms in the general population in Italy, the Netherlands and the United Kingdom. Brain Inj. 2019;33(8):1078–86.

46. Ruff RM, Weyer JC. Myths and mild traumatic brain injury. Psychol Inj Law. 2009;2(1):34–42.
47. Ruff RM. Mild traumatic brain injury and neural recovery: rethinking the debate. Gentry T, editor. NeuroRehabilitation. 2011;28(3):167–80.
48. Hiploylee C, Dufort PA, Davis HS, Wennberg RA, Tartaglia MC, Mikulis D, et al. Longitudinal study of postconcussion syndrome: not everyone recovers. J Neurotrauma. 2017;34(8):1511–23.
49. Pinner M, Børgensen SE, Jensen R, Birket-Smith M, Gade A, Riis JO. Consensus-driven guidelines regarding commotio cerebri (Konsensusrapport om commotio cerebri (hjernerystelse) og det postcommotionelle syndrom). 2003:1–82.
50. Hou R, Moss-Morris R, Peveler R, Mogg K, Bradley BP, Belli A. When a minor head injury results in enduring symptoms: a prospective investigation of risk factors for postconcussional syndrome after mild traumatic brain injury. J Neurol Neurosurg Psychiatry. 2012;83(2):217–23.
51. Nolin P, Villemure R, Heroux L. Determining long-term symptoms following mild traumatic brain injury: method of interview affects self-report. Brain Inj. 2006;20(11):1147–54.
52. Sawchyn JM, Brulot MM, Strauss E. Note on the use of the postconcussion syndrome checklist. Arch Clin Neuropsychol Off J Natl Acad Neuropsychol. 2000;15(1):1–8.
53. Silverberg ND, Panenka WJ, Iverson GL. Work productivity loss after mild traumatic brain injury. Arch Phys Med Rehabil. 2018;99(2):250–6.
54. Toronto Rehabilitation Institute [Internet]. Available from: https://www.uhn.ca/TorontoRehab.
55. Gladstone J. From psychoneurosis to ICHD-2: an overview of the state of the art in post-traumatic headache. Headache. 2009;49(7):1097–111.
56. Dikmen S, Machamer J, Fann JR, Temkin NR. Rates of symptom reporting following traumatic brain injury. J Int Neuropsychol Soc JINS. 2010;16(3):401–11.
57. Lew HL, Lin P-H, Fuh J-L, Wang S-J, Clark DJ, Walker WC. Characteristics and treatment of headache after traumatic brain injury: a focused review. Am J Phys Med Rehabil. 2006;85(7):619–27.
58. Lew HL, Poole JH, Guillory SB, Salerno RM, Leskin G, Sigford B. Persistent problems after traumatic brain injury: the need for long-term follow-up and coordinated care. J Rehabil Res Dev. 2006;43(2):vii–x.
59. International Headache Society. International classification of headache disorders 3rd edition [Internet]. [cited 2021 Aug 17]. Available from: https://www.ichd-3.org/.
60. Parachute. Concussion guide for athletes [Internet]. [cited 2021 Aug 17]. Available from: https://parachute.ca/wp-content/uploads/2019/06/Concussion-Guide-for-Athletes.pdf.
61. Diener H-C, Limmroth V. Medication-overuse headache: a worldwide problem. Lancet Neurol. 2004;3(8):475–83.
62. Ontario Neurotrauma Foundation. Guideline for concussion/mild traumatic brain injury & prolonged symptoms 3rd Edition, for adults over 18 years of age - Appendix 6.7 - Prophylactic Therapy [Internet]. [cited 2021 Aug 17]. Available from: https://braininjuryguidelines.org/concussion/.
63. Wiseman-Hakes C, Colantonio A, Gargaro J. Sleep and wake disorders following traumatic brain injury: a systematic review of the literature. Crit Rev Phys Rehabil Med. 2009;21(3–4):317–74.
64. Castriotta RJ, Wilde MC, Lai JM, Atanasov S, Masel BE, Kuna ST. Prevalence and consequences of sleep disorders in traumatic brain injury. J Clin Sleep Med JCSM Off Publ Am Acad Sleep Med. 2007;3(4):349–56.
65. Ouellet M-C, Beaulieu-Bonneau S, Morin CM. Insomnia in patients with traumatic brain injury: frequency, characteristics, and risk factors. J Head Trauma Rehabil. 2006;21(3):199–212.
66. Wiseman-Hakes C, Murray B, Moineddin R, Rochon E, Cullen N, Gargaro J, et al. Evaluating the impact of treatment for sleep/wake disorders on recovery of cognition and communication in adults with chronic TBI. Brain Inj. 2013;27(12):1364–76.

67. Chaput G, Giguère J-F, Chauny J-M, Denis R, Lavigne G. Relationship among subjective sleep complaints, headaches, and mood alterations following a mild traumatic brain injury. Sleep Med. 2009;10(7):713–6.
68. Chan LG, Feinstein A. Persistent sleep disturbances independently predict poorer functional and social outcomes 1 year after mild traumatic brain injury. J Head Trauma Rehabil. 2015;30(6):E67–75.
69. Ontario Neurotrauma Foundation. Guideline for concussion/mild traumatic brain injury & persistent symptoms 3rd Edition, for adults over 18 years of age - sleep-wake disturbances [Internet]. [cited 2021 Aug 17]. Available from: https://braininjuryguidelines.org/concussion/.
70. Ouellet M-C, Morin CM. Cognitive behavioral therapy for insomnia associated with traumatic brain injury: a single-case study. Arch Phys Med Rehabil. 2004;85(8):1298–302.
71. Ouellet M-C, Morin CM. Efficacy of cognitive-behavioral therapy for insomnia associated with traumatic brain injury: a single-case experimental design. Arch Phys Med Rehabil. 2007;88(12):1581–92.
72. Espie CA, MacMahon KMA, Kelly H-L, Broomfield NM, Douglas NJ, Engleman HM, et al. Randomized clinical effectiveness trial of nurse-administered small-group cognitive behavior therapy for persistent insomnia in general practice. Sleep. 2007;30(5):574–84.
73. Meares S, Shores EA, Batchelor J, Baguley IJ, Chapman J, Gurka J, et al. The relationship of psychological and cognitive factors and opioids in the development of the postconcussion syndrome in general trauma patients with mild traumatic brain injury. J Int Neuropsychol Soc JINS. 2006;12(6):792–801.
74. Ponsford J, Willmott C, Rothwell A, Cameron P, Kelly AM, Nelms R, et al. Factors influencing outcome following mild traumatic brain injury in adults. J Int Neuropsychol Soc JINS. 2000;6(5):568–79.
75. Ontario Neurotrauma Foundation. Guideline for concussion/mild traumatic brain injury & persistent symptoms 3rd Edition, for adults over 18 years of age - mental health disorders [Internet]. [cited 2021 Aug 17]. Available from: https://braininjuryguidelines.org/concussion/.
76. Mollayeva T, Kendzerska T, Mollayeva S, Shapiro CM, Colantonio A, Cassidy JD. A systematic review of fatigue in patients with traumatic brain injury: the course, predictors and consequences. Neurosci Biobehav Rev. 2014;47:684–716.
77. Cantor JB, Ashman T, Gordon W, Ginsberg A, Engmann C, Egan M, et al. Fatigue after traumatic brain injury and its impact on participation and quality of life. J Head Trauma Rehabil. 2008;23(1):41–51.
78. Bay E, de-Leon MB. Chronic stress and fatigue-related quality of life after mild to moderate traumatic brain injury. J Head Trauma Rehabil. 2011;26(5):355–63.
79. Englander J, Bushnik T, Oggins J, Katznelson L. Fatigue after traumatic brain injury: association with neuroendocrine, sleep, depression and other factors. Brain Inj. 2010;24(12):1379–88.
80. Bushnik T, Englander J, Katznelson L. Fatigue after TBI: association with neuroendocrine abnormalities. Brain Inj. 2007;21(6):559–66.
81. Schnieders J, Willemsen D, de Boer H. Factors contributing to chronic fatigue after traumatic brain injury. J Head Trauma Rehabil. 2012;27(6):404–12.
82. Sullivan KA, Blaine H, Kaye S-A, Theadom A, Haden C, Smith SS. A systematic review of psychological interventions for sleep and fatigue after mild traumatic brain injury. J Neurotrauma. 2018;35(2):195–209.
83. Northern Lincolnshire and Goole National Health Institute. Fatigue: "Pacing" as a Strategy [Internet]. [cited 2021 Aug 17]. Available from: https://www.nlg.nhs.uk/content/uploads/2015/09/IFP-0875-Fatigue-Pacing-as-a-Strategy.pdf.
84. Zhang W-T, Wang Y-F. Efficacy of methylphenidate for the treatment of mental sequelae after traumatic brain injury. Medicine (Baltimore). 2017;96(25):e6960.
85. Johansson B, Wentzel A-P, Andréll P, Rönnbäck L, Mannheimer C. Long-term treatment with methylphenidate for fatigue after traumatic brain injury. Acta Neurol Scand. 2017;135(1):100–7.
86. Iaccarino MA, Philpotts LL, Zafonte R, Biederman J. Stimulant use in the management of mild traumatic brain injury: a qualitative literature review. J Atten Disord. 2020;24(2):309–17.

87. Wylie GR, Freeman K, Thomas A, Shpaner M, OKeefe M, Watts R, et al. Cognitive improvement after mild traumatic brain injury measured with functional neuroimaging during the acute period. PLoS One. 2015;10(5):e0126110.
88. Rabinowitz AR, Levin HS. Cognitive sequelae of traumatic brain injury. Psychiatr Clin North Am. 2014;37(1):1–11.
89. Metting Z, Rödiger LA, Stewart RE, Oudkerk M, De Keyser J, van der Naalt J. Perfusion computed tomography in the acute phase of mild head injury: regional dysfunction and prognostic value. Ann Neurol. 2009;66(6):809–16.
90. McInnes K, Friesen CL, MacKenzie DE, Westwood DA, Boe SG. Mild traumatic brain injury (mTBI) and chronic cognitive impairment: a scoping review. PLoS One. 2017;12(4):e0174847.
91. Nordström A, Edin BB, Lindström S, Nordström P. Cognitive function and other risk factors for mild traumatic brain injury in young men: nationwide cohort study. BMJ. 2013;346:f723.
92. Binder LM, Rohling ML, Larrabee GJ. A review of mild head trauma. Part I: meta-analytic review of neuropsychological studies. J Clin Exp Neuropsychol. 1997;19(3):421–31.
93. Schretlen DJ, Shapiro AM. A quantitative review of the effects of traumatic brain injury on cognitive functioning. Int Rev Psychiatry Abingdon Engl. 2003;15(4):341–9.
94. Frencham KAR, Fox AM, Maybery MT. Neuropsychological studies of mild traumatic brain injury: a meta-analytic review of research since 1995. J Clin Exp Neuropsychol. 2005;27(3):334–51.
95. Belanger HG, Curtiss G, Demery JA, Lebowitz BK, Vanderploeg RD. Factors moderating neuropsychological outcomes following mild traumatic brain injury: a meta-analysis. J Int Neuropsychol Soc JINS. 2005;11(3):215–27.
96. Oldenburg C, Lundin A, Edman G, Nygren-de Boussard C, Bartfai A. Cognitive reserve and persistent post-concussion symptoms—a prospective mild traumatic brain injury (mTBI) cohort study. Brain Inj. 2016;30(2):146–55.
97. Bigler ED. Neuropsychology and clinical neuroscience of persistent post-concussive syndrome. J Int Neuropsychol Soc. 2008;14(1):1–22.
98. Cooper DB, Bunner AE, Kennedy JE, Balldin V, Tate DF, Eapen BC, et al. Treatment of persistent post-concussive symptoms after mild traumatic brain injury: a systematic review of cognitive rehabilitation and behavioral health interventions in military service members and veterans. Brain Imaging Behav. 2015;9(3):403–20.
99. Brunger H, Ogden J, Malia K, Eldred C, Terblanche R, Mistlin A. Adjusting to persistent post-concussive symptoms following mild traumatic brain injury and subsequent psycho-educational intervention: a qualitative analysis in military personnel. Brain Inj. 2014;28(1):71–80.
100. Broshek DK, De Marco AP, Freeman JR. A review of post-concussion syndrome and psychological factors associated with concussion. Brain Inj. 2015;29(2):228–37.
101. Ontario Neurotrauma Foundation. Guideline for concussion/mild traumatic brain injury & persistent symptoms 3rd edition, for adults over 18 years of age - cognitive difficulties [Internet]. [cited 2021 Aug 17]. Available from: https://braininjuryguidelines.org/concussion/.
102. Cullen KE. The vestibular system: multimodal integration and encoding of self-motion for motor control. Trends Neurosci. 2012;35(3):185–96.
103. Brodsky JR, Shoshany TN, Lipson S, Zhou G. Peripheral vestibular disorders in children and adolescents with concussion. Otolaryngol--Head Neck Surg Off J Am Acad Otolaryngol-Head Neck Surg. 2018;159(2):365–70.
104. Haripriya GR, Mary P, Dominic M, Goyal R, Sahadevan A. Incidence and treatment outcomes of post traumatic BPPV in traumatic brain injury patients. Indian J Otolaryngol Head Neck Surg. 2018;70(3):337–41.
105. Hoffer ME, Gottshall KR, Moore R, Balough BJ, Wester D. Characterizing and treating dizziness after mild head trauma. Otol Neurotol Off Publ Am Otol Soc Am Neurotol Soc Eur Acad Otol Neurotol. 2004;25(2):135–8.
106. Wang A, Zhou G, Kawai K, O'Brien M, Shearer AE, Brodsky JR. Benign paroxysmal positional vertigo in children and adolescents with concussion. Sports Health. 2021;13(4):380–6.

107. Staab JP, Eckhardt-Henn A, Horii A, Jacob R, Strupp M, Brandt T, et al. Diagnostic criteria for persistent postural-perceptual dizziness (PPPD): consensus document of the committee for the classification of vestibular disorders of the Bárány Society. J Vestib Res Equilib Orientat. 2017;27(4):191–208.
108. Popkirov S, Stone J, Holle-Lee D. Treatment of persistent postural-perceptual dizziness (PPPD) and related disorders. Curr Treat Options Neurol. 2018;20(12):50.
109. Herdman S, Clendaniel RA, editors. Vestibular rehabilitation [Internet]. 4th edn. Philadelphia: F.A.DavisCompany;2014.p.630.(Contemporaryperspectivesinrehabilitation).Availablefrom: https://www.fadavis.com/product/physical-therapy-vestibular-rehabilitation-herdman-4.
110. Corwin DJ, Wiebe DJ, Zonfrillo MR, Grady MF, Robinson RL, Goodman AM, et al. Vestibular deficits following youth concussion. J Pediatr. 2015;166(5):1221–5.
111. Jacobson GP, Newman CW. The development of the dizziness handicap inventory. Arch Otolaryngol Head Neck Surg. 1990;116(4):424–7.
112. Bell DR, Guskiewicz KM, Clark MA, Padua DA. Systematic review of the balance error scoring system. Sports Health Multidiscip Approach. 2011;3(3):287–95.
113. Takacs J, Garland SJ, Carpenter MG, Hunt MA. Validity and reliability of the community balance and mobility scale in individuals with knee osteoarthritis. Phys Ther. 2014;94(6):866–74.
114. Schubert MC, Minor LB. Vestibulo-ocular physiology underlying vestibular hypofunction. Phys Ther. 2004;84(4):373–85.
115. Murray DA, Meldrum D, Lennon O. Can vestibular rehabilitation exercises help patients with concussion? A systematic review of efficacy, prescription and progression patterns. Br J Sports Med. 2017;51(5):442–51.
116. Bhattacharyya N, Gubbels SP, Schwartz SR, Edlow JA, El-Kashlan H, Fife T, et al. Clinical practice guideline: benign paroxysmal positional vertigo (Update). Otolaryngol--Head Neck Surg Off J Am Acad Otolaryngol-Head Neck Surg. 2017;156(3_suppl):S1–47.
117. Prokopakis E, Vlastos IM, Tsagournisakis M, Christodoulou P, Kawauchi H, Velegrakis G. Canalith repositioning procedures among 965 patients with benign paroxysmal positional vertigo. Audiol Neurootol. 2013;18(2):83–8.
118. McDonnell MN, Hillier SL. Vestibular rehabilitation for unilateral peripheral vestibular dysfunction. Cochrane ENT Group, editor. Cochrane Database Syst Rev [Internet]. 2015 Jan 13 [cited 2021 Aug 17]. Available from: https://doi.org/10.1002/14651858.CD005397.pub4.
119. Alsalaheen BA, Whitney SL, Mucha A, Morris LO, Furman JM, Sparto PJ. Exercise prescription patterns in patients treated with vestibular rehabilitation after concussion. Physiother Res Int J Res Clin Phys Ther. 2013;18(2):100–8.
120. Aligene K, Lin E. Vestibular and balance treatment of the concussed athlete. NeuroRehabilitation. 2013;32(3):543–53.
121. Akin FW, Davenport MJ. Validity and reliability of the Motion Sensitivity Test. J Rehabil Res Dev. 2003;40(5):415–21.
122. Alsalaheen BA, Mucha A, Morris LO, Whitney SL, Furman JM, Camiolo-Reddy CE, et al. Vestibular rehabilitation for dizziness and balance disorders after concussion. J Neurol Phys Ther JNPT. 2010;34(2):87–93.
123. Master CL, Master SR, Wiebe DJ, Storey EP, Lockyer JE, Podolak OE, et al. Vision and vestibular system dysfunction predicts prolonged concussion recovery in children. Clin J Sport Med Off J Can Acad Sport Med. 2018;28(2):139–45.
124. Cockerham GC, Goodrich GL, Weichel LED, Orcutt JC, Rizzo JF, Bower CKS, et al. Eye and visual function in traumatic brain injury. J Rehabil Res Dev. 2009;46(6):811.
125. Lara F, Cacho P, García A, Megías R. General binocular disorders: prevalence in a clinic population. Ophthalmic Physiol Opt J Br Coll Ophthalmic Opt Optom. 2001;21(1):70–4.
126. Master CL, Scheiman M, Gallaway M, Goodman A, Robinson RL, Master SR, et al. Vision diagnoses are common after concussion in adolescents. Clin Pediatr (Phila). 2016;55(3):260–7.
127. Hunt AW, Mah K, Reed N, Engel L, Keightley M. Oculomotor-based vision assessment in mild traumatic brain injury: a systematic review. J Head Trauma Rehabil. 2016;31(4):252–61.

128. Kapoor N, Ciuffreda KJ. Vision disturbances following traumatic brain injury. Curr Treat Options Neurol. 2002;4(4):271–80.
129. Simpson-Jones ME, Hunt AW. Vision rehabilitation interventions following mild traumatic brain injury: a scoping review. Disabil Rehabil. 2019;41(18):2206–22.
130. Ciufredda KJ, Ludlam DP. Conceptual model of optometric vision care in mild traumatic brain injury. J Behav Optom. 2011;22:10–2.
131. Military Health System. Assessment and management of visual dysfunction associated with mild traumatic brain injury - Department of Defense Clinical Recommendation [Internet]. [cited 2021 Aug 17]. Available from: https://health.mil/Reference-Center/Publications/2020/07/31/Assessment-and-Management-of-Visual-Dysfunction-Associated-with-Mild-TBI-Clinical-Recommendation.
132. Rouse MW, Borsting EJ, Mitchell GL, Scheiman M, Cotter SA, Cooper J, et al. Validity and reliability of the revised convergence insufficiency symptom survey in adults. Ophthalmic Physiol Opt J Br Coll Ophthalmic Opt Optom. 2004;24(5):384–90.
133. Borsting EJ, Rouse MW, Mitchell GL, Scheiman M, Cotter SA, Cooper J, et al. Validity and reliability of the revised convergence insufficiency symptom survey in children aged 9 to 18 years. Optom Vis Sci Off Publ Am Acad Optom. 2003;80(12):832–8.
134. Pavlou M, Whitney SL, Alkathiry AA, Huett M, Luxon LM, Raglan E, et al. Visually induced dizziness in children and validation of the pediatric visually induced dizziness questionnaire. Front Neurol. 2017;8:656.
135. Mucha A, Collins MW, Elbin RJ, Furman JM, Troutman-Enseki C, DeWolf RM, et al. A brief vestibular/ocular motor screening (VOMS) assessment to evaluate concussions: preliminary findings. Am J Sports Med. 2014;42(10):2479–86.
136. Anzalone AJ, Blueitt D, Case T, McGuffin T, Pollard K, Garrison JC, et al. A positive vestibular/ocular motor screening (VOMS) is associated with increased recovery time after sports-related concussion in youth and adolescent athletes. Am J Sports Med. 2017;45(2):474–9.
137. Snegireva N, Derman W, Patricios J, Welman KE. Eye tracking technology in sports-related concussion: a systematic review and meta-analysis. Physiol Meas. 2018;39(12):12TR01.
138. Barton JJS, Ranalli PJ. Vision therapy: ocular motor training in mild traumatic brain injury. Ann Neurol. 2020;88(3):453–61.
139. Dwyer B, Katz DI. Postconcussion syndrome. Handb Clin Neurol. 2018;158:163–78. https://doi.org/10.1016/B978-0-444-63954-7.00017-3.

Chapter 9
Pharmacological Therapies for Concussions

Edward D. Hall, Eugene Park, and Andrew J. Baker

Introduction

Pharmacotherapy in broad terms may target underlying biological processes or target symptom complexes such as headache. This chapter explores potential therapies aimed at the underlying biological processes. There is a spectrum of symptoms, signs, and imaging findings that co-varies with force intensity in traumatic brain injury (TBI). It is therefore reasonable to consider that the pathophysiology of concussion includes processes that are seen in moderate and severe TBI – but to a lesser extent. Therefore, the first section of this chapter explores the recent history of the inroads into the pharmacological therapies aimed at moderate and severe TBI as well as mild TBI and concussion with the assumption of relevance related to common mechanisms between severity categories.

The pharmacological approach to concussion can be framed as targeting secondary injury mechanisms and/or targeting symptoms associated with concussion and persistent concussion syndrome (PCS). As we understand concussion and PCS better, and the role of comorbidities that are either pre-existent or precipitated, the approach also includes the active identification of clinical entities that are amenable to evidence-based interventions. Of course, the goal is to achieve a clinical pathological correlation, as well as an understanding of the pathophysiology and its

E. D. Hall (✉)
Spinal Cord & Brain Injury Research Center (SCoBIRC), University of Kentucky College of Medicine, Lexington, KY, USA
e-mail: edhall@uky.edu

E. Park
Keenan Research Centre, Li Ka Shing Knowledge Institute at St. Michael's Hospital, Toronto, ON, Canada

A. J. Baker
Departments of Critical Care and Anesthesia, St. Michael's Hospital, Unity Health Toronto and University of Toronto, Toronto, ON, Canada

T. A. Schweizer, A. J. Baker (eds.), *Tackling the Concussion Epidemic*,
https://doi.org/10.1007/978-3-030-93813-0_9

precision mitigation in concussion. Unfortunately, there is generally a lack of evidence-based direction to guide treatments [1–3]. Furthermore, while there is evidence following concussions/mild traumatic brain injury (mTBI) of persisting cellular changes from laboratory-supported research, this is largely in the context of the absence of evident anatomical injury or lesions by basic clinical imaging [4–6] adding to the challenges of clinical research. This chapter, therefore, provides a description of recent history of the development of a pharmacologic approach to the secondary cellular mechanisms, largely in the context of moderate or severe TBI. This narrative arises from the assumption that the traumatic etiology in concussion points to a shared pathophysiology and pharmacologic potential. The chapter ends with a brief survey of the pharmacological approaches for treating symptoms of neurotransmitter dysfunction and concussion-related comorbidities, while recognizing the importance of a multi-dimensional approach, as discussed in many chapters of this book.

Pharmacological Approaches to Limit Neurodegenerative Mechanisms: Antioxidant Therapies for Concussions/Mild TBI and Moderate/Severe TBI

Traumatic brain injury (TBI) involves a "primary" mechanical insult to the brain, which initiates a rapidly evolving, "secondary" biochemical injury. This biochemical "cascade" that is responsible for the post-TBI neurodegenerative events that take place during the first minutes, hours, and days after injury may initiate progressive vascular, neuronal, and glial degeneration resulting in permanent mild, moderate, or severe neurological disability or death. TBI researchers have documented that the "secondary" injury process begins with the injury-induced depolarization of excitatory glutamate-releasing neurons, which causes excessive release of the brain's principal excitatory neurotransmitter glutamate. This results in overactivation of the N-methyl-D-aspartate (NMDA) receptor on downstream neurons, which causes "excitotoxic" intra-neuronal accumulation of calcium (Ca^{++}) and sodium (Na^{+}) [7, 8]. The increases in intracellular Ca^{++} activate the proteolytic enzymes calpain I and II, leading to progressive degradation of cytoskeletal neurofilament proteins (e.g., α-spectrin). Brain cellular mitochondria, in addition to their principal bioenergetic functions (electron transport and ATP synthesis) perform mitochondrial "buffering" of intracellular Ca^{++}. However, as the intra-mitochondrial matrix Ca^{++} accumulation builds to an extreme level, where it is above the cytoplasmic Ca^{++} "set point", it causes mitochondrial membrane potential ($\Delta\psi$)-dependent electron transport, ATP synthesis, and mitochondrial Ca^{++} buffering to fail, resulting release of previously buffered intra-mitochondrial Ca^{++} [9, 10].

Pathophysiological studies in male CF-I mice [11, 12], and Sprague-Dawley rats [13, 14] using controlled cortical impact (CCI)-induced TBI paradigms, demonstrated a gradually increasing formation of reactive oxygen species (ROS) and reactive nitrogen species (RNS) which initiate progressive lipid peroxidation (LP) that begins during the first post-TBI minutes and slowly increases over the first 4–12 hours. Beyond that time window, any opportunity for effective pharmacological LP inhibitory neuroprotection has closed. Nevertheless, our neuroprotective drug studies have shown that certain LP targeting antioxidants possess a clinically practical therapeutic window that ranges between 4 and 12 hours in both moderate and severe rodent and human TBI studies.

While ROS-/RNS-induced LP-mediated brain damage may be less intense in Glasgow Coma Score (GCS 13–15) "mild" TBI/concussion patients, recent studies have shown that ROS-/RNS-induced LP occurs in single "mild" TBI models (concussions) and more so in "repetitive concussions" that are allowed to occur within a "return-to-play" period limited 3 days since. Thus, pharmacological inhibition of LP appears to be a promising neuroprotective strategy in concussive TBIs which constitute ~75–80% of the acute TBI patient population.

Biochemistry of Free Radical-Induced Lipid Peroxidation, Protein Oxidation, and Carbonylation

One of the most extensively validated "secondary injury" mechanisms revealed in experimental TBI studies is the early post-traumatic increase in reactive oxygen species (ROS) and reactive nitrogen species (RNS) that cause oxygen radical-induced oxidative damage to brain cellular lipids and proteins [15, 16]. This chapter outlines the key sources of ROS, RNS, and their highly reactive (i.e., rapidly oxidizing) free radicals, the pathophysiological mechanisms associated with oxidative neural damage, and, most importantly, pharmacological antioxidants that have been shown to produce neuroprotective actions that limit ROS-/RNS-initiated neurodegeneration.

Superoxide Radical

The primordial oxygen free radical that comes from several pathophysiological sources involves the single electron (e^-) reduction of an oxygen molecule (O_2) to produce the superoxide radical (O_2^-). Superoxide can be generated from several sources; one of the main sources is $O_2^{\bullet-}$ leakage from complex I of the mitochondrial electron transport chain in Ca^{++}-overloaded brain mitochondria. However, $O_2^{\bullet-}$ is considered by many free radical chemists to be a modestly reactive radical, but nevertheless one that can react with other molecules to give rise to more reactive, and thus more damaging, radical species. The reason that $O_2^{\bullet-}$ is only modestly reactive is that it can act as either an oxidant that is capable of stealing an electron from another oxidizable molecule or as a reductant by which it donates its unpaired electron to another radical species, thus acting as an antioxidant. However, if $O_2^{\bullet-}$ reacts with a proton (H^+) to form a hydroperoxyl radical ($HO^{\bullet}_2$), this results in a superoxide form that is much more likely to trigger LP (i.e., to act as an electron thief). This is more likely to occur in injured tissue where tissue acidosis is present that favors the predominance of $HO^{\bullet}_2$.

Superoxide Dismutase, Pro-oxidant Effects of Iron, and Tissue Lactic Acidosis

One of the most important endogenous antioxidants is the enzyme superoxide dismutase (SOD) which rapidly catalyzes the dismutation of O_{2-} into H_2O_2 and oxygen. At low pH, O_{2-} can dismutate spontaneously. The formation of highly reactive oxygen radicals, which have unpaired electron(s) in their outer molecular orbitals, and the propagation of LP chain reactions are fueled by non-radical ROS, which do not have unpaired electron(s) but are chemically reactive. For example, OH· radicals are generated in the iron-catalyzed Fenton reaction where ferrous iron (Fe^{2+}) is oxidized to form the highly reactive OH· in the presence of H_2O_2 ($Fe^{2+} + H_2O_2 \rightarrow Fe^{3+} + OH\cdot + OH-$). Superoxide, acting as a reducing agent (i.e., an electron-donating antioxidant), can donate its unpaired electron to ferric iron (Fe^{3+}), cycling it back to the ferrous state (Fe^{2+}, via the Haber-Weiss reaction $O_{2-} + Fe^{3+} \rightarrow Fe^{2+} + O_2$). This sets up additional Fe^{2+}-catalyzed Fenton reactions and increased production of OH·.

Under physiological conditions, iron is tightly regulated by its transport protein transferrin and the storage protein, ferritin, both of which bind the ferric (Fe^{3+}) form of iron. However, this reversible bond of transferrin and ferritin with ferric iron decreases with declining pH (below pH 7). Tissue acidosis is known to occur in the traumatized CNS and to cause the release of iron and initiation of iron-dependent oxygen radical production and LP. A second source of iron comes from hemoglobin released during injury-induced brain tissue hemorrhage.

Peroxynitrite and Its Highly Reactive Radicals

Although $O_2^{\bullet-}$ is much less reactive than OH· radical, its reaction with nitric oxide (NO.) radical forms the highly reactive RNS peroxynitrite (PN: $ONOO_-$). This reaction (O_{2-} + NO. → ONOO-) occurs with a very fast rate constant which out-competes mitochondrial manganese SOD's ability to convert $O_2^{\bullet-}$ into H_2O_2. Subsequently, at physiological pH, $ONOO^-$ will largely undergo protonation to form peroxynitrous acid (ONOOH) or it can react with carbon dioxide (CO_2) to form nitrosoperoxycarbonate ($ONOOCO_3^-$). The ONOOH can break down to form highly reactive nitrogen dioxide ($NO^{\bullet}_2$) and hydroxyl radical ($ONOOH \rightarrow NO^{\bullet}_2 + OH\cdot$). Alternatively, the $ONOOCO_3^-$ can decompose into $NO^{\bullet}_2$ and carbonate radical ($CO^{\bullet}_{-3}$)

($ONOOCO_3^- \rightarrow NO^\bullet_2 + CO^\bullet_{-3}$). Each of these PN-derived radicals ($NO^\bullet_2$, OH·, and $CO^\bullet_{-3}$) are highly reactive and able to initiate and propagate LP neurodegeneration with diffusion rate-limited speed [15, 16].

Lipid Peroxidation and the Formation of Highly Reactive Protein Carbonyls

Increased production of reactive free radicals (i.e., "oxidative stress") in the injured brain has been shown to cause oxidative damage to cellular lipids and proteins leading to functional compromise and cell death in both the cerebral microvascular and brain parenchymal compartments. Extensive study has confirmed that a major form of radical-induced oxidative damage involves ROS/RNS radical attack on brain cell membrane polyunsaturated fatty acids (PUFAs) that triggers the process of LP that is characterized by three distinct steps: *initiation*, *propagation*, and *termination and carbonylation* [17].

Initiation

LP is initiated when one of the highly reactive oxygen radicals (hydroxyl radical, OH·; nitrogen dioxide radical, $NO^\bullet_2$; and carbonate radical, $CO^\bullet_{-3}$) reacts with polyunsaturated fatty acids (PUFAs) such as arachidonic acid (AA), linoleic acid (LA), eicosapentaenoic acid (EPA), or docosahexaenoic acid (DHA), all of which have multiple allylic carbons that are susceptible to LP. Initiation of LP begins when the reactive radicals steal a hydrogen atom of a PUFA from its associated electrons from one of their allylic carbons. The basis for the susceptibility of the allylic carbon(s) of the PUFAs to having of its allylic carbon electrons stolen by a highly electrophilic free radical is that the carbon is surrounded by two relatively electrophilic double bonds which act to pull one of the electrons from the allylic carbon. Consequently, a reactive free radical can easily pull the hydrogen electron off the allylic carbon because its commitment to pairing with the carbon electron has been weakened by the surrounding electronegative double bonds. This results in the original radical being quenched while the polyunsaturated fatty acid (L) becomes a lipid radical ($L^\bullet$).

Propagation

Subsequently, LP propagation begins when the unstable $L^\bullet$ reacts with O_2 to form a lipid peroxyl radical ($LOO^\bullet$). The $LOO^\bullet$ in turn abstracts a hydrogen atom from an adjacent PUFA, yielding a lipid hydroperoxide (LOOH) and a second $L^\bullet$, which sets off a series of LP propagating "chain" reactions in the brain microvascular, neuronal, and glial cells.

Termination and Carbonylation

The LP propagation reactions are terminated in the third step when the substrates (i.e., peroxidizable PUFAs) become depleted and a lipid radical reacts with another radical to yield potentially neurotoxic non-radical, but highly reactive, aldehydic end products referred to as carbonyls. Two highly reactive and neurotoxic carbonyls of LP are 4-hydroxynonenal (4-HNE) and acrolein, both of which have been well characterized in both TBI and spinal cord injury (SCI) experimental models [13–16, 18]. The LP-derived 4-HNE and acrolein covalently bind to proteins, mainly the basic amino acids (lysine, histidine, arginine) by either Schiff base or Michael addition reactions, which alter the structural and functional properties of brain proteins.

Contribution of Lipid Peroxidative Damage to Mitochondrial Failure, Intracellular Calcium Overload, and Activation of Calcium-Dependent Proteolytic Enzymes and Neurodegeneration

The impact of ROS/RNS production is heightened when oxygen radicals amplify other secondary injury pathways creating a continuous cycle of neuronal ion imbalance, Ca^{++} buffering impairment, mitochondrial dysfunction, glutamate-induced excitotoxicity, and microvascular disruption. One example of ROS-induced ionic disruption arises from LP-induced damage to the plasma membrane ATP-driven Ca^{2+} pump (Ca^{++}-ATPase) and Na^+ pump (Na^+/K^+-ATPase), which contributes to increases in intracellular Ca^{++} concentrations, mitochondrial dysfunction, and additional ROS production. Both Ca^{++}-ATPase and Na^+-/K^+-ATPase disruptions result in further increases in intracellular Ca^{++} and Na^+ accumulation, respectively [15], the latter causing reversal of the Na^+/Ca^{++} exchanger which further exacerbates intracellular Ca^{++} [19, 20]. As already noted above, PN, formed from mitochondrial Ca^{++} overload, also contributes to post-TBI mitochondrial dysfunction. Specifically, nitric oxide (NO·), formed from mitochondrial nitric oxide synthase (mNOS), in turn reacts with $O_2^{\bullet-}$ to produce the highly toxic PN, which impairs mitochondrial respiratory function (electron transport and ATP synthesis) and Ca^{2+} buffering capacity via its derived free radicals. Indeed, increased PN-derived 3-NT and 4-HNE have been detected during the time of mitochondrial dysfunction and correlated with respiratory [13, 14] and Ca^{2+} buffering impairment [21].

Lipid peroxidation-derived neurotoxic aldehydes (carbonyls) 4-HNE or acrolein have been shown in neuronal or astrocytic cultures or in synaptosomes to impair glutamate uptake and to inhibit mitochondrial function [22–25]. On the other hand, glutamate-induced excitotoxic damage in synaptosomal or neuronal cultures is attenuated by pharmacological LP inhibition, confirming that oxidative damage is a promoter of glutamate excitotoxicity [7, 8].

Phase II and III Clinical Trial Results of PEG-SOD and Tirilazad Mesylate: Initial Validation of Antioxidant Neuroprotection in TBI Animal Models and Moderately Severe Human TBI

Polyethylene Glycol (PEG)-Conjugated Superoxide Dismutase (SOD): Scavenging the Primordial Superoxide Radical

The initial studies of free radical scavenging compounds in TBI models were carried out with Cu^{++}/Zn^{++} SOD in the collaborative work of Kontos, Wei, and Povlishock at the Medical College of Virginia. These investigators showed that post-traumatic cerebral microvascular dysfunction was initiated by $O_2^{\bullet-}$ generated as a by-product of TBI-triggered activation of the arachidonic acid cascade which begins during the first minutes and hours after TBI [26–28]. Their pioneering experimental work in rodent and feline TBI models demonstrated that administration of Cu^{++}/Zn^{++} SOD prevented post-traumatic free radical–induced disruption of cerebrovascular autoregulatory dysfunction.

This work led to phase II and III clinical trials in which the more metabolically stable polyethylene glycol (PEG)-covalently conjugated bovine Cu^{++}/Zn^{++} SOD (PEG-SOD; generic name: pegorgotein; trade name Dismutec®) was examined in moderate and/or severe TBI patients. The PEG modification of SOD had been shown by the pharmaceutical sponsor Sterling-Winthrop to not modify the SOD activity, but rather to dramatically increase its in vivo metabolic stability, which increased the half-life of SOD activity in rats from 8 minutes to over 30 hours with PEG-SOD [29].

An initial phase II double-blinded dose-response study, conducted at two centers (the Medical College of Virginia and the University of Maryland Shock Trauma Center), showed a positive trend in humans as well [30]. Specifically, 104 patients with severe TBI were randomized to either PEG alone or one of three PEG-SOD dose groups (2000, 5000, or 10,000 Units/kg administered as a single i.v. bolus) within 4 hours after TBI. Glasgow Outcome Scale (GOS)-assessed outcome at 3 and again at 6 months, in 91 and 93 patients, respectively, showed that at 3 months, 44% of the PEG-treated patients were either in a persistent vegetative state or had died, while only 20% of the patients who received the 10,000 Units/kg PEG-SOD dose were in those bad outcome groups ($p < 0.03$) compared to placebo. At 6 months post-TBI, these figures were reduced from 36% for PEG-treated patients to only 21% for PEG-SOD, respectively ($p < 0.04$). Differences in outcome between PEG and either of the two lower PEG-SOD dose groups were not statistically significant.

These encouraging phase II placebo-controlled clinical results carried out in two highly experienced trauma centers (Medical College of Virginia and the University of Maryland) inspired a subsequent phase III, 29 trauma center study that randomized 463 "severe" TBI patients randomized to either PEG or 10,000 or 20,000 PEG-SOD Units/kg i.v. administered within 8 hours after severe TBI. Disappointingly, neither PEG-SOD dose, administered within the 8-hour post-TBI time window, showed a significant benefit in terms of increased survival or improved neurological outcomes in "severe" TBI. This implies that the PEG-SOD antioxidant neuroprotective therapeutic efficacy window may be limited to the 4-hour post-TBI time frame. However, in the 10,000 Units/kg PEG-SOD-treated patients, there was a significantly lower incidence of adult respiratory distress syndrome (ARDS) in the 10,000 Units/kg PEG-SOD patients compared to the PEG-treated TBI patients ($p < 0.015$) [31]. Unfortunately, PEG-SOD trials in TBI patients were discontinued by the corporate sponsor Sterling-Winthrop, and questions of whether repeated, rather than single PEG-SOD, dosing would be more effective were never resolved and PEG-SOD passed into pharmacological history.

Despite the failure of PEG-SOD in human TBI, experimental studies have shown that transgenic mice that overexpress Cu^{++}/Zn^{++} SOD are significantly protected against post-TBI pathophysiology and neurodegeneration [32–36]. This fully supports the importance of post-traumatic $O_2^{\bullet-}$ in post-traumatic secondary injury, despite the fact that targeting this primordial radical, which is only at its highest level during the first few hours after TBI, may not be the best antioxidant target for severe TBI compared to interrupting the ROS-/RNS-initiated LP process, which, as our recent rodent TBI studies show, does not peak in intensity until 4–12 hours after TBI [13, 14].

21-Aminosteroid Tirilazad Mesylate – Membrane Stabilization + Scavenging of Lipid Peroxyl Radicals

Consistent with targeting LP as probably the dominant mechanism involved in post-TBI oxidative damage, the 21-aminosteroid LP inhibitor tirilazad mesylate, trade name Freedox®, was discovered to potently inhibit free radical-induced, iron-catalyzed LP by a combination of catalytic LOO· scavenging along with a membrane-stabilizing action that limits the propagation of LP reactions between an LOO· and an adjacent polyunsaturated fatty acid [37]. The protective efficacy of tirilazad was demonstrated in multiple animal models of acute TBI in mice [38], rats [39], and cats [40]. While this highly lipophilic compound is largely localized in the microvascular endothelium, the early post-traumatic disruption of the BBB allows the penetration of tirilazad into the brain parenchyma [41]. Nevertheless, experimental data derived from the rat controlled cortical impact, and the mouse diffuse concussive head injury models have shown that a major effect of tirilazad is to lessen post-traumatic microvascular damage, as evidenced by attenuation of blood–brain barrier (BBB) opening [41, 42].

Nearly coincident with the PEG-SOD phase II and III TBI trials, the LP inhibitor tirilazad mesylate was taken into clinical development in the early 1990s. Following a phase II dose-escalation study that demonstrated the drug's safety in TBI patients, in two phase III multi-center clinical trials, the ability of tirilazad mesylate to improve neurological recovery in moderately and severely injured TBI patients was evaluated. One trial was conducted in North America and the other in Europe, Australia, and South Africa. In both trials, TBI patients were treated randomly within 4 hours after TBI with tirilazad (10 mg/kg i.v. q6h for 5 days) or its aqueous vehicle (as a placebo). However, the North American trial was never published due to a major confounding imbalance in the blinded randomization of the moderate and severe patients to placebo or tirilazad with regard to injury severity and pre-treatment neurological status.

In contrast, the parallel European-Australasian-South African phase III moderate/severe TBI trial that enrolled 1120 moderate (GCS 9–12) and/or severe (GCS 4–8) TBI patients showed much better randomization balance between the placebo- and tirilazad-treated patients. Additionally, the principal investigator of this study, Dr. Larry Marshall, Chairman of the Department of Neurosurgery at UC San Diego and a leading TBI expert, urged us to include computerized axial tomography (CT) confirmation of traumatic SAH (49.7% of placebo treated and 50.4% of tirilazad treated), which was published in the *Journal of Neurosurgery* [43].

A post-hoc analysis showed that tirilazad-treated, moderately injured (GOS 9–12) male TBI patients with traumatic subarachnoid hemorrhage (tSAH) had a significantly lower incidence of 6-month mortality after treatment with tirilazad (7.1%) compared to placebo (25.0%, $p < 0.042$). Also in severely injured males with tSAH, tirilazad also lessened mortality from 42.5% in placebo-treated to 33.3% ($p < 0.026$). Additionally, 6-month post-TBI GOS favorable outcome was increased by 21%, from 60% in placebo-treated to 81.5% in tirilazad-treated moderate tSAH patients, albeit not significantly ($p < 0.13$).

The borderline significance of this tirilazad mesylate improvement in favorable outcome in tSAH patients would have needed to be replicated, had Pharmacia & Upjohn and tirilazad mesylate (Freedox®) survived the rampant "merger mania" of the decade of the 1990s. Nevertheless, from a scientific point of view, this result is consistent with the fact that tirilazad is also highly effective in reducing SAH-induced brain edema and cerebral vasospasm in multiple animal models of aneurysmal SAH [37] and in humans with aneurysmal SAH [44, 45] and traumatic SAH [43].

Current Enthusiasm for Antioxidant Neuroprotective Drug Discovery for Concussion (GCS 13–15)

Two recently published review articles have strongly encouraged mild TBI/concussion preclinical and clinical investigators toward an increased focus on the pharmaceutical investigation of various antioxidants for their neuroprotective utility. One of those reviews has stated that *"Of the several biochemical changes that occur in a patient's brain following a concussion, an increase in reactive oxygen species (ROS) is of particular concern"* [46].

A prominent and highly productive group of Italian TBI investigators state in their recent review the following evaluation concerning the history of "*Antioxidant Therapies in Traumatic Brain Injury*" research and development.

> A large number of studies have evaluated the efficacy of antioxidant administration to decrease TBI-associated damage in various animal models and in a limited number of clinical trials. Points of weakness of preclinical studies are represented by the large variability in the TBI model adopted, in the antioxidant tested, in the timing, dosages and routes of administration used, and in the variety of molecular or neurocognitive parameters evaluation. The analysis of the very few clinical studies does not allow strong conclusions to be drawn on the real effectiveness of antioxidant administration to TBI patients [47].

Later in their review, the authors state their view that the post-concussion antioxidant neuroprotective therapeutic window, which they believe, is limited to only 3 hours post-concussion and believe that antioxidant administration for concussions should be started shortly following admission.

> According to what is stated above, sports-related concussions are a type of TBI in which prevention might effectively be applied either by modifying rules of those sports disciplines at higher risk of concussion, or in preventively treating athletes with drugs capable of inhibiting specific molecular pathways activated by concussions. It should also be taken into account that drug treatments might be helpful in allowing safer return of athletes to play. In this light, few studies have been carried out to evaluate the effects of the administration of antioxidants prior to concussion in reducing molecular changes and symptoms associated with concussion [47].

The views of Di Pietro et al. [47] leave the reader initially uncertain about their enthusiasm for post-concussion antioxidant efficacy and for the therapeutic practicality of post-TBI antioxidant neuroprotection. However, their view is certainly reasonable about trying to pharmacologically intercept the highly reactive, hydrogen peroxide (H_2O_2)-derived hydroxyl radical ($OH\cdot$) or the peroxynitrite-generated nitrogen dioxide ($NO^{\bullet}_2$), both of which peak within the concussed mouse brain within first 15 minutes post-injury as evidenced by 4-HNE or 3-NT immunostaining [48]. Logically, for prophylactic pharmacological neuroprotective treatment to be widely accepted for pre-treatment of competitive athletes at risk for concussions, antioxidant pre-treatments need to be safe and devoid of stimulant or depressant neuropharmacology.

Accordingly, in their review, Di Pietro et al. [47] highlight several "nutraceuticals," including ascorbic acid (vitamin C), N-acetyl-cysteine (a glutathione (GSH) congener), flavonoids, resveratrol, α-tocopherol (vitamin E), coenzyme $Q_{10,}$ carotenoids (natural products that possess antioxidant and anti-inflammatory properties), and omega 3 fatty acids including docosahexaenoic acid (DHA).

Concerning DHA as a potentially approvable prophylactic antioxidant neuroprotective agent, a recent study evaluated the effects of pretreatment of 81 Division I American football athletes who were recruited and randomly administered 2.4 or 6 g/day of DHA/day. The football players, during the 189-day season of the study, were randomly serum-sampled for neurofilament light (NFL) levels, as a measure of concussion-induced axonal injury. Surprisingly, the lowest DHA daily dose (2 g/day) produced the best effect in serum NFL levels suggesting that DHA administered at higher doses may possess a biphasic/U-shaped dose-response curve in regard to the axonal protective effects of DHA [49].

Newer Multi-mechanistic Pharmacological Approaches for Antioxidant Neuroprotection That May Be Parenterally or Orally Administrable for Early Treatment of Mild TBI/Concussion (GCS 13–15)

Pharmacological Nrf2-Antioxidant Response Element (ARE) Activation Enhancement

The body's endogenous antioxidant defense system is largely regulated by the nuclear factor E2-related factor 2/antioxidant response element (Nrf2/ARE) signaling pathway at the transcriptional level [50, 51]. Recent work has revealed that following controlled cortical impact TBI in mice there is a progressive activation of the Nrf2-ARE system in the traumatically-injured brain, as evidenced by an increase in HO-1 mRNA and protein that peaks at 72 hours after TBI. However, this effect does not precede, but rather is coincident with the post-injury increase in LP-related 4-HNE [52]. Therefore, it is apparent that this endogenous neuroprotective antioxidant response needs to be pharmacologically sped up and increased in magnitude if it is to be capable of exerting meaningful acute post-TBI neuroprotection. Two nutraceutical Nrf2 activators that have been shown to speed up Nrf2/ARE expression and provide effective neuroprotective actions in TBI models are sulforaphane and carnosic acid.

Sulforaphane

Administration of the natural product sulforaphane, a Nrf2/ARE signaling activator found in high concentrations in broccoli, significantly reduced contusion volume and increased post-SCI coordination. These positive outcomes were a result of sulforaphane-induced increases in Nrf2, glutamine, and decreases in inflammatory cytokines, IL-1b and TNFα [53]. The mRNA levels of Nrf2-regulated antioxidant enzymes heme oxygenase (HO-1) and NADPH:quinone oxidoreductase-1 (NQO1) are upregulated 24 hours post-TBI [54]. In contrast, Nrf2 knockout mice were found to be susceptible to increased oxidative stress and neurologic deficits following TBI compared to their wild-type counterparts [55].

Sulforaphane is also neuroprotective in various animal models of TBI specifically reducing cerebral edema and oxidative stress and improving BBB function and cognitive deficits [56]. Studies by Chen and coworkers [57] demonstrated increased cortical expression of Nrf2 and HO-1 in the rat SAH model. However, treatment with sulforaphane further increased the expression of Nrf2, HO-1, NQ01, and glutathione S-transferase-α1 (GST-α1), resulting in the reduction of brain edema, cortical neuronal death, and motor deficits.

Carnosic Acid in Moderately Severe TBI

Another Nrf2-ARE activating natural product, carnosic acid, from the herb rosemary, has been shown to more effectively induce this antioxidant defense system than the prototype Nrf2-ARE activator sulforaphane [58]. This is because the parent carnosic acid, which contains a di-phenolic catechol moiety, is capable of scavenging LOO· radicals, making it in part, a typical electron-donating LP-inhibiting antioxidant. Additionally, the catechol of carnosic acid is metabolically converted to a carnosic acid quinone which is responsible for activating the Nrf2/ARE signaling pathway. Thus, carnosic acid is a dual mechanism antioxidant with combined electron-donating properties and Nrf2/ARE-activating activity.

It has been shown that administration of carnosic acid to non-TBI mice is able to significantly increase the resistance of isolated cortical mitochondria, harvested 48 hours later, to the respiratory depressant effects of 4-HNE applied in vitro together with a decrease in 4-HNE modification of mitochondrial proteins [11] Subsequent studies, with a single 1 mg/kg i.p. dose of carnosic acid administered to mice at 15 minutes after controlled cortical impact TBI, demonstrated preserved mitochondrial respiratory function along with a reduction in the level of LP-mediated damage in mitochondria harvested from the injured cortex at 24 hours after TBI [59]. Furthermore, carnosic acid's antioxidant effects were still apparent when its post-TBI administration was delayed until 8 hours post-injury in terms of an attenuation of the neurotoxic LP-derived 4-HNE and 3-NT in the injured cortical tissue together with a decrease in Ca^{2+}-activated, calpain-mediated, neuronal cytoskeletal degradation. Regarding the latter neuronal protective effect, a decrease in 48-hour cytoskeletal degradation was also shown to occur, even with a post-TBI treatment delay of 8-hours post-TBI (Fig. 9.1).

Carnosic Acid in Repetitive Mild TBI (rmTBI)

Recent studies by Dash and colleagues [60] have demonstrated that carnosic acid is able to improve neurological recovery in a mouse repetitive mild concussion TBI (rmTBI) model. Each mouse received 3 concussions, each 72 hours apart. At 30 minutes after each TBI, the mice received a 1 mg/kg i.p. dose of carnosic acid. In the rm-TBI paradigm, carnosic acid (1 mg/kg i.p dose) was shown to improve spatial learning and memory beginning at 15 days after the last injury [60].

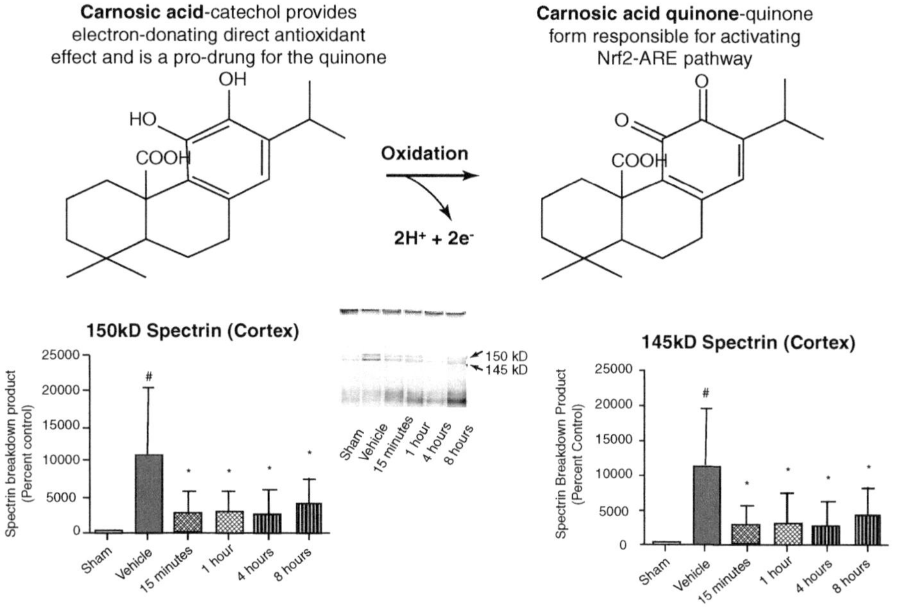

Fig. 9.1 *Top*: Multi-mechanistic antioxidant carnosic acid (CA): Although the parent drug CA (left) possesses direct radical scavenging activity due to its electron-donating diphenolic hydroxyls, its ability to activate the Nrf2-ARE pathway requires its oxidation to the more electrophilic CA ortho-quinone (right) species which is facilitated by conditions of oxidative stress. Both CA and the CA quinone are highly brain penetrable. *Bottom*: Delayed administration of CA provides a clinically relevant therapeutic window. The administration of CA (1 mg/kg i.p.) was delayed to either 15 minutes, 1 hour, 4 hours, or 8 hours post-injury for the initial i.p. injection followed by a booster injection at 24 hours post-injury. At 48 hours post-injury, ipsilateral cortical tissue was collected for Western blot analysis of α-spectrin breakdown products as an indication of cellular cytoskeletal degradation. All delayed time points (15 minutes, 1 hour, 4 hours, and 8 hours) were significantly decreased ($p < 0.05$) as compared to vehicle for both the 150 kD (caspase and calpain derived) and 145 kD (calpain specific) breakdown products. Analyzed by one-way ANOVA followed by Student Newman-Keuls post-hoc test. * = $p < 0.05$. Error bars represent +/− SD. $n = 8$–10 per group. (Reprinted from Miller et al. [53]. With permission from Elsevier)

Mitochondrial Protection by Scavenging of Lipid Peroxidation-Derived Protein Carbonyls 4-Hydroxynonenal (4-HNE) and Acrolein

Penicillamine

We have previously demonstrated that D-penicillamine is able to scavenge the RNS peroxynitrite (PN) [61] and to protect brain mitochondria from PN-induced respiratory dysfunction in isolated rat brain mitochondria [62]. D-penicillamine has also been documented to form an irreversible (covalent) bond to primary aldehydes enabling the drug to scavenge neurotoxic LP-derived carbonyl compounds such as 4-HNE and acrolein [63]. Consistent with that mechanism of action, D-penicillamine has been shown to attenuate the levels of 4-HNE-modified mitochondrial proteins following exposure of isolated brain mitochondria to 4-HNE [12]. The PN scavenging action of D-penicillamine, along with its carbonyl scavenging capability, may jointly explain our previous findings that acutely administered D-penicillamine can improve early neurological recovery of mice subjected to moderately severe concussive TBI [64].

Phenelzine

Another FDA-approved hydrazine-containing drug phenelzine, long used for certain depressive patients, does not seem to compromise arterial blood pressure as readily as hydralazine. Accordingly, a very recently published paper has shown that phenelzine administration to rats subjected to acute contusion SCI mitigated post-SCI neuropathic pain, reduced motor deficits, and improved spinal cord tissue sparing [65]. Earlier studies have demonstrated neuroprotective efficacy in a rodent ischemia-reperfusion stroke model, which was attributed to reducing "aldehyde load" in the stroke-injured brain [66]. In vitro studies have documented the ability of phenelzine to protect isolated rat brain mitochondria from the respiratory depressant effects of 4-HNE together with a concentration-related attenuation of the accumulation of 4-HNE-modified mitochondrial proteins. Subsequent in vivo studies in the rat controlled cortical impact TBI model have found that a single 10 mg/kg subcutaneous dose of phenelzine can also reduce early (3 hours) post-traumatic mitochondrial respiratory failure, as well as cortical lesion volume at 14 days post-injury [67]. More recently, we have observed that phenelzine is able to protect isolated mitochondria from respiratory functional depression and carbonyl modification of mitochondrial proteins following application of the more highly reactive aldehyde acrolein [68].

To better define the optimal neuroprotective use of phenelzine, additional in vivo TBI studies shown in Fig. 9.2 have demonstrated that repeated dosing with phenelzine over a 60-hour post-TBI period is able to reduce LP-derived mitochondrial carbonylation and bioenergetic failure at its 72-hour peak, along with a reduction in cortical lesion volume that is greater than that seen with only a single early dose. This makes sense since the adequate carbonyl-scavenging drug levels logically need to be maintained during the 72-hour long time course of post-traumatic generation of LP-derived neurotoxic aldehydes [68]. Subsequent in vivo TBI experiments revealed that in addition to preserving mitochondrial bioenergetics out to 72-hours post-TBI, phenelzine administration was able to significantly improve intraneuronal calcium homeostasis, to maintain mitochondrial membrane potential, and, thereby, to partially protect neuronal cytoskeletal integrity [69].

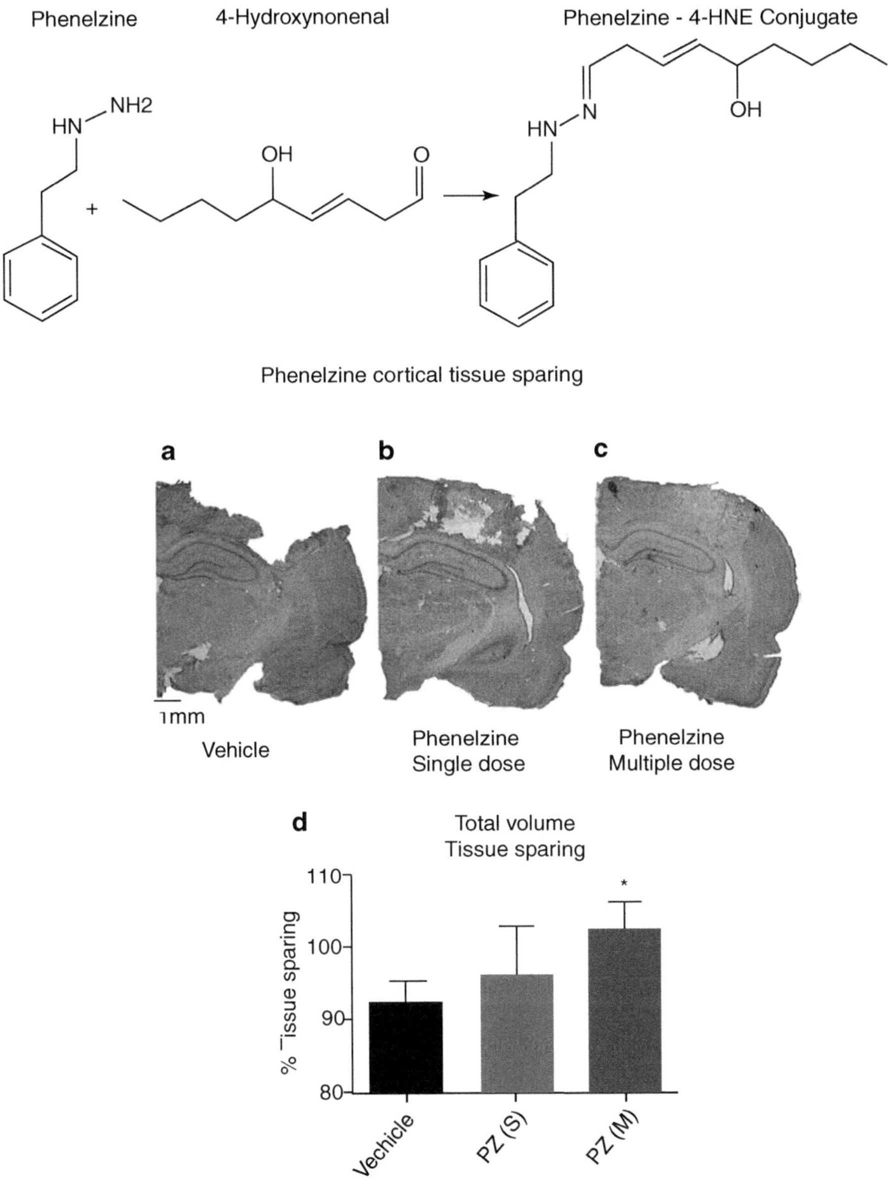

Fig. 9.2 Covalent reactivity of the hydrazine side chain of phenelzine with the lipid peroxidation-derived carbonyl 4-hydroxynonenal. (**a**) Vehicle (0.9% saline)-treated rat brain injected 15 minutes after TBI. All groups (vehicle, PZ(s), PZ(m)) were euthanized 72 hours after first injection. (**b**) Phenelzine single dose (PZs)-treated animal, injected with a single dose of PZ, 15 minutes after injury at 10 mg/kg. (**c**) Brain of PZ-treated rat with a multiple dosing paradigm (PZm): single subcutaneous injection of PZ 15 minutes after injury, followed by maintenance dosing of 5 mg/kg every 12 hours thereafter. (**d**) Percent of tissue sparing followed by either vehicle (saline), PZ(s), or PZ(m) treatment did not exhibit a statistically significant amount of cortical tissue sparing when compared to vehicle. However, PZm significantly increased the total volume of spared cortical tissue. One-way ANOVA ($F = 8.5$, $df = 2{,}20$, $p < 0.002$) followed by Dunnett's post-hoc test. * = $p < 0.05$ compared to vehicle. Error bars represent mean ± SD; $n = 7$–8 rats per group. (Reprinted from Cebak et al. [68]. With permission from Mary Ann Liebert, Inc.)

Hydralazine

More recently, it has been documented that a variety of FDA-approved hydrazine ($NH-NH_2$)-containing compounds, including the anti-hypertensive agent hydralazine and the anti-depressant phenelzine, can covalently react with the carbonyl moieties of 4-HNE or acrolein which prevents their covalent binding to susceptible amino acids in proteins [70]. Most impressive is the fact that the application of hydrazines can rescue cultured cells from 4-HNE toxicity even when administered after the 4-HNE has already covalently bound to cellular proteins [70]. Consistent with this effect being neuroprotective, Shi and colleagues at Purdue University have shown that hydralazine inhibits either compression- or acrolein-mediated injuries to rat spinal cord tissue [71]. However, a concern about hydralazine is that it is a potent arterial vasodilator that has long been used for interruption of hypertensive crises. Thus, hydralazine might potentially worsen post-TBI hypotension and decrease arterial perfusion in the injured brain and/or spinal cord that might offset hydralazine carbonyl scavenging neuroprotective effects. However, Shi and coworkers, in their rat SCI studies with hydralazine, have documented that the neuroprotective dose they are using in SCI models (5 mg/kg i.p.) does not have significant hypotensive effects, and the achieved levels in spinal cord tissue are sufficient to reduce the post-SCI accumulation of the neurotoxic aldehyde acrolein [72].

Combinatorial Antioxidant Neuroprotection

Antioxidant neuroprotective therapeutic discovery directed at acute TBI has consistently been focused upon attempting to inhibit the secondary injury cascade by pharmacological targeting of a single oxidative damage mechanism. As presented above, these efforts have included either enzymatic scavenging of superoxide radicals with SOD [29–31] or inhibition of LP with tirilazad [43]. While each of these strategies alone has shown protective efficacy in multiple animal models of TBI, phase III clinical trials with either compound failed to demonstrate a statistically significant positive effect, although post-hoc subgroup analysis showed that the microvascularly localized tirilazad may have efficacy in moderate and severe TBI patients with tSAH [43]. While many reasons have been identified as possible contributors to the failure, one logical explanation has to do with the possible need to interfere at multiple points in the oxidative damage portion of the secondary injury cascade, either simultaneously or in a phased manner, to achieve a clinically demonstrable level of neuroprotection. To begin to address this hypothesis, we are currently exploring the possibility that reducing posttraumatic oxidative damage more completely and less variably might be achievable by combined treatment with two mechanistically complimentary antioxidant compounds. Figure 9.3 summarizes the theoretical rationale for a multi-mechanistic antioxidant therapy for TBI, whether in concussions and mild, moderate, or severe TBI. It is anticipated that the combination of two or more antioxidant mechanistic strategies may improve the extent of

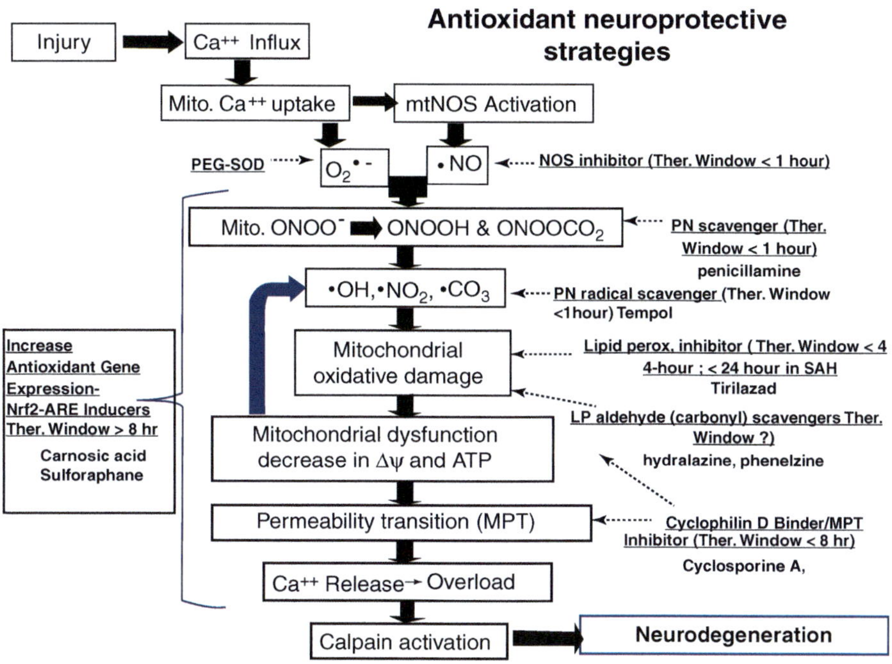

Fig. 9.3 Rationale for the combination of two or more antioxidant strategies to achieve a more effective and consistent (i.e., less variable) neuroprotective effect in the injured brain

neuroprotective efficacy and lessen the variability of the effect obtained with single antioxidant strategies.

Preliminary data, shown in Fig. 9.4 (top), suggests that a combination treatment that includes a peroxynitrite radical scavenger Tempol with an LP inhibitor U-83836E in mice subjected to controlled cortical impact TBI is more effective in reducing 48-hour calpain-mediated neurofilament damage (i.e., α-spectrin breakdown). In parallel, experiments (Fig. 9.4 (bottom)) showed that the same treatment combination reduced 7-day post-TBI cortical tissue damage. In the case of both parameters, the variability of the data is reduced to approximately half of that seen in the parallel groups treated with either of the two drugs alone.

Temporal Window of Metabolic Brain Vulnerability to Repeat Concussions

Regarding concussions (GCS 13–15), arguably one of most important considerations concerning recovery is whether this victim is part of a TBI group that is likely to have frequent repetitive concussions. This scenario most commonly applies to high school, college, or professional athletes who participate (particularly as

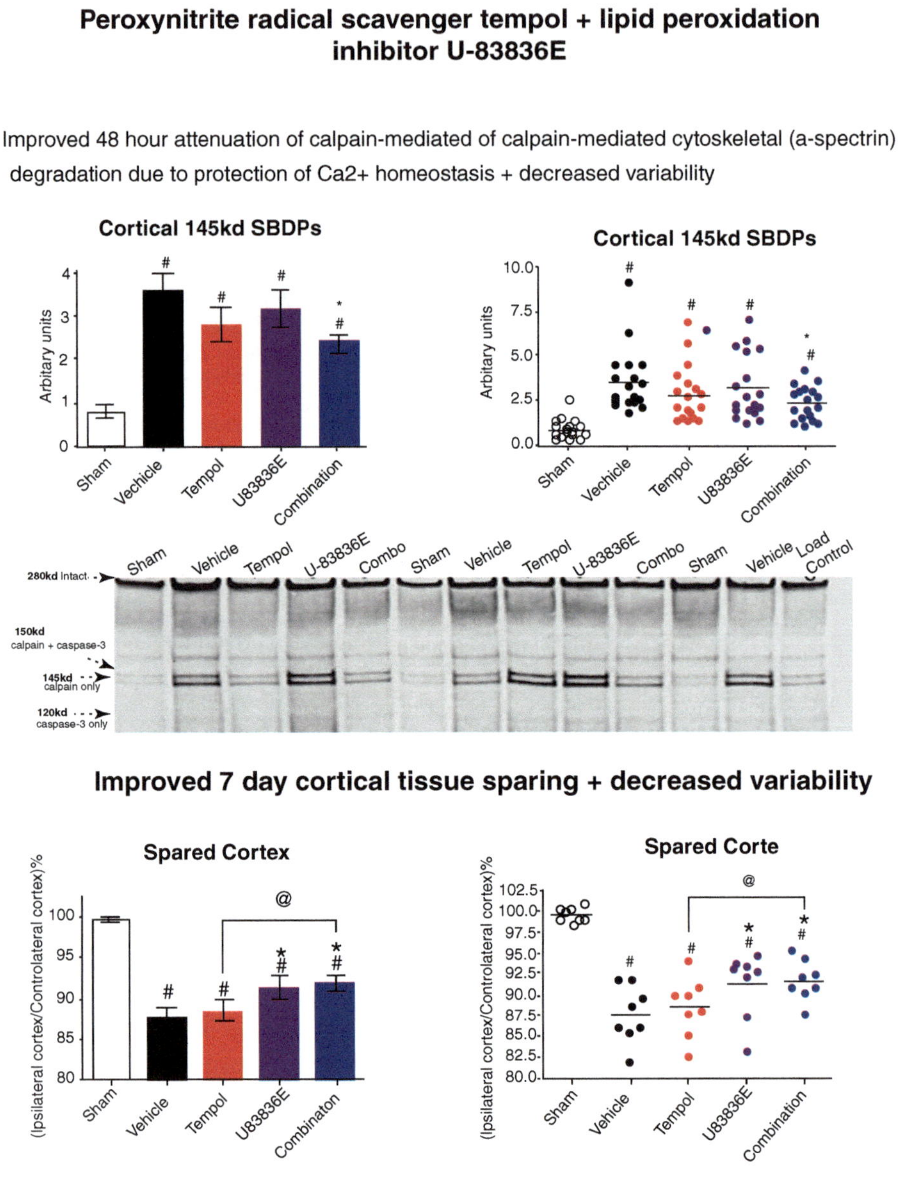

Fig. 9.4 *Top*: Comparison of the effects of the PN radical scavenger Tempol and the lipid peroxidation inhibitor U-83836E alone and in combination (each administered at 15 minutes post-TBI) on 48 hours calpain-mediated cortical neuronal cytoskeletal α-spectrin degradation (SBDP = spectrin breakdown products), measured by Western blot, at its 48 hours post-injury peak in the mouse controlled cortical impact TBI model ($N = 18$ male mice/group; # = $p < 0.05$ compared to sham, uninjured mice) and on 7 day cortical tissue sparing ($N = 8$/group; # = $p < 0.05$ compared to sham, *$p < 0.05$ compared to vehicle group). Values = mean ± standard deviation. *Bottom*: Comparison of the effects of the PN radical scavenger Tempol and the lipid peroxidation inhibitor U-83836E alone and in combination on 7 day post-TBI cortical tissue sparing (@ = $p < 0.05$ compared to Tempol). For both neuroprotective parameters, it is apparent that Tempol and U- 83836E result in as much as a 50% decrease in variability. (Adapted from Hall et al. [123]. With permission from Elsevier)

"starters") in rough and tumble contact sports (e.g., football, rugby, ice hockey, soccer), and are likely to sustain repeat concussions during one or more seasons, or in frontline military personnel who are at risk for repetitive concussions from frequent exposure to explosions from mortar or artillery fire during their deployments.

Accordingly, the important question concerns how frequently athletes or military personnel sustain concussions, and how long should they be removed from athletic competition, or combat, to allow for adequate rest and recovery between their repetitive concussions. Fortunately, we currently live in a time when the prevailing opinion is that if an athlete, or frontline soldier, sustains a concussion, they need to be relieved from duty for a period of time before they are allowed to return to either athletic competition or combat. While it is uncertain whether the needed rest period to enable adequate post-concussion recovery is different between rodents and humans, it seems likely that post-TBI oxidative and nitrosative damage time courses in acutely brain-injured rodents and probably humans are similar.

Vagnozzi et al. (2007) determined that there is a temporal window of brain vulnerability recovery in rats undergoing repeat mTBIs with the optimal timing for the second impact being delayed until after at least 5 days following the last impact. Specifically, radical-induced lipid peroxidation (LP), measured by accumulation of malondialdehyde (MDA), does not die down completely until oxidative and nitrosative-induction of LP has returned to its nearly normal level at least 5 days post-TBI. This is the first study to show the existence of oxidative and nitrosative stresses in repeat TBIs, and their neuroprotective modulation by the time interval between two concussive episodes separated by a post-concussion recovery period to allow time for repair from post-TBI free radical-induced LP.

Pharmacological Approaches for Treating Post-concussion Symptoms

Both neurocognitive and neuropsychiatric sequelae following concussion and mild TBI (mTBI) likely arise with alterations to the usually physiologically controlled balances in neurotransmitters [73–75]. Common targets in the treatment of concussion include addressing neurotransmitter imbalances in dopamine, serotonin, the cholinergic system, and the noradrenergic system. Dopamine (DA) plays numerous roles in cognition, attention, executive function, and memory. The dopaminergic response to TBI is complex owing to the numerous areas of DA activity including the prefrontal cortex [76], hippocampus[77], and striatum [78]. DA receptors, D1 and D2, are differentially expressed throughout these brain structures [79, 80] and also fluctuate in expression levels in response to injury [81]. There is evidence that genetic variability in DA receptors can influence outcome after concussion [82]. Several animal studies support a role for targeting DA dysregulation as a potential pharmacotherapeutic target following TBI [83]. DA receptor agonists have been evaluated clinically in the treatment of concussion with varying measures of success. Bromocriptine, a D2 agonist, has demonstrated some benefit in working memory in concussed patients, while Pergolide, a D1/D2 agonist, had even greater effect [84]. Other clinically evaluated DA agonists also include amantadine, which has shown to improve cognitive processing and functional improvement [85]. Amantadine facilitates the presynaptic release of DA while inhibiting its uptake, effectively increasing the concentration and duration of its neurotransmitter effects [86, 87]. Amantadine is also a weak NMDA receptor antagonist that may confer some limited neuroprotection following trauma [87].

Post-traumatic headache is one of the most frequently reported symptoms following concussion and can persist for months after injury [88–90]. Metoclopramide is a dopamine and serotonin receptor antagonist that has been shown to reduce headaches [91, 92], but there is no evidence to support a rationale for its use in mTBI. The antagonism of DA runs counter to several studies in which agonism of DA improves symptom reporting following concussions.

The cholinergic system drives sensory processing [93], attention [94], sleep [95], arousal [96], and memory [97]. To date only one randomized control study [98] and one open-label study [99] have examined modulation of the cholinergic pathways with some success. In the RCT trial, galantamine did not confer an improvement in primary outcome (cognitive symptoms) but was associated with an improvement in secondary outcomes (episodic memory) relative to placebo [98]. Galantamine, a competitive inhibitor of acetylcholinesterase, prevents the breakdown of acetylcholine, thereby increasing the synaptic presence and duration of the neurotransmitter [100]. Given the extensive axonal projections of dopaminergic [101] and cholinergic neurons [102] to major structures throughout the brain, and the seemingly beneficial effects of pharmacological modulation of these pathways, it seems logical that future pharmacological initiatives should address concussion from the perspective of axonal injury in these neuronal subpopulations.

Extrapolation of potential drug effectiveness in concussion is based on the treatment of other neurodegenerative disorders, psychiatric conditions, or more severe forms of brain trauma. Many pharmacological agents have pleiotropic effects across a variety of receptors and signaling systems and in this regard target numerous pathways simultaneously. Methylphenidate, for example, is a compound used in the treatment of attention deficit hyperactivity disorder (ADHD) [103] through targeting dopaminergic and noradrenergic pathways by increasing the synaptic concentration. This is achieved by inhibiting uptake of serotonin and dopamine in the synaptic cleft [104]. Its application to treat impairments in executive function and depression demonstrates some promising potential in TBI [105]. However, this recent study by Al-Adawi et al. is hampered by similarly recurring pitfalls in many pharmacological TBI trials. These include open-label applications (i.e., no double-blinded experimental design or control group), small sample size, as well as a heterogeneity of injury severities. The multimodal activity of pharmacological compounds and the interpretation of their effects on concussion outcome are complicated by the complexity of the brain. Fluoxetine, a selective serotonin reuptake inhibitor (SSRI) which is commonly used in the treatment of depression, also demonstrates effectiveness in motor recovery after stroke, improved memory and cognition in Alzheimer's disease patients, a decrease in tremor severity in Parkinson's patients, and improved outcome in a host of other neurological disorders including multiple sclerosis and epilepsy [106]. Similarly, clinical trials with other SSRIs including sertraline [107, 108] and citalopram [109] have also demonstrated improvements in the treatment of depression. Animal studies point to numerous pathways of activation by these compounds including effects on plasticity through BDNF upregulation [110] and neurogenesis [111]. In animal studies, fluoxetine's effects on depression also suggest a role in reducing neuroinflammation [112]. While imaging of neuroinflammation in mBTI patients has recently been demonstrated [113], understanding the role of inflammation in development of post-concussive symptoms and neurodegenerative diseases such as chronic traumatic encephalopathy (CTE) is in its infancy. Collectively, pharmacotherapies may influence both cognitive and psychiatric outcomes simultaneously. Determining the interdependence of these systems remains a challenge in study design.

The use of cannabidiol (CBD) has gained recent attention as a potential treatment for concussion and mild TBI. Patients in the acute post-concussive phase report reduced symptom severity scores with cannabis use [114]. CBD receptors are expressed widely throughout the brain. Two receptors have been identified to date, cannabinoid receptor 1 and receptor 2 (CB1 and CB2, respectively) 115. CB1 is mainly expressed in axons and synaptic terminals while CB2 is highly expressed in microglia [116]. CBD has been implicated in several in vivo and in vitro studies to influence various aspects of cellular signaling including activity of the blood–brain barrier, dopaminergic agonism, neurogenesis, neuroprotection and immune modulatory effects [117]. It is also known to interact with a variety of receptors in the CNS with implications for modulation of symptoms following concussion. Among these, the serotonin receptor, 5HT1A, has a high affinity for CBD as an agonist and has potential roles in reducing pain, anxiety, and headaches [118]. Similarly, CBD is a weak agonist for the vanilloid receptor, TRPV1, a ligand-gated ion channel involved in nociception and is expressed pre-synaptically on afferent neurons and sensory ganglia [119].

Neuropsychiatric conditions have been studied through several large, randomized control trials of various drugs including amantadine [120], rivastigmine [121], and sertraline [108]. These compounds have demonstrated effects on irritability, cognitive impairment, and depression, respectively. In all cases, however, the placebo treatment groups also demonstrate a considerable effect on outcome. In the amantadine trial, an improvement in irritability outcomes with drug treatment was also paralleled by a significant effect in the placebo group. This potentially masked the true biological effects of amantadine [120]. The rivastigmine trial demonstrated some benefit on cognitive outcome. However, inclusion criteria for TBI patients with GCS ranges from 3 to 15 imply heterogeneous pathophysiologies across injury severities which further complicates the interpretation of rivastigmine's actions at a mechanistic level. In all these studies, the significant contribution of placebo effects on improving outcome compared to medicated groups demonstrates the importance of psychological and neurobiological input on pharmacotherapeutic activity [122].

References

1. McCrory P, Meeuwisse W, Dvorak J, et al. Consensus statement on concussion in sport-the 5(th) international conference on concussion in sport held in Berlin, October 2016. Br J Sports Med. 2017;51:838–47.
2. Comper P, Bisschop SM, Carnide N, Tricco A. A systematic review of treatments for mild traumatic brain injury. Brain Inj. 2005;19:863–80.
3. Marshall S, Bayley M, McCullagh S, et al. Updated clinical practice guidelines for concussion/mild traumatic brain injury and persistent symptoms. Brain Inj. 2015;29:688–700.
4. Kamins J, Giza CC. Concussion-mild traumatic brain injury: recoverable injury with potential for serious sequelae. Neurosurg Clin N Am. 2016;27:441–52.
5. Giza C, Greco T, Prins ML. Concussion: pathophysiology and clinical translation. Handb Clin Neurol. 2018;158:51–61.
6. Barkhoudarian G, Hovda DA, Giza CC. The molecular pathophysiology of concussive brain injury – an update. Phys Med Rehabil Clin N Am. 2016;27:373–93.
7. Monyer H, Hartley DM, Choi DW. 21-Aminosteroids attenuate excitotoxic neuronal injury in cortical cell cultures. Neuron. 1990;5:121–6.
8. Pellegrini-Giampietro DE, Cherici G, Alesiani M, Carla V, Moroni F. Excitatory amino acid release and free radical formation may cooperate in the genesis of ischemia-induced neuronal damage. J Neurosci. 1990;10:1035–41.
9. Nicholls DG. Mitochondrial calcium function and dysfunction in the central nervous system. Biochim Biophys Acta. 2009;1787:1416–24.
10. Nicholls DG. Brain mitochondrial calcium transport: origins of the set-point concept and its application to physiology and pathology. Neurochem Int. 2017;109:5–12.
11. Miller DM, Singh IN, Wang JA, Hall ED. Administration of the Nrf2-ARE activators sulforaphane and carnosic acid attenuates 4-hydroxy-2-nonenal-induced mitochondrial dysfunction ex vivo. Free Radic Biol Med. 2013;57:1–9.
12. Singh IN, Sullivan PG, Deng Y, Mbye LH, Hall ED. Time course of post-traumatic mitochondrial oxidative damage and dysfunction in a mouse model of focal traumatic brain injury: implications for neuroprotective therapy. J Cereb Blood Flow Metab. 2006;26:1407–18.
13. Hill RL, Kulbe JR, Singh IN, Wang JA, Hall ED. Synaptic mitochondria are more susceptible to traumatic brain injury-induced oxidative damage and respiratory dysfunction than non-synaptic mitochondria. Neuroscience. 2018;386:265–83.
14. Hill RL, Singh IN, Wang JA, Hall ED. Time courses of post-injury mitochondrial oxidative damage and respiratory dysfunction and neuronal cytoskeletal degradation in a rat model of focal traumatic brain injury. Neurochem Int. 2017;111:45–56.
15. Bains M, Hall ED. Antioxidant therapies in traumatic brain and spinal cord injury. Biochim Biophys Acta. 2012;1822:675–84.
16. Hall ED, Vaishnav RA, Mustafa AG. Antioxidant therapies for traumatic brain injury. Neurotherapeutics. 2010;7:51–61.
17. Gutteridge JM. Lipid peroxidation and antioxidants as biomarkers of tissue damage. Clin Chem. 1995;41:1819–28.
18. Hamann K, Shi R. Acrolein scavenging: a potential novel mechanism of attenuating oxidative stress following spinal cord injury. J Neurochem. 2009;111:1348–56.
19. Rohn TT, Hinds TR, Vincenzi FF. Ion transport ATPases as targets for free radical damage. Protection by an aminosteroid of the Ca2+ pump ATPase and Na+/K+ pump ATPase of human red blood cell membranes. Biochem Pharmacol. 1993;46:525–34.
20. Rohn TT, Hinds TR, Vincenzi FF. Inhibition of Ca2+-pump ATPase and the Na+/K+-pump ATPase by iron-generated free radicals. Protection by 6,7-dimethyl-2,4-DI-1-pyrrolidinyl-7H-pyrrolo[2,3-d] pyrimidine sulfate (U-89843D), a potent, novel, antioxidant/free radical scavenger. Biochem Pharmacol. 1996;51:471–6.
21. Sullivan PG, Krishnamurthy S, Patel SP, Pandya JD, Rabchevsky AG. Temporal characterization of mitochondrial bioenergetics after spinal cord injury. J Neurotrauma. 2007;24:991–9.

22. Keller JN, Mark RJ, Bruce AJ, et al. 4-Hydroxynonenal, an aldehydic product of membrane lipid peroxidation, impairs glutamate transport and mitochondrial function in synaptosomes. Neuroscience. 1997;80:685–96.
23. Keller JN, Pang Z, Geddes JW, et al. Impairment of glucose and glutamate transport and induction of mitochondrial oxidative stress and dysfunction in synaptosomes by amyloid beta-peptide: role of the lipid peroxidation product 4-hydroxynonenal. J Neurochem. 1997;69:273–84.
24. Lovell MA, Xie C, Markesbery WR. Acrolein, a product of lipid peroxidation, inhibits glucose and glutamate uptake in primary neuronal cultures. Free Radic Biol Med. 2000;29:714–20.
25. Springer JE, Azbill RD, Mark RJ, Begley JG, Waeg G, Mattson MP. 4-hydroxynonenal, a lipid peroxidation product, rapidly accumulates following traumatic spinal cord injury and inhibits glutamate uptake. J Neurochem. 1997;68:2469–76.
26. Kontos HA. Oxygen radicals in CNS damage. Chem Biol Interact. 1989;72:229–55.
27. Kontos HA, Povlishock JT. Oxygen radicals in brain injury. Cent Nerv Syst Trauma. 1986;3:257–63.
28. Kontos HA, Wei EP. Superoxide production in experimental brain injury. J Neurosurg. 1986;64:803–7.
29. Muizelaar JP. Clinical trials with Dismutec (pegorgotein; polyethylene glycol-conjugated superoxide dismutase; PEG-SOD) in the treatment of severe closed head injury. Adv Exp Med Biol. 1994;366:389–400.
30. Muizelaar JP, Marmarou A, Young HF, et al. Improving the outcome of severe head injury with the oxygen radical scavenger polyethylene glycol-conjugated superoxide dismutase: a phase II trial. J Neurosurg. 1993;78:375–82.
31. Muizelaar JP, Kupiec JW, Rapp LA. PEG-SOD after head injury. J Neurosurg. 1995;83:942.
32. Chan PH, Epstein CJ, Li Y, et al. Transgenic mice and knockout mutants in the study of oxidative stress in brain injury. J Neurotrauma. 1995;12:815–24.
33. Gladstone DJ, Black SE, Hakim AM, Heart, Stroke Foundation of Ontario Centre of Excellence in Stroke Recovery. Toward wisdom from failure: lessons from neuroprotective stroke trials and new therapeutic directions. Stroke. 2002;33:2123–36.
34. Lewen A, Matz P, Chan PH. Free radical pathways in CNS injury. J Neurotrauma. 2000;17:871–90.
35. Mikawa S, Kinouchi H, Kamii H, et al. Attenuation of acute and chronic damage following traumatic brain injury in copper, zinc-superoxide dismutase transgenic mice. J Neurosurg. 1996;85:885–91.
36. Xiong Y, Shie FS, Zhang J, Lee CP, Ho YS. Prevention of mitochondrial dysfunction in post-traumatic mouse brain by superoxide dismutase. J Neurochem. 2005;95:732–44.
37. Hall ED, McCall JM, Means ED. Therapeutic potential of the lazaroids (21-aminosteroids) in acute central nervous system trauma, ischemia and subarachnoid hemorrhage. Adv Pharmacol. 1994;28:221–68.
38. Hall ED, Yonkers PA, McCall JM, Braughler JM. Effects of the 21-aminosteroid U74006F on experimental head injury in mice. J Neurosurg. 1988;68:456–61.
39. McIntosh TK, Thomas M, Smith D, Banbury M. The novel 21-aminosteroid U74006F attenuates cerebral edema and improves survival after brain injury in the rat. J Neurotrauma. 1992;9:33–46.
40. Dimlich RV, Tornheim PA, Kindel RM, Hall ED, Braughler JM, McCall JM. Effects of a 21-aminosteroid (U-74006F) on cerebral metabolites and edema after severe experimental head trauma. Adv Neurol. 1990;52:365–75.
41. Hall ED, Yonkers PA, Andrus PK, Cox JW, Anderson DK. Biochemistry and pharmacology of lipid antioxidants in acute brain and spinal cord injury. J Neurotrauma. 1992;9(Suppl 2):S425–42.
42. Smith SL, Andrus PK, Zhang JR, Hall ED. Direct measurement of hydroxyl radicals, lipid peroxidation, and blood-brain barrier disruption following unilateral cortical impact head injury in the rat. J Neurotrauma. 1994;11:393–404.
43. Marshall LF, Maas AI, Marshall SB, et al. A multicenter trial on the efficacy of using tirilazad mesylate in cases of head injury. J Neurosurg. 1998;89:519–25.

44. Kassell NF, Haley EC Jr, Apperson-Hansen C, Alves WM. Randomized, double-blind, vehicle-controlled trial of tirilazad mesylate in patients with aneurysmal subarachnoid hemorrhage: a cooperative study in Europe, Australia, and New Zealand. J Neurosurg. 1996;84:221–8.
45. Lanzino G, Kassell NF. Double-blind, randomized, vehicle-controlled study of high-dose tirilazad mesylate in women with aneurysmal subarachnoid hemorrhage. Part II. A cooperative study in North America. J Neurosurg. 1999;90:1018–24.
46. Kim K, Priefer R. Evaluation of current post-concussion protocols. Biomed Pharmacother. 2020;129:110406.
47. Di Pietro V, Yakoub KM, Caruso G, et al. Antioxidant therapies in traumatic brain injury. Antioxidants (Basel). 2020;9(3):260.
48. Hall ED, Detloff MR, Johnson K, Kupina NC. Peroxynitrite-mediated protein nitration and lipid peroxidation in a mouse model of traumatic brain injury. J Neurotrauma. 2004;21:9–20.
49. Oliver JM, Jones MT, Kirk KM, et al. Effect of docosahexaenoic acid on a biomarker of head trauma in American football. Med Sci Sports Exerc. 2016;48:974–82.
50. Kensler TW, Wakabayashi N, Biswal S. Cell survival responses to environmental stresses via the Keap1-Nrf2-ARE pathway. Annu Rev Pharmacol Toxicol. 2007;47:89–116.
51. Zhang DD. Mechanistic studies of the Nrf2-Keap1 signaling pathway. Drug Metab Rev. 2006;38:769–89.
52. Miller DM, Wang JA, Buchanan AK, Hall ED. Temporal and spatial dynamics of nrf2-antioxidant response elements mediated gene targets in cortex and hippocampus after controlled cortical impact traumatic brain injury in mice. J Neurotrauma. 2014;31:1194–201.
53. Wang X, de Rivero Vaccari JP, Wang H, et al. Activation of the nuclear factor E2-related factor 2/antioxidant response element pathway is neuroprotective after spinal cord injury. J Neurotrauma. 2012;29:936–45.
54. Yan W, Wang HD, Hu ZG, Wang QF, Yin HX. Activation of Nrf2-ARE pathway in brain after traumatic brain injury. Neurosci Lett. 2008;431:150–4.
55. Hong SC, Goto Y, Lanzino G, Soleau S, Kassell NF, Lee KS. Neuroprotection with a calpain inhibitor in a model of focal cerebral ischemia. Stroke. 1994;25:663–9.
56. Dash PK, Zhao J, Orsi SA, Zhang M, Moore AN. Sulforaphane improves cognitive function administered following traumatic brain injury. Neurosci Lett. 2009;460:103–7.
57. Chen G, Fang Q, Zhang J, Zhou D, Wang Z. Role of the Nrf2-ARE pathway in early brain injury after experimental subarachnoid hemorrhage. J Neurosci Res. 2011;89:515–23.
58. Satoh T, Kosaka K, Itoh K, et al. Carnosic acid, a catechol-type electrophilic compound, protects neurons both in vitro and in vivo through activation of the Keap1/Nrf2 pathway via S-alkylation of targeted cysteines on Keap1. J Neurochem. 2008;104:1116–31.
59. Miller DM, Singh IN, Wang JA, Hall ED. Nrf2-ARE activator carnosic acid decreases mitochondrial dysfunction, oxidative damage and neuronal cytoskeletal degradation following traumatic brain injury in mice. Exp Neurol. 2015;264:103–10.
60. Maynard ME, Underwood EL, Redell JB, et al. Carnosic acid improves outcome after repetitive mild traumatic brain injury. J Neurotrauma. 2019;36:2147–52.
61. Althaus JS, Oien TT, Fici GJ, Scherch HM, Sethy VH, VonVoigtlander PF. Structure activity relationships of peroxynitrite scavengers an approach to nitric oxide neurotoxicity. Res Commun Chem Pathol Pharmacol. 1994;83:243–54.
62. Singh IN, Sullivan PG, Hall ED. Peroxynitrite-mediated oxidative damage to brain mitochondria: protective effects of peroxynitrite scavengers. J Neurosci Res. 2007;85:2216–23.
63. Wood PL, Khan MA, Moskal JR. Mechanism of action of the disease-modifying anti-arthritic thiol agents D-penicillamine and sodium aurothiomalate: restoration of cellular free thiols and sequestration of reactive aldehydes. Eur J Pharmacol. 2008;580:48–54.
64. Hall ED, Kupina NC, Althaus JS. Peroxynitrite scavengers for the acute treatment of traumatic brain injury. Ann N Y Acad Sci. 1999;890:462–8.
65. Chen Z, Park J, Butler B, et al. Mitigation of sensory and motor deficits by acrolein scavenger phenelzine in a rat model of spinal cord contusive injury. J Neurochem. 2016;138:328–38.

66. Wood PL, Khan MA, Moskal JR, Todd KG, Tanay VA, Baker G. Aldehyde load in ischemia-reperfusion brain injury: neuroprotection by neutralization of reactive aldehydes with phenelzine. Brain Res. 2006;1122:184–90.
67. Singh IN, Gilmer LK, Miller DM, Cebak JE, Wang JA, Hall ED. Phenelzine mitochondrial functional preservation and neuroprotection after traumatic brain injury related to scavenging of the lipid peroxidation-derived aldehyde 4-hydroxy-2-nonenal. J Cereb Blood Flow Metab. 2013;33:593–9.
68. Cebak JE, Singh IN, Hill RL, Wang JA, Hall ED. Phenelzine protects brain mitochondrial function in vitro and in vivo following traumatic brain injury by scavenging the reactive carbonyls 4-hydroxynonenal and acrolein leading to cortical histological neuroprotection. J Neurotrauma. 2017;34:1302–17.
69. Hill RL, Singh IN, Wang JA, Hall ED. Effects of phenelzine administration on mitochondrial function, calcium handling, and cytoskeletal degradation after experimental traumatic brain injury. J Neurotrauma. 2019;36:1231–51.
70. Galvani S, Coatrieux C, Elbaz M, et al. Carbonyl scavenger and antiatherogenic effects of hydrazine derivatives. Free Radic Biol Med. 2008;45:1457–67.
71. Hamann K, Nehrt G, Ouyang H, Duerstock B, Shi R. Hydralazine inhibits compression and acrolein-mediated injuries in ex vivo spinal cord. J Neurochem. 2008;104:708–18.
72. Park J, Zheng L, Marquis A, et al. Neuroprotective role of hydralazine in rat spinal cord injury-attenuation of acrolein-mediated damage. J Neurochem. 2014;129:339–49.
73. Kmieciak-Kolada K, Felinska W, Stachura Z, Majchrzak H, Herman ZS. Concentration of biogenic amines and their metabolites in different parts of brain after experimental cerebral concussion. Pol J Pharmacol Pharm. 1987;39:47–53.
74. Kline AE, Yu J, Massucci JL, Zafonte RD, Dixon CE. Protective effects of the 5-HT1A receptor agonist 8-hydroxy-2-(di-n-propylamino)tetralin against traumatic brain injury-induced cognitive deficits and neuropathology in adult male rats. Neurosci Lett. 2002;333:179–82.
75. Kosari-Nasab M, Shokouhi G, Azarfarin M, Bannazadeh Amirkhiz M, Mesgari Abbasi M, Salari AA. Serotonin 5-HT1A receptors modulate depression-related symptoms following mild traumatic brain injury in male adult mice. Metab Brain Dis. 2019;34:575–82.
76. Trujillo P, van Wouwe NC, Lin YC, et al. Dopamine effects on frontal cortical blood flow and motor inhibition in Parkinson's disease. Cortex. 2019;115:99–111.
77. McNamara CG, Dupret D. Two sources of dopamine for the hippocampus. Trends Neurosci. 2017;40:383–4.
78. Howe MW, Tierney PL, Sandberg SG, Phillips PE, Graybiel AM. Prolonged dopamine signalling in striatum signals proximity and value of distant rewards. Nature. 2013;500:575–9.
79. Sawaguchi T, Goldman-Rakic PS. D1 dopamine receptors in prefrontal cortex: involvement in working memory. Science. 1991;251:947–50.
80. Levey AI, Hersch SM, Rye DB, et al. Localization of D1 and D2 dopamine receptors in brain with subtype-specific antibodies. Proc Natl Acad Sci U S A. 1993;90:8861–5.
81. Lan YL, Li S, Lou JC, Ma XC, Zhang B. The potential roles of dopamine in traumatic brain injury: a preclinical and clinical update. Am J Transl Res. 2019;11:2616–31.
82. Abrahams S, McFie S, Lacerda M, et al. Unravelling the interaction between the DRD2 and DRD4 genes, personality traits and concussion risk. BMJ Open Sport Exerc Med. 2019;5(1):e000465.
83. Bales JW, Kline AE, Wagner AK, Dixon CE. Targeting dopamine in acute traumatic brain injury. Open Drug Discov J. 2010;2:119–28.
84. Flashman LA, McDonald BC, Ford JC, et al. Differential effects of pergolide and bromocriptine on working memory performance and brain activation after mild traumatic brain injury. J Neurotrauma. 2021;38:225–34.
85. Reddy CC, Collins M, Lovell M, Kontos AP. Efficacy of amantadine treatment on symptoms and neurocognitive performance among adolescents following sports-related concussion. J Head Trauma Rehabil. 2013;28:260–5.

86. Farnebo LO, Fuxe K, Goldstein M, Hamberger B, Ungerstedt U. Dopamine and noradrenaline releasing action of amantadine in the central and peripheral nervous system: a possible mode of action in Parkinson's disease. Eur J Pharmacol. 1971;16:27–38.
87. Mizoguchi K, Yokoo H, Yoshida M, Tanaka T, Tanaka M. Amantadine increases the extracellular dopamine levels in the striatum by re-uptake inhibition and by N-methyl-D-aspartate antagonism. Brain Res. 1994;662:255–8.
88. Seifert TD, Evans RW. Posttraumatic headache: a review. Curr Pain Headache Rep. 2010;14:292–8.
89. Blume HK, Vavilala MS, Jaffe KM, et al. Headache after pediatric traumatic brain injury: a cohort study. Pediatrics. 2012;129:e31–9.
90. Faux S, Sheedy J. A prospective controlled study in the prevalence of posttraumatic headache following mild traumatic brain injury. Pain Med. 2008;9:1001–11.
91. Bresee N, Aglipay M, Dubrovsky AS, et al. No association between metoclopramide treatment in ED and reduced risk of post-concussion headache. Am J Emerg Med. 2018;36:2225–31.
92. Friedman BW, Babbush K, Irizarry E, White D, John GE. An exploratory study of IV metoclopramide+diphenhydramine for acute post-traumatic headache. Am J Emerg Med. 2018;36:285–9.
93. Minces V, Pinto L, Dan Y, Chiba AA. Cholinergic shaping of neural correlations. Proc Natl Acad Sci U S A. 2017;114:5725–30.
94. Sarter M, Hasselmo ME, Bruno JP, Givens B. Unraveling the attentional functions of cortical cholinergic inputs: interactions between signal-driven and cognitive modulation of signal detection. Brain Res Brain Res Rev. 2005;48:98–111.
95. Xu M, Chung S, Zhang S, et al. Basal forebrain circuit for sleep-wake control. Nat Neurosci. 2015;18:1641–7.
96. Szymusiak R. Magnocellular nuclei of the basal forebrain: substrates of sleep and arousal regulation. Sleep. 1995;18:478–500.
97. Bertrand D, Wallace TL. A review of the cholinergic system and therapeutic approaches to treat brain disorders. Curr Top Behav Neurosci. 2020;45:1–28.
98. McAllister TW, Zafonte R, Jain S, et al. Randomized placebo-controlled trial of methylphenidate or galantamine for persistent emotional and cognitive symptoms associated with PTSD and/or traumatic brain injury. Neuropsychopharmacology. 2016;41:1191–8.
99. Tenovuo O. Central acetylcholinesterase inhibitors in the treatment of chronic traumatic brain injury-clinical experience in 111 patients. Prog Neuro-Psychopharmacol Biol Psychiatry. 2005;29:61–7.
100. Razay G, Wilcock GK. Galantamine in Alzheimer's disease. Expert Rev Neurother. 2008;8:9–17.
101. Bjorklund A, Dunnett SB. Dopamine neuron systems in the brain: an update. Trends Neurosci. 2007;30:194–202.
102. Li X, Yu B, Sun Q, et al. Generation of a whole-brain atlas for the cholinergic system and mesoscopic projectome analysis of basal forebrain cholinergic neurons. Proc Natl Acad Sci U S A. 2018;115:415–20.
103. Abikoff H, Hechtman L, Klein RG, et al. Symptomatic improvement in children with ADHD treated with long-term methylphenidate and multimodal psychosocial treatment. J Am Acad Child Adolesc Psychiatry. 2004;43:802–11.
104. Volkow ND, Ding YS, Fowler JS, et al. Is methylphenidate like cocaine? Studies on their pharmacokinetics and distribution in the human brain. Arch Gen Psychiatry. 1995;52:456–63.
105. Al-Adawi S, Al-Naamani A, Jaju S, et al. Methylphenidate improves executive functions in patients with traumatic brain injuries: a feasibility trial via the idiographic approach. BMC Neurol. 2020;20:103.
106. Mostert JP, Koch MW, Heerings M, Heersema DJ, De Keyser J. Therapeutic potential of fluoxetine in neurological disorders. CNS Neurosci Ther. 2008;14:153–64.
107. Fann JR, Uomoto JM, Katon WJ. Sertraline in the treatment of major depression following mild traumatic brain injury. J Neuropsychiatry Clin Neurosci. 2000;12:226–32.

108. Ashman TA, Cantor JB, Gordon WA, et al. A randomized controlled trial of sertraline for the treatment of depression in persons with traumatic brain injury. Arch Phys Med Rehabil. 2009;90:733–40.
109. Rapoport MJ, Chan F, Lanctot K, Herrmann N, McCullagh S, Feinstein A. An open-label study of citalopram for major depression following traumatic brain injury. J Psychopharmacol. 2008;22:860–4.
110. Yan L, Xu X, He Z, et al. Antidepressant-like effects and cognitive enhancement of coadministration of Chaihu Shugan San and fluoxetine: dependent on the BDNF-ERK-CREB signaling pathway in the hippocampus and frontal cortex. Biomed Res Int. 2020;2020:2794263.
111. Levy MJF, Boulle F, Emerit MB, et al. 5-HTT independent effects of fluoxetine on neuroplasticity. Sci Rep. 2019;9:6311.
112. Kosari-Nasab M, Shokouhi G, Ghorbanihaghjo A, Abbasi MM, Salari AA. Anxiolytic- and antidepressant-like effects of Silymarin compared to diazepam and fluoxetine in a mouse model of mild traumatic brain injury. Toxicol Appl Pharmacol. 2018;338:159–73.
113. Ebert SE, Jensen P, Ozenne B, et al. Molecular imaging of neuroinflammation in patients after mild traumatic brain injury: a longitudinal (123) I-CLINDE single photon emission computed tomography study. Eur J Neurol. 2019;26:1426–32.
114. Lawrence DW, Foster E, Comper P, et al. Cannabis, alcohol and cigarette use during the acute post-concussion period. Brain Inj. 2020;34:42–51.
115. Mackie K. Cannabinoid receptors: where they are and what they do. J Neuroendocrinol. 2008;20(Suppl 1):10–4.
116. Cabral GA, Raborn ES, Griffin L, Dennis J, Marciano-Cabral F. CB2 receptors in the brain: role in central immune function. Br J Pharmacol. 2008;153:240–51.
117. Singh J, Neary JP. Neuroprotection following concussion: the potential role for cannabidiol. Can J Neurol Sci. 2020;47:289–300.
118. Elliott MB, Ward SJ, Abood ME, Tuma RF, Jallo JI. Understanding the endocannabinoid system as a modulator of the trigeminal pain response to concussion. Concussion. 2017;2:CNC49.
119. Benitez-Angeles M, Morales-Lazaro SL, Juarez-Gonzalez E, Rosenbaum T. TRPV1: structure, endogenous agonists, and mechanisms. Int J Mol Sci. 2020;21:3421.
120. Hammond FM, Sherer M, Malec JF, et al. Amantadine effect on perceptions of irritability after traumatic brain injury: results of the amantadine irritability multisite study. J Neurotrauma. 2015;32:1230–8.
121. Silver JM, Koumaras B, Chen M, et al. Effects of rivastigmine on cognitive function in patients with traumatic brain injury. Neurology. 2006;67:748–55.
122. Polich G, Iaccarino MA, Kaptchuk TJ, Morales-Quezada L, Zafonte R. Placebo effects in traumatic brain injury. J Neurotrauma. 2018;35:1205–12.
123. Hall ED, Wang JA, Miller DM, et al. Newer pharmacological approaches for antioxidant neuroprotection in traumatic brain injury. Neuropharmacology. 2019;145(B):247–58.

Chapter 10
Sex, Gender, and Concussion

Angela Colantonio

Introduction

Concussion is the most common form of "mild traumatic brain injury" (mTBI). It is estimated that 42–45 million people sustain a mild traumatic brain injury (mTBI) in the world annually. These injuries account for up to 75 to 90% of all traumatic brain injuries (TBI) [1]. There is a growing body of literature documenting both sex and gender influences relevant to concussion that has historically not been given sufficient attention. The lack of literature in this area affects the ability to inform clinical practice. To date, clinical guidelines on concussion have not systematically addressed any considerations of sex and gender [2].

This chapter outlines examples of sex and gender considerations with respect to concussion, rather than an exhaustive review of the literature. This article utilizes the Canadian Institutes of Health Research [3] definitions for sex and gender whereby sex refers to the biological and physiological characteristics that distinguish males from females and gender refers to the "socially constructed roles, behaviors, expressions and identities of girls, women, boys, men, and gender diverse people." Although sex (male/female) and gender (men/women) are both commonly discussed as discrete and binary concepts, they have been acknowledged as both fluid and dynamic and are interrelated [3]. These constructs, however, are often used interchangeably in the literature. At the same time, their influence on health status is often difficult to deconstruct [4].

Preclinical data have been dominated by research on male rodents/animals. While similarities exist, these data may not necessarily be generalizable to females. Further, because historically TBI in general differentially affects men, research has been largely based on male-dominated samples. There are, however, known sex

A. Colantonio (✉)
Department of Occupational Science and Occupational Therapy, University of Toronto, Toronto, ON, Canada
e-mail: angela.colantonio@utoronto.ca

T. A. Schweizer, A. J. Baker (eds.), *Tackling the Concussion Epidemic*,
https://doi.org/10.1007/978-3-030-93813-0_10

differences throughout the brain in both neural anatomy and function and this research is rapidly growing [5]. Further, the World Health Organization (WHO) has identified gender as a determinant of health, given the unequal opportunities that girls and women experience globally [6]. Concussion rates vary by gender and have disproportionately affected boys and men. At the same time, gender considerations are typically not systematically integrated into research studies of prevention and treatment [7].

To fully appreciate the complexity of considering sex and gender in concussion, it is important to understand epidemiological trends at the population level. There are important differences in concussion rates across the lifespan that vary by sex and context. This chapter provides some general estimates and touch upon the influence of sex in concussion in the context of sports, vulnerable populations, violence, and workplace injury.

Concussion Rates

Research has found that, overall, rates of concussion/mild traumatic brain injury are higher among males, when considering the entire population across the life span. Rates based on emergency room visits have been found to be highest in males 4 years old and younger – the youngest age group – who are primarily injured due to falls. Brain injury rates among males also spike in adolescence and young adulthood, when motor vehicle crashes are more common. This is largely attributed to risk-taking behavior, particularly among boys/men at this stage in life [8].

There is an increase in concussion incidence rates again in older adults, mainly attributable to falls. However, unlike the younger age group, the highest rates for mild TBI have been found among older women [9]. Overall, half of Medicare populations in the United States were female. A study of home care recipients with a TBI diagnosis in Canada also showed that clients were predominantly female [10]. In North America, older adults are the fastest growing segment of the population with larger numbers of women affected by concussion, making sex and gender differences in TBI a growing concern. To date, there are no concussion guidelines specifically for older adults, unlike pediatric and adult populations overall.

The rate of reported concussion and mild traumatic brain injury has increased dramatically over the last decades, with the greatest increases among young women. One Canadian study of emergency room and physician visits showed a 6.3-fold increase in rates for female children and adolescents versus a 3.6-fold increase among boys. This may be because girls are participating more frequently in higher-risk activities like sports [11]. A population-based study of over 90,000 patients with a first concussion-related emergency department (ED) visit in Ontario, Canada, over a 10-year period (March 2003 to December 2011) also showed a dramatic increase among females particularly through sports-related injury. There was an 83% increase versus 19% increase in numbers of concussions among males over this 10-year period [12]. Clearly, this increase in rates of reported concussions needs to be investigated further, which involves sex and gender considerations.

Concussions in the Workplace

In the labor context, reported concussion rates have increased dramatically among persons injured at the workplace and involve more female workers sustaining moderate to severe injuries. Some sectors such as education, healthcare, and government-related occupations have higher rates of injury for women [13–15]. Additionally, there is an emerging body of literature on sex differences in military populations, given the increasing numbers of women obtaining employment in this workforce sector. Post-concussion syndrome is more commonly reported among women employed in the military, and it is reported that these women exhibit a different constellation of symptoms than their male counterparts [16]. Female veterans with mild TBI report more somatosensory or vestibular symptoms and experience a broad range of nonspecific symptoms (e.g., nausea, fatigue, disturbances in sleep and appetite) that are less frequently reported by male veterans. Differences in chronic pain reporting also exist; male veterans tend to report low back pain, whereas female veterans are more likely to report headache and depression following TBI. In terms of psychiatric sequelae, a higher incidence of post-traumatic stress disorders (PTSDs) has been found among female veterans [16].

Concussions in Vulnerable Populations

Among marginalized populations, such as groups who are incarcerated, are involved with the justice system, or are homeless, rates of TBI overall tend to be higher in males [17]. Assault as a mechanism of injury is the most likely cause of TBI within marginalized groups, when compared to the general population [18]. The literature has primarily focused on TBI across all levels of severity without a specific focus on concussion. However, mTBI was the most common form of TBI overall in these studies. Sex differences in these studies include women reporting more adverse early-life experiences than males or incarcerated persons without TBI and being more likely to have had a TBI prior to contact with the justice system [19].

Assault as a mechanism of injury is more common in marginalized populations, which includes instances of intimate partner violence (IPV). This mechanism of injury and context has been overlooked in the concussion literature. It is estimated that 1 in 4 North American women have been affected by IPV, and the majority of injuries are hits to the head, face, and neck. These injuries tend to be repetitive and occur over a prolonged period of time. While there have been no national studies on this topic, commonly cited rates of injury are from 30 to 74% or higher [20, 21]. We may assume these are underestimates, given that most of these injuries are likely not reported. For example, one study found that none of the women with identified brain injury sought out care [22]. Most of the studies examining prevalence of TBI related to IPV rely on smaller samples [23]. Because these studies are frequently based on shelter populations, less is known of concussion in the general population experiencing IPV. Screening instruments for TBI are typically not validated for this population, which requires a more tailored approach due to the nature of these injuries. Further, TBI has been identified as a considerable knowledge gap among frontline providers working with abused women [20].

More generally, it is difficult to get accurate numbers of persons who sustain concussions for many reasons. Because loss of consciousness is not required for the diagnosis of concussion, some individuals may not realize that they have sustained one. Not everyone affected by concussion seeks care, and as such, data sources that rely on contact with the health care system will miss these injuries. Further, concussion can be underdiagnosed, or missed in emergency rooms or other contexts, as some of the symptoms may overlap with conditions such as post-traumatic stress disorder and neck injuries. Sex- and gender-specific biases, such as concussion education biases, add to the complexity in diagnosing concussions. These biases are discussed in detail in a later section on sex differences in concussions.

Other Sex Differences

Epidemiological studies have also identified comorbid conditions that can vary by sex at time of injury. For instance, a large population-based study of concussions reported in the emergency room identified comorbid neck injury to be significantly higher among females across mechanisms of injury. This difference was not found in the youngest and oldest age groups, suggesting co-occurring injury is reported mostly during a woman's childbearing years. Hence, the interaction with age in the evaluation of many sex differences reported needs to be seriously considered when making assessments of similarities and differences [12].

The paucity of sex-stratified data has also extended to research on prognosis after mTBI/concussion. Based on a review of over 200 studies from 1980 to 2012, it was shown that only 7% of studies that recruit both sexes report sex-stratified results regarding mTBI prognosis [24]. This review, which reanalyzed results from a previously published report without sex stratification, showed that no sex differences were found for risk of dementia, primary brain tumor, return to work, or post-traumatic stress syndrome. Females were found to have a higher risk of suicide and epilepsy in the children and young adult populations, while males were found to be at a higher risk of schizophrenia.

Concussions in Sports

A large number of studies addressing sex differences in concussion have looked at sports-related injury. In some games with similar rules, the rates of concussion are higher among girls and women. One large study of college athletes found that females had a 1.4 times higher overall concussion injury rate than males in sex-comparable sports such as soccer, basketball, and baseball [25]. A review of 101 articles published by June 2016 found some evidence to support that females in teenage years may be more vulnerable for symptoms that persist more than one month than boys based on large-scale observational studies after sports-related concussion [26]. Other reviews have concluded that females report a greater number and more severe and persistent symptoms compared to males [27, 28]. Female athletes have been found to score better on verbal memory and motor processing speed. However, females reported more sleep and emotional and cognitive symptoms compared to male athletes [29]. However, there are inconsistent findings with respect to the distribution of symptomatology by sex. Sex differences have not been found consistently with respect to other functional outcomes such as disability and community integration in other non-sports-related populations after mild traumatic brain injury [30–32].

Sex Differences in Concussions Explained

Biological Factors

Studies have provided both biological and socio-cultural explanations to explain differences in incidence and outcomes by sex – largely in sports-related populations. Differences in neck musculature, for instance, have been proposed to explain the higher incidence with similar exposures through sports [33]. Girls and women generally have less isometric neck strength, neck girth, and head mass, resulting in lower levels of head/neck segment stiffness. This may make them more susceptible to rotational forces after a blow to the head. One study found that for every pound increase in neck strength, odds of concussion decreased by 5% [34].

Females experiencing concussions have also been found to experience more comorbid mental health conditions such as anxiety and depression including pre-injury [35]. Pre-existing mental health conditions such as anxiety and depression are associated with poorer outcomes [2]. The higher prevalence of these conditions in females overall could be an important factor in influencing post-concussion outcomes.

Hormonal differences among men and women may also play a role. Progesterone – known as a female reproductive hormone – varies throughout a female's reproductive cycle with higher levels associated with better concussion recovery [36, 37].

There is some evidence to suggest that menstrual phase at the time of injury may affect outcomes. Because women exhibit poorer outcomes when they are injured during the luteal phase of their cycle, when progesterone is highest, it has been hypothesized that this is the result of progesterone withdrawal [37]. Women injured during the luteal phase also report poorer outcomes one month after injury. A more recent study, however, did not find that menstrual phase was related to symptom scores in the acute injury phase, at 1 to 2 weeks post-injury [13].

Menstrual cycle disruption has been reported as common after moderate to severe traumatic brain injury [38], and more recently this has been reported after concussion [39, 40]. Females between the ages of 12 and 21 had a significant increase in the risk of multiple abnormal menstrual patterns after a sports-related concussion compared with orthopedic injury. This disruption can be an indicator for hypothalamic pituitary-ovarian axis functioning that is affected by concussion and impacts progesterone levels [39].

Researchers have also begun to examine to what degree medications such as birth control may account for sex differences. One study of 90 collegiate varsity athletes found that non-contraceptive users had higher symptom severity than contraceptive users. Those using contraceptives may experience less fluctuation in progesterone levels [41]. However, one study found that reported use of birth control was associated with poorer symptoms scores 2 weeks post-injury [13].

With respect to structural differences in the brain, female axons have been found to be smaller and more susceptible to injury [42, 43]. More recently, Dolle et al. [43] have reported that female axons consistently have fewer microtubules than male axons and as such are at a greater risk of failure during trauma.

Structural differences in the brain by sex may also influence function and concussion outcomes. For instance, compared to females, the male corpus callosum has larger fibers and a larger cross-sectional area. Females tend to utilize both hemispheres of the brain for a majority of tasks, while there are more intra-hemispheric connections among males. Because women are more likely to utilize inter-hemispheric connections, strain in the corpus callosum area, which is the channel for inter-hemispheric connections, may have a greater impact on females [44, 45].

Gender or Socio-Cultural Factors

Other explanations for differences in outcomes by sex are related to gender or socio-cultural factors. These include proposed differences in symptom reporting which show that overall, females are more likely to seek out care [46]. It is thought that girls and women are more likely to be honest and to report symptoms more frequently than boys, although this has not been found to be consistent [47]. Conversely, it is believed that boys and men underreport symptoms due to gender norms related to masculinity, which dictate that it is less socially acceptable for men and boys to admit symptomatology and weakness [48, 49].

One study reported that the identification with male gender norms by both males and females was most associated with playing through injury. It also identified gender differences in concussion education, showing that female athletes were more knowledgeable about concussion which is likely to affect reporting [50].

It has also been well documented that male depression is poorly understood. Current measures of depression are more likely to reflect what is most expressed by girls and women, which may not fully capture depressive symptoms in men [51]. Instruments developed to assess male depression have been developed but are not routinely used. Treatment options based on what may be most acceptable for women such as psychotherapy may not be preferred by men [49]. In a study of patient preferences for care among more severely injured youth, treatment preferences were found whereby males expressed a preference for improving their health through physical activity and, as such, this may be a preferred option for recovery in this cohort. In contrast, girls were more likely to report preferences for same-sex providers and appreciate the support of family [52].

Other environmental factors that relate to gender that should be considered are whether girls receive differential coaching that may put them at risk of injury. In addition, there is a question as to whether sports equipment and rules could be tailored to prevent injuries that differentially affect girls and women. One example is whether a different soccer ball should be considered given that the ball/head ratio is larger for girls and women.

In studies of workplace injury, injured workers with persistent symptoms after mild traumatic brain injury report that return to work was easier in more "feminine" environments versus ones that were "masculine" or male dominated. The support of coworkers was important for all injured workers [53]. In a study that modeled pain after mild traumatic brain injury occurring at the workplace, different sociodemographic, injury related, behavioral and clinical variables emerged when sex stratified analyses were employed which explained more variance separately than when data were combined. Although this study reported no sex differences in pain frequencies or severity, the study highlights the importance of the contribution of sex- and gender-related variables in understanding the multi-dimensional construct of pain which is frequently associated with concussion and post-concussion syndrome [54].

In focus groups conducted among women with brain injury, women reported feeling dismissed when being seen by clinicians [55]. Research indicates that expectations of recovery are predominantly based on the experience of males, and this may influence the expectations of healthcare providers serving women. It may be that females express themselves differently than males [56]. A better understanding of the gendered experience of concussion may lead to better care and a more tailored approach [55].

There are many questions that arise from the literature to date that can inform clinical practice. For instance, if neck musculature is a factor in concussions, would greater attention to neck strengthening in girls and women be associated with fewer concussions? If risk-taking behavior is associated with male gender norms and masculinity, should prevention and treatment messages aim to influence culture, reduce injuries, and improve men's and boy's health-seeking behavior? Finally, there may be sex- and gender-specific preferences for care which have yet to be thoroughly investigated. Testing the effect of integrating patient preferences which are sensitive to sex and gender is another important area of investigation.

Moving Forward

Currently, no interventional studies have evaluated concussion management approaches which explicitly consider sex and gender. The Canadian Institutes for Health Research is funding an ongoing interventional study that aims to assess how the use of sex- and gender-focused educational materials may result in better TBI outcomes, including concussion, and to subsequently inform guidelines. This ongoing, integrated knowledge transfer study benefits from the perspectives of patients, caregivers, and service providers across the gender spectrum, who are informing the development of knowledge transfer materials [57].

It is very important for clinicians to recognize that there is increasing demand from end-users for concussion treatment that considers sex and gender. An example of this demand for gender-specific information and care was the establishment of the international task force on girls and women with acquired brain injury. After the first international workshop on women and TBI in 2010, at the request of Marilyn Spivack, founder of the Brain Injury Association of America, this task force was formed through the American Congress of Rehabilitation Medicine [58]. This international group of researchers, clinicians, patient ambassadors, and advocates has been meeting monthly for over 8 years. The task force produced a first-of-its-kind special issue in the Archives of Physical Medicine and Rehabilitation, entitled "Sex, Gender and Traumatic Brain Injury." This issue was guest-edited by Dr. Angela Colantonio [7]. Some leading advocacy groups such as Pink Concussions, the first non-profit addressing research and education in brain injury, have raised awareness internationally for the need for this research. This organization has mounted numerous summits and conferences and hosts support groups for thousands of women who are seeking sex and gender information (www.pinkconcussions.com).

There are a number of other encouraging developments. Since this first workshop on women and TBI in 2010, there are now several conferences a year that address this topic internationally. The National Institutes of Health recently funded a 2-day free workshop entitled *Understanding Traumatic Brain Injury in Women*, with leading experts. Results of this 2017 workshop are freely available online. Federal funding agencies in North America and Europe now call for mandatory items on grant applications that address sex and gender, and an increasing number of journals are indicating that manuscripts should address sex and gender [59]. In

Canada, the Status of Women Canada has recently launched a concussion video, *Sex and Gender Considerations in Concussion Care*. This video takes an intersectional approach by extending sex and gender to include other factors such as age, ethnicity, and socioeconomic status [60].

Sex and gender are seminal factors to include in research studies, but the intersection of other factors is also clearly important. Nature Reviews Neurology recently published a review relevant to our ongoing research entitled *Traumatic brain injury: Sex, gender and intersecting vulnerabilities* [30]. This review addresses sex and gender considerations across all levels of severity with a focus on marginalized populations and is a resource for the reader. This review offers a comprehensive model by which to understand risks and protective factors related to sex and gender, which considers a multitude of factors. As the diagram below demonstrates, it is virtually impossible to delineate all the factors, yet it remains important to consider a broad range of potential contributors (see Fig. 10.1).

Some considerations for clinicians are captured in the following questions, which can be asked by those involved in concussion care. To what extent has concussion research included sex and gender and what are the study contexts? Are studies biased toward one sex/gender or is there equitable representation? Do the analyses and discussion reflect sex and gender considerations? Are there biological factors that inform care such as propensity for comorbid conditions by sex? What gendered considerations are important? To what extent can treatment plans address gender identities? Are prevention and management communications presented in ways that are gender inclusive? Also important is to consider concussion in specific

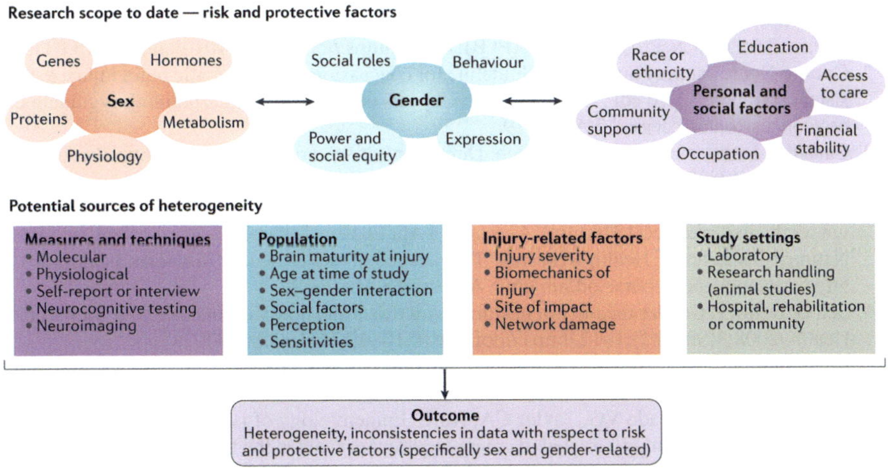

Fig. 10.1 Scopes and limitations of TBI studies. The figure summarizes the scope of studies to date that have explored sex, gender, and intersecting vulnerabilities in the context of traumatic brain injury (TBI). The figure also lists potential sources of heterogeneity of study results, which must be considered in future studies to help us better understand the protective and risk factors associated with sex and gender. (Reprinted from Mollayeva et al. [30]. With permission from Springer Nature)

contexts. For example, to what extent are concussions captured in contexts that are prevalent across genders? How can screening measures be made appropriate for the intimate partner context or other contexts, for instance, versus that of sports?

Much of our current literature is based on sports and therefore based on a younger and more fit population. More research must be conducted on a broader range of populations affected, including more marginalized populations, and across the age continuum. It is critical to note that there is virtually no information on concussion in gender diverse individuals. Clinical intake and research studies more commonly provide opportunities to report non-binary gender identities. The extent to which research on persons who identify as men or women is generalizable to non-binary gender identities needs further exploration.

Overall, it is important for clinicians to keep informed of the rapidly growing area of sex and gender considerations in TBI research. This information should be integrated in health care curricula and continuing education materials. Such resources are available for instance on both the CIHR and NIH websites. Understanding sex and gender-based differences, as well as similarities, is foundational to good science. It helps us inform more tailored approaches to concussion prevention, management, and education. Further, in Canada, this approach is consistent with Health Canada's Sex and Gender Action Plan which aims to integrate sex, and diversity considerations into all the work of Health Canada [61–63].

References

1. Gardner RC, Yaffe K. Epidemiology of mild traumatic brain injury and neurodegenerative disease. Mol Cell Neurosci. 2015;66(Pt B):75–80. https://doi.org/10.1016/j.mcn.2015.03.001.
2. Ontario Neurotrauma Foundation. Guideline for concussion/mild traumatic brain injury & persistent stymptoms, 3rd edn, for adults over 18 years of age. 2018. http://braininjuryguidelines.org/concussion/. Accessed 3 Apr 2019.
3. Canadian Institutes of Health Research. Definitions of sex and gender. 2015. http://www.cihr-irsc.gc.ca/e/47830.html. Accessed 3 Apr 2019.
4. Canadian Institutes of Health Research. Science is better with sex and gender. 2018. http://www.cihr-irsc.gc.ca/e/51310.html. Accessed 3 Apr 2019.
5. National Institutes of Health Research. The science of sex & gender in human health. 2017. https://sexandgendercourse.od.nih.gov/. Accessed 3 Apr 2019.
6. Sen G, Ostlin P. Gender inequity in health: why it exists and how we can change it. Glob Public Health. 2008;3(Suppl 1):1–12. https://doi.org/10.1080/17441690801900795.
7. Colantonio A. Sex, gender, and traumatic brain injury [special issue]. Arch Phys Med Rehabil. 2016;97(2 Suppl):S1–S70.
8. Cancelliere C, Coronado VG, Taylor CA, Xu L. Epidemiology of isolated versus nonisolated mild traumatic brain injury treated in emergency departments in the United States, 2006-2012: sociodemographic characteristics. J Head Trauma Rehabil. 2017;32(4):E37–46. https://doi.org/10.1097/htr.0000000000000260.
9. Albrecht JS, Hirshon JM, McCunn M, Bechtold KT, Rao V, Simoni-Wastila L, Smith GS. Increased rates of mild traumatic brain injury among older adults in US emergency departments, 2009-2010. J Head Trauma Rehabil. 2016;31(5):E1–7. https://doi.org/10.1097/htr.0000000000000190.

10. McGuire C, Kristman VL, Martin L, Bedard M. Characteristics and incidence of traumatic brain injury in older adults using home Care in Ontario from 2003-2013. Can Geriatr J. 2017;20(1):2–9. https://doi.org/10.5770/cgj.20.228.
11. Zemek RL, Grool AM, Rodriguez Duque D, DeMatteo C, Rothman L, Benchimol EI, Guttmann A, Macpherson AK. Annual and seasonal trends in ambulatory visits for pediatric concussion in Ontario between 2003 and 2013. J Pediatr. 2017;181(222–228):e222. https://doi.org/10.1016/j.jpeds.2016.10.067.
12. Sutton M, Chan V, Escobar M, Mollayeva T, Hu Z, Colantonio A. Neck injury comorbidity in concussion-related emergency department visits: a population-based study of sex differences across the life span. J Womens Health (Larchmt). 2019;28(4):473–82.
13. El-Khechen Richandi G, Comper P, Bayley M, Colantonio A. The impact of menstrual phase on outcomes of females with concussion. Poster presented at the International Brain Injury Association 13th World Congress on Brain Injury, Toronto, 13–16 March 2019; abstract published in Brain Injury (2019).
14. Colantonio A, Mroczek D, Patel J, Lewko J, Fergenbaum J, Brison R. Examining occupational traumatic brain injury in Ontario. Can J Public Health. 2010;101(Suppl 1):S58–62.
15. Gray J, Mollayeva T, Bayley M, Mihailidis A, Liss GM, Gibson B, Lewko J, Sharma B, Nowrouzi-Kia B, Colantonio A. Concussion in the workplace. 2016. https://www.youtube.com/watch?v=4Jpqw1sA2d8.
16. Kim LH, Quon JL, Sun FW, Wortman KM, Adamson MM, Harris OA. Traumatic brain injury among female veterans: a review of sex differences in military neurosurgery. Neurosurg Focus. 2018;45(6):E16. https://doi.org/10.3171/2018.9.focus18369.
17. Durand E, Chevignard M, Ruet A, Dereix A, Jourdan C, Pradat-Diehl P. History of traumatic brain injury in prison populations: a systematic review. Ann Phys Rehabil Med. 2017;60(2):95–101. https://doi.org/10.1016/j.rehab.2017.02.003.
18. Topolovec-Vranic J, Ennis N, Colantonio A, Cusimano MD, Hwang SW, Kontos P, Ouchterlony D, Stergiopoulos V. Traumatic brain injury among people who are homeless: a systematic review. BMC Public Health. 2012;12(1):1059. https://doi.org/10.1186/1471-2458-12-1059.
19. Colantonio A, Kim H, Allen S, Asbridge M, Petgrave J, Brochu S. Traumatic brain injury and early life experiences among men and women in a prison population. J Correct Health Care. 2014;20(4):271–9. https://doi.org/10.1177/1078345814541529.
20. Haag HL, Sokoloff S, MacGregor N, Broekstra S, Cullen N, Colantonio A. Battered and brain injured: assessing knowledge of traumatic brain injury among intimate partner violence service providers. J Women's Health. 2019;28(7):990–6. https://doi.org/10.1089/jwh.2018.7299.
21. Kwako LE, Glass N, Campbell J, Melvin KC, Barr T, Gill JM. Traumatic brain injury in intimate partner violence: a critical review of outcomes and mechanisms. Trauma Violence Abuse. 2011;12(3):115–26. https://doi.org/10.1177/1524838011404251.
22. Jackson H, Philp E, Nuttall RL, Diller L. Traumatic brain injury: a hidden consequence for battered women. Prof Psychol Res Pract. 2002;33(1):39–45. https://doi.org/10.1037/0735-7028.33.1.39.
23. Haag HL, Jones D, Joseph T, Colantonio A. Battered & brain injured: traumatic brain injury among women survivors of intimate partner violence - a scoping review. Trauma Violence Abuse. 2019:1–18. https://doi.org/10.1177/1524838019850623.
24. Cancelliere C, Donovan J, Cassidy JD. Is sex an indicator of prognosis after mild traumatic brain injury: a systematic analysis of the findings of the World Health Organization collaborating Centre task force on mild traumatic brain injury and the international collaboration on mild traumatic brain injury prognosis. Arch Phys Med Rehabil. 2016;97(2 Suppl):S5–18. https://doi.org/10.1016/j.apmr.2014.11.028.
25. Covassin T, Moran R, Elbin RJ. Sex differences in reported concussion injury rates and time loss from participation: an update of the National Collegiate Athletic Association Injury Surveillance Program from 2004-2005 through 2008-2009. J Athl Train. 2016;51(3):189–94. https://doi.org/10.4085/1062-6050-51.3.05.

26. Iverson GL, Gardner AJ, Terry DP, Ponsford JL, Sills AK, Broshek DK, Solomon GS. Predictors of clinical recovery from concussion: a systematic review. Br J Sports Med. 2017;51(12):941–8. https://doi.org/10.1136/bjsports-2017-097729.
27. Covassin T, Savage JL, Bretzin AC, Fox ME. Sex differences in sport-related concussion long-term outcomes. Int J Psychophysiol. 2018;132(Pt A):9–13. https://doi.org/10.1016/j.ijpsycho.2017.09.010.
28. King NS. A systematic review of age and gender factors in prolonged post-concussion symptoms after mild head injury. Brain Inj. 2014;28(13–14):1639–45. https://doi.org/10.3109/02699052.2014.954271.
29. Covassin T, Elbin RJ, Larson E, Kontos A. Sex and age differences in depression and baseline sport-related concussion neurocognitive performance and symptoms. Clin J Sport Med. 2012;22(2):98–104. https://doi.org/10.1097/JSM.0b013e31823403d2.
30. Mollayeva T, Mollayeva S, Colantonio A. Traumatic brain injury: sex, gender and intersecting vulnerabilities. Nat Rev Neurol. 2018;14(12):711–22. https://doi.org/10.1038/s41582-018-0091-y.
31. Mollayeva T, Shapiro CM, Mollayeva S, Cassidy JD, Colantonio A. Modeling community integreation in workers with delayed recovery from mild traumatic brain injury. BMC Neurol. 2015;15:194.
32. Mollayeva T, Pratt B, Mollayeva S, Shapiro CM, Cassidy JD, Colantonio A. The relationship between insomina and disability inworkers with mild traumatic brain injury/concussion: insomnia and disaibity in chronic mild traumatic brain injury. Sleep Med. 2016;20:157–66.
33. Tierney RT, Higgins M, Caswell SV, Brady J, McHardy K, Driban JB, Darvish K. Sex differences in head acceleration during heading while wearing soccer headgear. J Athl Train. 2008;43(6):578–84. https://doi.org/10.4085/1062-6050-43.6.578.
34. Collins CL, Fletcher EN, Fields SK, Kluchurosky L, Rohrkemper MK, Comstock RD, Cantu RC. Neck strength: a protective factor reducing risk for concussion in high school sports. J Prim Prev. 2014;35(5):309–19. https://doi.org/10.1007/s10935-014-0355-2.
35. Brown DA, Elsass JA, Miller AJ, Reed LE, Reneker JC. Differences in symptom reporting between males and females at baseline and after a sports-related concussion: a systematic review and meta-analysis. Sports Med. 2015;45(7):1027–40. https://doi.org/10.1007/s40279-015-0335-6.
36. Wright DK, O'Brien TJ, Shultz SR, Mychasiuk R. Sex matters: repetitive mild traumatic brain injury in adolescent rats. Ann Clin Transl Neurol. 2017;4(9):640–54. https://doi.org/10.1002/acn3.441.
37. Wunderle K, Hoeger KM, Wasserman E, Bazarian JJ. Menstrual phase as predictor of outcome after mild traumatic brain injury in women. J Head Trauma Rehabil. 2014;29(5):E1–8. https://doi.org/10.1097/htr.0000000000000006.
38. Colantonio A, Mar W, Escobar M, Yoshida K, Velikonja D, Rizoli S, Cusimano M, Cullen N. Women's health outcomes after traumatic brain injury. J Womens Health (Larchmt). 2010;19(6):1109–16. https://doi.org/10.1089/jwh.2009.1740.
39. Snook ML, Henry LC, Sanfilippo JS, Zeleznik AJ, Kontos AP. Association of Concussion with Abnormal Menstrual Patterns in adolescent and young women. JAMA Pediatr. 2017;171(9):879–86. https://doi.org/10.1001/jamapediatrics.2017.1140.
40. Biscardi M, Shafi R, Einstein G, Cullen N, Colantonio A. Menopause, anti- Müllerian hormone and cognition in a cohort of women with persistent symptoms following TBI: a case for future research. Brain Inj. 2020;35(8):934–42.
41. Gallagher V, Kramer N, Abbott K, Alexander J, Breiter H, Herrold A, Lindley T, Mjaanes J, Reilly J. The effects of sex differences and hormonal contraception on outcomes after collegiate sports-related concussion. J Neurotrauma. 2018;35(11):1242–7. https://doi.org/10.1089/neu.2017.5453.
42. Babikian T, Marion SD, Copeland S, Alger JR, O'Neill J, Cazalis F, Mink R, Giza CC, Vu JA, Hilleary SM, Kernan CL, Newman N, Asarnow RF. Metabolic levels in the corpus callosum and their structural and behavioral correlates after moderate to severe pediatric TBI. J Neurotrauma. 2010;27(3):473–81. https://doi.org/10.1089/neu.2009.1058.

43. Dolle JP, Jaye A, Anderson SA, Ahmadzadeh H, Shenoy VB, Smith DH. Newfound sex differences in axonal structure underlie differential outcomes from in vitro traumatic axonal injury. Exp Neurol. 2018;300:121–34. https://doi.org/10.1016/j.expneurol.2017.11.001.
44. Solomito MJ, Reuman H, Wang DH. Sex differences in concussion: a review of brain anatomy, function, and biomechanical response to impact. Brain Inj. 2019;33(2):105–10. https://doi.org/10.1080/02699052.2018.1542507.
45. Shafi R, Crawley AP, Tartaglia MC, Tator CH, Green RE, Mikulis DJ, Colantonio A. Sex-specific differences in resting-state functional connectivity of large-scale networks in postconcussion syndrome. Sci Rep. 2020;10:21982. https://doi.org/10.1038/s41598-020-77137-4.
46. Stergiou-Kita M, Mansfield E, Sokoloff S, Colantonio A. Gender influences on return to work after mild traumatic brain injury. Arch Phys Med Rehabil. 2016;97(2 Suppl):S40–5. https://doi.org/10.1016/j.apmr.2015.04.008.
47. Demakis GJ, Rimland CA. Untreated mild traumatic brain injury in a young adult population. Arch Clin Neuropsychol. 2010;25(3):191–6. https://doi.org/10.1093/arclin/acq004.
48. Stergiou-Kita M, Mansfield E, Bezo R, Colantonio A, Garritano E, Lafrance M, Lewko J, Mantis S, Moody J, Power N, Theberge N, Westwood E, Travers K. Danger zone: men, masculinity and occupational health and safety in high risk occupations. Saf Sci. 2015;80:213–20. https://doi.org/10.1016/j.ssci.2015.07.029.
49. American Psychological Association Boys and Men Guidelines Group. APA guidelines for psychological practice with boys and men. American Psychological Association. 2018. https://www.apa.org/about/policy/boys-men-practice-guidelines.pdf. Accessed 3 Apr 2019.
50. Kroshus E, Baugh CM, Stein CJ, Austin SB, Calzo JP. Concussion reporting, sex, and conformity to traditional gender norms in young adults. J Adolesc. 2017;54:110–9. https://doi.org/10.1016/j.adolescence.2016.11.002.
51. Oliffe JL, Phillips MJ. Men, depression and masculinities: a review and recommendations. J Men's Health. 2008;5(3):194–202. https://doi.org/10.1016/j.jomh.2008.03.016.
52. Lindsay S, Proulx M, Maxwell J, Hamdani Y, Bayley M, Macarthur C, Colantonio A. Gender and transition from pediatric to adult health care among youth with acquired brain injury: experiences in a transition model. Arch Phys Med Rehabil. 2016;97(2 Suppl):S33–9. https://doi.org/10.1016/j.apmr.2014.04.032.
53. Colantonio A, Salehi S, Kristman V, Cassidy JD, Carter A, Vartanian O, Bayley M, Kirsh B, Hebert D, Lewko J, Kubrak O, Mantis S, Vernich L. Return to work after work-related traumatic brain injury. NeuroRehabilitation. 2016;39(3):389–99. https://doi.org/10.3233/nre-161370.
54. Mollayeva T, Cassidy JD, Shapiro CM, Mollayeva S, Colantonio A. Concussion/mild traumatic brain injury related chronic pain in males and females: a diagnostic modelling study. Medicine. 2017;96(7):e5917.
55. Haag HL, Caringal M, Sokoloff S, Kontos P, Yoshida K, Colantonio A. Being a woman with acquired brain injury: challenges and implications for practice. Arch Phys Med Rehabil. 2016;97(2 Suppl):S64–70. https://doi.org/10.1016/j.apmr.2014.12.018.
56. Despins EH, Turkstra LS, Struchen MA, Clark AN. Sex-based differences in perceived pragmatic communication ability of adults with traumatic brain injury. Arch Phys Med Rehabil. 2016;97(2 Suppl):S26–32. https://doi.org/10.1016/j.apmr.2014.06.023.
57. Mollayeva T, Amodio V, Mollayeva S, D'Souza A, Colquhoun H, Haag HL, Quilico E, Colantonio A. A gender transformative approach to improve outcomes and equity among persons with traumatic brain injury. BMJ Open. 2019;9:e024674. https://doi.org/10.1136/bmjopen-2018-024674.
58. Harris JE, Colantonio A, Bushnik T, Constantinidou F, Dawson D, Goldin-Lauretta Y, Swaine B, Warren H. Advancing the health and quality-of-life of girls and women after traumatic brain injury: workshop summary and recommendations. Brain Inj. 2012;26(2):177–82. https://doi.org/10.3109/02699052.2011.635361.

59. Heidari S, Babor TF, De Castro P, Tort S, Curno M. Sex and gender equity in research: rationale for the SAGER guidelines and recommended use. Res Integrity Peer Rev. 2016;1(1):2. https://doi.org/10.1186/s41073-016-0007-6.
60. Status of Women Canada. Appling GBA to concussion prevention & treatment [Video]. 2018. https://cfc-swc.gc.ca/med/multimedia/videos/index-en.html.
61. Health Canada. Sex and gender implementation report. 2018. Retrieved from: https://www.canada.ca/en/health-canada/corporate/transparency/corporate-management-reporting/sex-gender-based-analysis-action-implementation-report.html.
62. Canadian Institutes of Health Research. Online training modules: integrating sex & gender in health research. 2017. http://www.cihr-irsc.gc.ca/e/49347.html. Accessed 3 Apr 2019.
63. Valera EM, Joseph AC, Snedaker K, Breiding MJ, Robertson CL, Colantonio A, Levin H, Pugh MJ, Yurgelun-Todd D, Mannix R, Bazarian JJ, Turtzo C, Turkstra LS, Begg L, Cummings DM, Bellgowan PSF. Understanding traumatic brain injury in females: a state-of-the-art summary and future directions. J Head Trauma Rehabil. 2021;36(1):E1–E17. https://doi.org/10.1097/HTR.0000000000000652.

Chapter 11
Chronic Traumatic Encephalopathy

David G. Munoz and Ian R. Mackenzie

Historical Introduction

The concept that repeated trauma to the brain can lead to progressive neurological deterioration was a late development in medicine, considering that the clinical observation was probably available since classical times. In 1928, Markland described in a letter to the *Journal of the American Medical Association* (*JAMA*) what he called punch-drunk syndrome in boxers [1]. Several names were provided for the condition. Probably the most common at the time was dementia pugilistica [2], but it was Critchley [3] who coined the term chronic traumatic encephalopathy (CTE) that eventually became the standard nomenclature. The pathological basis of the condition was not understood until 1973 when Corsellis and his colleagues [4], working in Wickford, Essex, North London, UK, provided a detailed description of the findings in the brains of 15 elderly retired boxers. The gross findings were not surprising and included old contusions, widespread atrophy with enlarged ventricles and thinning of the corpus callosum, and cavum septum pellucidum. Neuronal loss and gliosis were identified on microscopic exam, as expected in the affected areas. Somewhat less expected was the observed depigmentation (loss of the normal black color) of the substantia nigra. However, the main surprise was the finding of abundant

D. G. Munoz (✉)
Keenan Research Centre for Biomedical Research, The Li Ka Shing Knowledge Institute, and Division of Pathology, St. Michael's Hospital, Toronto, ON, Canada

Department of Laboratory Medicine and Pathobiology, University of Toronto, Toronto, ON, Canada
e-mail: David.Munoz@unityhealth.to

I. R. Mackenzie
Department of Pathology, Vancouver General Hospital, Vancouver, BC, Canada

Department of Pathology and Laboratory Medicine, University of British Columbia, Vancouver, BC, Canada
e-mail: ian.mackenzie@vch.ca

T. A. Schweizer, A. J. Baker (eds.), *Tackling the Concussion Epidemic*,
https://doi.org/10.1007/978-3-030-93813-0_11

neurofibrillary tangles in the hippocampus, basal temporal cortex, and also the frontal and lateral neocortices. Neurofibrillary tangles are intracytoplasmic flame-shaped inclusions in neurons, originally demonstrated using silver stains. They were known to occur in Alzheimer's disease and other neurodegenerative diseases, including progressive supranuclear palsy, post-encephalitic parkinsonism, and peculiar parkinsonism dementia complex affecting the native population of the island of Guam. The detailed morphological studies of Patrick Hof, a Swiss immigrant to the USA (first San Diego and then the New York City), and his colleagues established in 1992 that the distribution of the neurofibrillary tangles in dementia pugilistica was different from that of Alzheimer's disease [5]. The tangles predominantly affect the deep layers of the cortex in Alzheimer's disease, whereas the superficial layers were preferentially involved in dementia pugilistica, a pattern similar to that seen in post-encephalitic parkinsonism and Guam parkinsonism dementia. The discovery that the microtubule-associated protein tau was the main component of the neurofibrillary tangles of Alzheimer's disease opened the door to the use of immunohistochemistry, which provided a much greater resolution for the study of distribution. Using tau immunohistochemistry, Geddes and colleagues made three additional critical observations: the clustering of neurofibrillary tangles around blood vessels; the predominant involvement of the depth of sulci; and the fact that involvement of the cerebral neocortex preceded the involvement of the hippocampus [6, 7].

During this period, it became clear that the condition was not limited to professional boxers, as it was described in a woman who was beaten frequently for much of her life [8], as well as a young woman with head-banging behaviors secondary to autism [9]. Nevertheless, CTE remained an obscure medical condition until public interest was awakened by the finding of the condition in former players in the National Football League reported by Omalu and his colleagues [10, 11]. However, it was not until the careful systematic studies of the Boston neuropathologist Ann McKee starting in 2009 that an understanding of CTE as a distinct entity could develop [12]. A critical tool for the advance was the systematic use of whole hemibrain sections pioneered by Heiko and Eva Braak [13], which combined with free-floating immunostaining of sections much thicker than the 5 μm histological standard allowed the visual detection of the distribution of lesions.

At present, CTE is a condition defined by its neuropathology in combination with a history of repeated head trauma. The neurological manifestations are diverse. In this chapter, we will examine first the features of the neuropathology that characterize CTE, followed by the associated clinical presentations and the possible contribution of cerebral trauma to common neurodegenerative conditions.

Neuropathology of Chronic Traumatic Encephalopathy

The contemporary concept of CTE is based on the abnormalities detected at the light microscopic level, in contrast to the gross abnormalities emphasized in the original reports of Corsellis et al. in former boxers [4]. The critical finding of CTE is the deposition of insoluble tau in neurons and astrocytes in specific patterns. The tau proteins in neurons consist of a mixture of four repeat and three repeat isoforms (4R and 3R tau, respectively) similar to that seen in Alzheimer's disease [14]. Such mixture of 3R/4R tau isoforms is also observed in several other conditions, including parkinsonism dementia of Guam, but not in other diseases with neurofibrillary tangles such as progressive supranuclear palsy or cortical basal degeneration, (where 4R tau predominates) or classical Pick's disease (3R tau). Detailed analysis of posttranslational modifications of tau has shown not only several additional similarities with Alzheimer's disease but also some differences, including the abundance of the tauC3 epitope, corresponding to tau truncated at D421 [15], and a novel tau filament fold [16]. In contrast to the neuronal tau, a marked predominance of 4R isoforms is present in the astrocytic tau deposits in CTE, a finding common to all conditions with glial tau deposits, including multiple forms of frontotemporal lobar degeneration, as well as aging-related tau astrogliopathy (ARTAG) [17, 18].

The picture that emerges from these observations is that the distinct features of CTE as currently understood are found neither in the gross pathology nor in the molecular characteristic of the proteins involved, but in the distribution of the lesions. The tau deposits take the morphology of neurofibrillary tangles accompanied by neuritic threads and dots, and astrocytic tangles (Fig. 11.1). A study

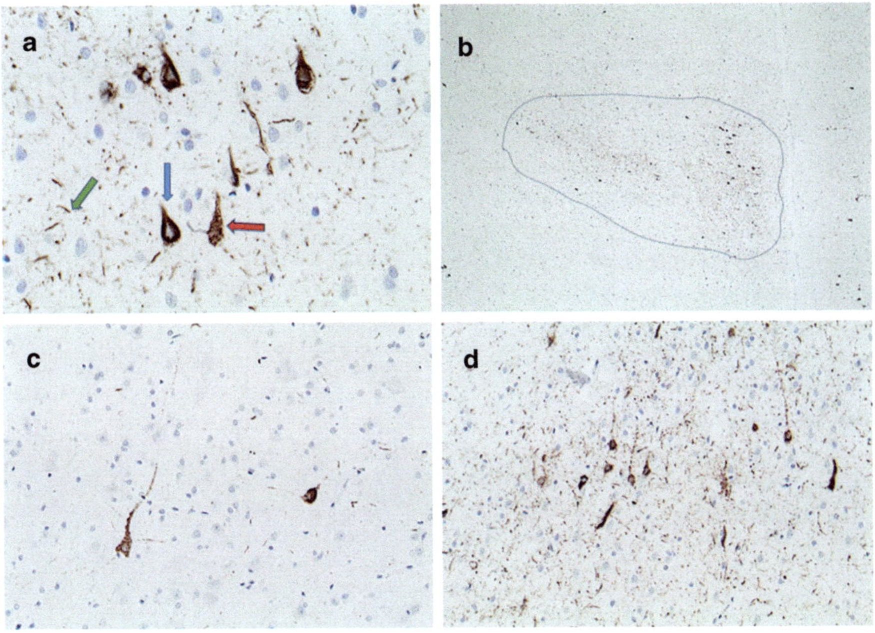

Fig. 11.1 Cerebral cortex of a patient with CTE immunostains for tau protein using antibody AT8. (**a**) The main lesions demonstrated are neurofibrillary tangles (blue arrow), pre-tangles (red arrow), and dystrophic neurites (green arrow). (**b**) The distribution of lesions is patchy. In the low power, an area of high density has been outlined by the blue dotted line. (**c, d**) Under medium microscope power, an area of low lesion density (**c**) is seen in contrast to an area of high lesion density (**d**)

involving 68 cases established the hierarchical pattern of development of lesions, following the approach initially used by Braak and Braak for the staging of Alzheimer's disease [19]. The lesions of CTE develop first at the depth of sulci in

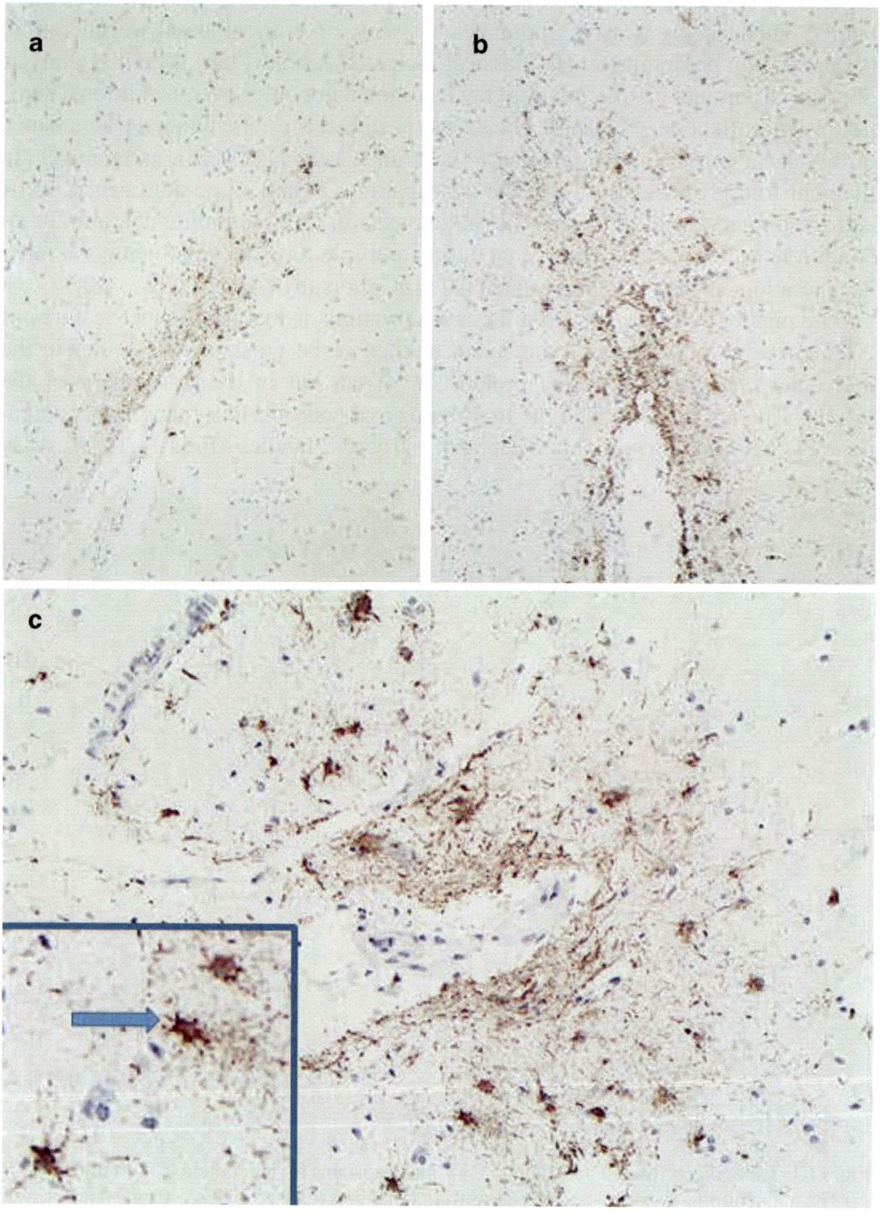

Fig. 11.2 The bottoms of sulci are a major point of accumulation of lesions (**a** and **b**). Under high power (**c**) many of the immunostained cells can be identified as astrocytes by their radiating processes morphology (inset, blue arrow)

the cerebral cortex, usually including the frontal, parietal, or temporal, but not the occipital, lobes (Fig. 11.2a, b). Such neocortical involvement in the absence of hippocampal pathology is in sharp contrast with the usual development of tau pathology in aging and Alzheimer's disease, where the hippocampus and entorhinal cortex are affected well in advance of the neocortex. The cortical lesions first appear at the depth of sulci and are centered by small blood vessels (stage I). They may be accompanied by separate glial clusters of thorny astrocytes [20] (Fig. 11.2c). In stage II not only are three or more foci present as described above, but also neurofibrillary tangles are located both along sulcal walls and at the crest of gyri, always occupying the superficial cortical laminae. In addition, neurofibrillary tangles are observed in the nucleus basalis of Meynert [21], just ventral to the globus pallidus and the anterior commissure, and the noradrenergic locus ceruleus in the dorsal pons. Stage III is characterized by the confluence of cortical lesions accompanied by the appearance of neurofibrillary tangles in the hippocampus, in a pattern somewhat different from that of aging and Alzheimer's disease, in that the CA4 and CA2 sectors are involved along with the CA1 sector [22] (Fig. 11.3). Other medial temporal lobe structures, including the entorhinal cortex and the amygdala, also become affected. Neurofibrillary tangles also appear in the olfactory bulbs, the dopaminergic substantia nigra, and the serotoninergic medial raphe. Finally, in stage IV, the neuronal and glial tau pathology is widespread and dense throughout the cerebral cortex, and neurofibrillary tangles appear in other locations, including the thalamus, dentate nucleus of the cerebellum, and spinal cord (Fig. 11.4).

Other pathologies develop concurrently with the tauopathy. Axonal degeneration is manifested by varicosities demonstrable with neurofilament stains and found in subcortical and deep white matter, as well as the deep layers of the cerebral cortex. This is accompanied by myelin pallor and degeneration [23]. In many severe cases of CTE, TDP 43, the protein associated with most cases of amyotrophic lateral sclerosis (ALS) and approximately half of the cases of frontotemporal dementia, is present in the form of deposits in the cytoplasm of neurons and glial cells, and both dot-like and thread-like neurites involving the cerebral cortex, hippocampus, diencephalon, brainstem, and even spinal cord [24]. The inclusions in the granular cell layer of the dentate gyrus in the hippocampus, as well as the preferential involvement in layer 2 of the cerebral cortex, are reminiscent of the distribution of lesions in the subtypes of frontotemporal lobar degeneration associated with TDP 43 [25].

Amyloid deposits in the form of diffuse plaques are not part of the early stages of disease. While they have been observed in later stages, it is not completely clear to what degree amyloid deposits occur as part of the normal aging process [26]. As could be expected, they are more commonly observed in carriers of the APOE E4 allele. In any case, their presence is associated with the increased development of both tauopathy and Lewy body formation.

It is only in the severe stages (III and IV) that gross pathological features become evident. The most obvious ones are enlargement of the lateral ventricles and third ventricle, with cavum septum pellucidum. This is accompanied by relatively subtle cerebral atrophy predominantly affecting the frontal and temporal lobes. The other distinctive lesion is the depigmentation of the substantia nigra. Atrophy of the mammillary bodies is another recognizable feature.

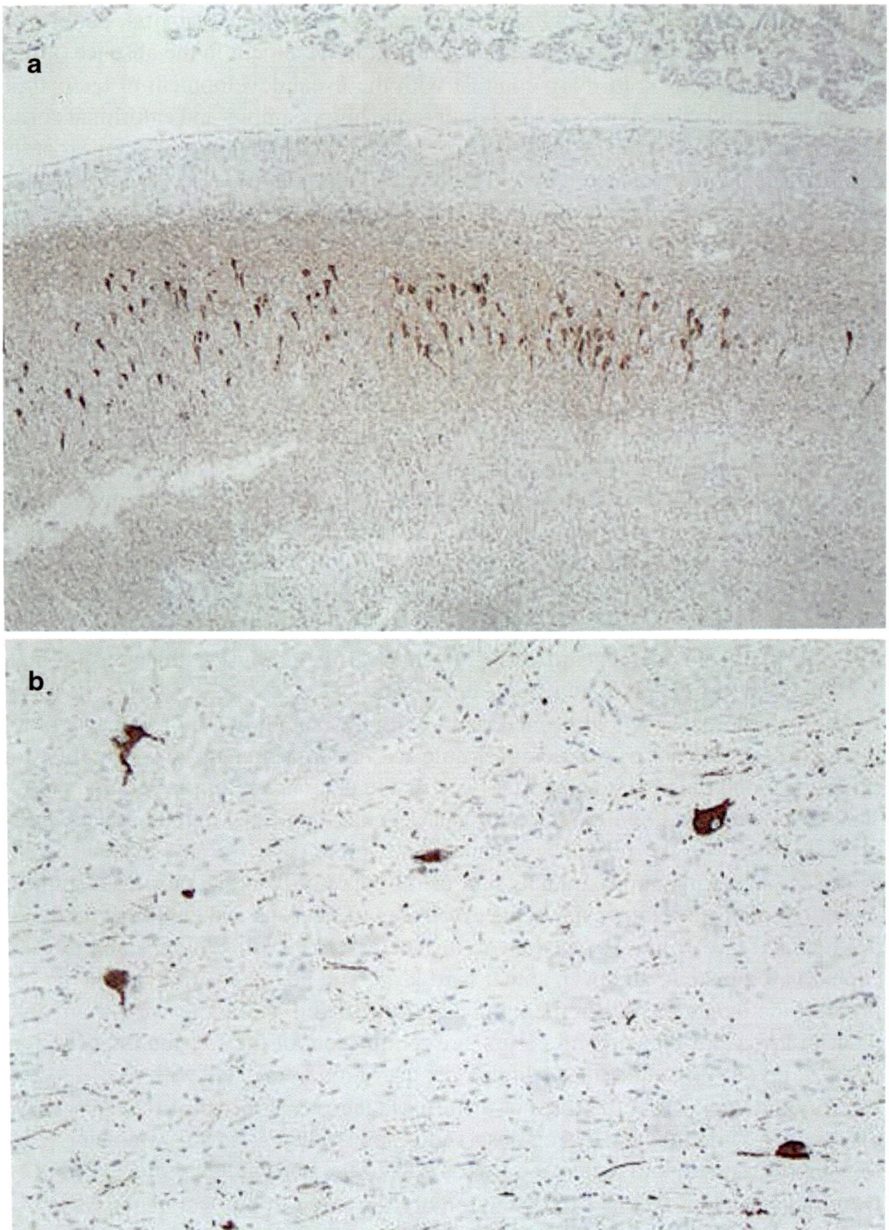

Fig. 11.3 (**a**) Tau immunostain of the hippocampus, demonstrating numerous neurofibrillary tangles in the CA2 sector. (**b**) Subcortical neurofibrillary tangles are identified by tau immunostaining in the oculomotor nucleus in the mid brain

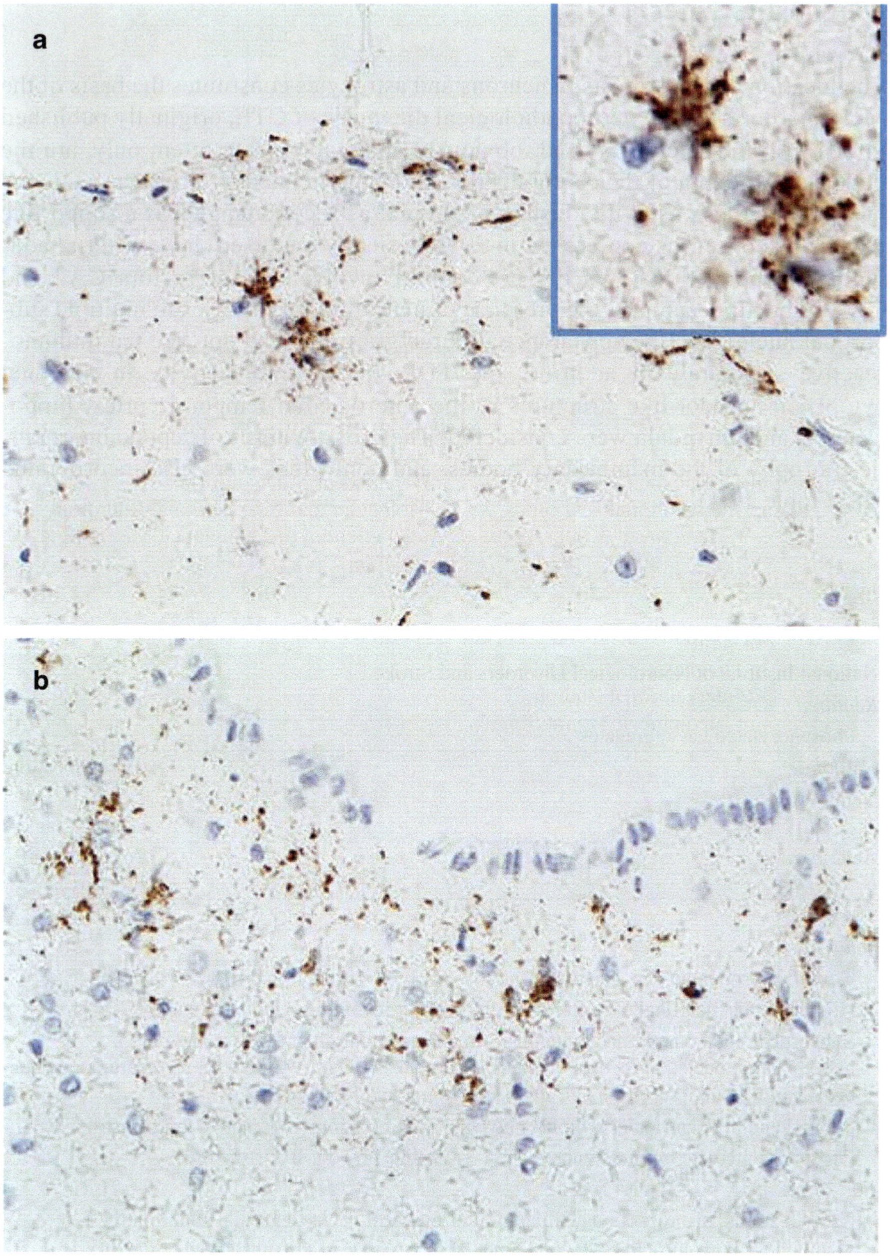

Fig. 11.4 Subpial (**a**) and subependymal (**b**) accumulation of tau immunoreactivity material, predominantly in astrocytes, best seen at higher magnification as shown in the inset

Diagnostic Criteria

The location of tau deposits in neurons and astrocytes constitutes the basis of the McKee criteria for the neuropathological diagnosis of CTE, originally published in 2013 [27]. The criteria used, obviously applicable postmortem only, are the perivascular, depth of the cerebral sulci, and superficial layer preferential distribution of the lesions, with clusters of subpial astrocytic tangles as a supportive feature. A consensus conference in 2016 basically endorsed these criteria, adding as supportive features the preferential involvement of sectors CA2 and CA4 in the hippocampus, the presence of neurofibrillary tangles in multiple subcortical nuclei, and the appearance of large grain-like and dot-like tau immunoreactive structures. In addition, the TDP 43 immunoreactivity in neuronal cytoplasm and dot-like structures in the anteromedial temporal cortex, hippocampus, and amygdala were considered. The gross features of septal abnormalities, atrophy of the mammillary bodies, and contusions were also incorporated [28] (Table 11.1).

Table 11.1 Criteria for the pathological diagnosis of CTE

National Institute of Neurological Disorders and Stroke	
Required	
Phosphorylated tau aggregates	
Cell types	Tissue location
Neurons	Depth of cortical sulci
Astrocytes	Around small blood vessels
Supportive	
Tau related	
Cortex: neuronal tau deposits preferentially involving layers II and III	
Hippocampus: neuronal tau deposits preferentially involving sectors CA2 and CA4	
Subcortical nuclei: neuronal tau deposits involving mammillary bodies and other hypothalamic nuclei, amygdala, nucleus accumbens, thalamus, midbrain tegmentum, nucleus basalis of Meynert, raphe nuclei, substantia nigra, and locus coeruleus	
Subpial and periventricular regions: phosphorylated tau immunoreactive thorny astrocytes	
Phosphorylated tau immunoreactive large grain-like and dot-like structures	
Non-tau related	
Gross features: disproportionate dilatation of the third ventricle, septal abnormalities, mammillary body atrophy, and contusions or other signs of previous traumatic injury	
Immunohistochemistry: TDP-43 immunoreactive neuronal cytoplasmic inclusions and dot-like structures in the hippocampus, anteromedial temporal cortex, and amygdala	

Clinical Syndrome

At this point there are no defined criteria for the clinical diagnosis of the manifestations of CTE. Traumatic encephalopathy syndrome has been the name proposed for this condition, for which only glimpses can be obtained by retrospective examination of the records of subjects who received a diagnosis of CTE at autopsy. However, this is a highly biased sample, constituted mostly by former athletes with a history of neurological impairment and the absence of comparable CTE negative cohort. With this caveat, the following picture emerges [27, 29, 30]. Most subjects have a history of repeated brain trauma, including concussions and sub-concussive injuries, in which consciousness was not lost. There is usually a long silent period of multiple years between the cessation of brain trauma and the appearance of the first symptoms. Characteristically this includes mood disturbances, such as depression and anxiety, and behavioral symptoms, such as poor emotional control, impulsivity, aggressiveness, and paranoia, which precede any cognitive changes. The latter affect functions traditionally associated with the frontal lobes, including not only executive function and processing speed but also memory. Motor symptoms occur in a subset of cases, and include extrapyramidal abnormalities, including parkinsonism, and a syndrome resembling amyotrophic lateral sclerosis, characterized by evidence of a combination of upper motor neuron (spasticity) and lower motor neuron (atrophy and fasciculations), resulting in severe weakness. The order of presentation is variable: in approximately two-thirds of cases, the amyotrophic lateral sclerosis symptoms precede mood, behavioral, or cognitive changes, whereas in the remaining third, the order is reversed.

The correlation between pathological findings and clinical symptoms has been studied by a few researchers [29]. In an examination of a neurodegenerative disorders brain bank, almost one-third of the individuals with a history of repeated brain trauma showed CTE, whereas not a single case was discovered among 198 individuals without a history of repeated brain trauma. However, in a study focused on multisystem atrophy, CTE was found in slightly more than 5% of the brains, but only half of them had a history of repeated brain trauma [31]. On the other hand, at least 10% of individuals with pathological evidence of early CTE (up to stage III) are asymptomatic [30]. Efforts are underway to obtain a diagnosis of CTE during life, using approaches similar to those utilized in Alzheimer's disease [32, 33]. One ongoing clinical trial utilizes cerebral tau imaging using positron emission tomography (*Tau Imaging of Chronic Traumatic Encephalopathy – Full Text View – ClinicalTrials.gov*, no date), and another one a combination of structural and functional imaging with biomarkers in cerebrospinal fluid, blood, and saliva (*The DIAGNOSE-CTE Research Project – Full Text View – ClinicalTrials.gov*, no date). In both trials, the subjects are current and former football players and controls, and detailed classical neurological exams with enhancements such as gait and ocular movement quantitative analysis are included. Presumably the physical signs will be less susceptible to confounding factors, such as the athletes' early retirement with the accompanying loss of privileged status, frequent lack of emphasis on education, and drug use. These studies should also be able to determine whether the traumatic encephalopathy syndrome is progressive and if so at what average rate.

Risk Factors

Contributing factors, other than repeated head trauma, must be important in determining who will develop CTE, since only some of the subjects receiving comparable amounts of head trauma, as in team sports, go on to develop any evidence of the disease. These factors are presently unknown, but most likely include genetic susceptibility and environmental influences. The question of whether a single severe head trauma can result in CTE is not completely resolved, but appears unlikely. In a pathological study of 33 individuals with a history of a single head injury, no cases of CTE were identified [34].

Mechanisms

Three mechanisms have emerged as likely contributors to the development of CTE: axonal injury, disruption of blood–brain barrier, and inflammation. The axonal injury, manifested by the presence of axonal swellings as described above, must be considered [23], but there are no clinical or experimental paradigms for the induction of tau deposition in response to cutting or crashing nerve fibers.

Disruption of the blood–brain barrier is suggested by the perivascular distribution of the lesions. Pathological evidence is provided by deposits of immunoglobulin G and fibrinogen in the cerebral cortex, predominantly in the perivascular distribution, observed in approximately half of the subjects after a single head injury years before death [35]. In live subjects, diffusion tensor image has provided evidence of blood–brain barrier disruption in college American football players, even without a history of concussion [36].

Inflammation associated with CTE is manifested by changes in the density and morphology of microglial cells, without infiltration by circulating inflammatory cells [37]. In addition, the cerebral tissue levels of cytokine, CCL 11, are elevated in CTE, but not in Alzheimer's disease. Moreover, the levels appear elevated in cerebrospinal fluid, raising the possibility of developing a biomarker for in vivo diagnosis of chronic traumatic encephalopathy [38].

Comorbidities

CTE's association with other neurodegenerative diseases increases with age and include Alzheimer's disease, Lewy body disease, frontotemporal degeneration, and amyotrophic lateral sclerosis. The critical question is whether these comorbidities are secondary to CTE, due to a shared risk factor (e.g., repeated head trauma) or simply coincidence.

The presence of beta amyloid diffuse plaques is considered compatible with the diagnosis of pure CTE, but the finding of a moderate to severe burden of neuritic plaques implies the additional diagnosis of Alzheimer's disease. In the presence of Alzheimer disease, it may not be possible to reach a diagnosis of CTE in the cerebral cortex because the Alzheimer tauopathy would obscure the specific locations of tau deposition in CTE (the depth of sulci and perivascular regions). In these cases, the diagnosis of CTE may still be possible by demonstrating tau accumulation in the form of neurofibrillary tangles in regions unaffected by Alzheimer's disease, such as the pulvinar nucleus, mammillary bodies, superior colliculi, basis pontis, and cerebellar dentate nucleus. Two important unresolved questions are whether CTE increases beta amyloid deposition exceeding normal age-specific prevalence and extent and whether repeated head injury may induce both CTE and, in susceptible individuals, Alzheimer's disease.

The association of repeated head injury with parkinsonism has been repeatedly documented [39, 40] and dramatically illustrated in the case of Muhammad Ali, who received a clinical diagnosis of Parkinson's disease 3 years after he retired, when he was a worldwide celebrity. How much of this clinical manifestation is due to deposition of alpha synuclein in the form of Lewy bodies as opposed to the subcortical tauopathy described earlier is less clear. There is some evidence that traumatic brain injury is associated with the formation of Lewy bodies, possibly in relation to the development of CTE pathology [41].

Finally, a syndrome consistent with amyotrophic lateral sclerosis, dominated by weakness and the combination of lower motor neuron (atrophy, fasciculations) and upper motor neuron (spasticity) signs, develops in 6–12% of subjects with CTE. These cases often have more severe TDP 43 pathology. The sequence of development of symptoms is not fixed; in some patients motor symptoms developed years before cognitive and behavioral symptoms typical of frontotemporal dementia, whereas in others the pattern is reversed.

Overall, the consensus is that, in addition to tau, other proteinopathies often contribute to the neurological deterioration in CTE [42].

Experimental Models

A mouse model of CTE has been described with the most important feature being progression after cessation of trauma. In this model, local deposits of insoluble tau appear locally within 24 hours of trauma but spread to distant sites by 6 months, a substantial proportion of a mouse life span. Thus, the mechanism of spread can be considered the central question for future research. Prion-like template spread has been experimentally demonstrated in a number of separate disease models, but alternative processes may be operative, including increased production of tau or its phosphorylated forms, as well as reduced clearing [43].

Conclusion

CTE has been established as the pathological substrate (or at least one of the substrates) of the neurological syndrome of progressive cognitive, behavioral, and motor deterioration developing years after suffering repeated head trauma. Although most commonly observed in professional athletes, it can also occur in other groups, including the military [44] and the victims of domestic abuse [8].

A major unresolved issue is how common CTE is in the general population and in particular among those participating in contact sports at the varsity or recreational level. CTE provides a unique insight into neurodegenerative disease because it is one of the few situations where the nature and timing of the environmental causative factors is known, and there is a long interval between injury and appearance of the lesions, providing excellent opportunities for further research.

References

1. Martland HS. Punch drunk. J Am Med Assoc. 1928;91(15):1103–7. https://doi.org/10.1001/jama.1928.02700150029009.
2. Millspaugh J. Dementia pugilistica. US Naval Med Bull. 1937;35:297–303.
3. Critchley M. Medical aspects of boxing, particularly from a neurological standpoint. Br Med J. 1957;1(5015):357. https://doi.org/10.1136/bmj.1.5015.357.
4. Corsellis JAN, Bruton CJ, Freeman-Browne D. The aftermath of boxing. Psychol Med. 1973;3(3):270–303. https://doi.org/10.1017/S0033291700049588.
5. Hof PR, et al. Differential distribution of neurofibrillary tangles in the cerebral cortex of dementia pugilistica and Alzheimer's disease cases. Acta Neuropathol. 1992;85(1):23–30. https://doi.org/10.1007/bf00304630.
6. Geddes JF, et al. Neurofibrillary tangles, but not Alzheimer-type pathology, in a young boxer. Neuropathol Appl Neurobiol. 1996;22(1):12–6. Available at: http://www.ncbi.nlm.nih.gov/pubmed/8866777. Accessed 29 Jan 2020.
7. Geddes JF, et al. Neuronal cytoskeletal changes are an early consequence of repetitive head injury. Acta Neuropathol. 1999;98(2):171–8. https://doi.org/10.1007/s004010051066.
8. Roberts GW, et al. Dementia in a punch-drunk wife. Lancet. 1990;335(8694):918–9. https://doi.org/10.1016/0140-6736(90)90520-F.
9. Hof PR, et al. Neuropathological observations in a case of autism presenting with self-injury behavior. Acta Neuropathol. 1991;82(4):321–6. https://doi.org/10.1007/BF00308819.
10. Omalu BI, et al. Chronic traumatic encephalopathy in a National Football League player. Neurosurgery. 2005;57(1):128–33. https://doi.org/10.1227/01.NEU.0000163407.92769.ED.
11. Omalu BI, et al. Chronic traumatic encephalopathy in a National Football League player: Part II. Neurosurgery. 2006;59(5):1086–92. https://doi.org/10.1227/01.NEU.0000245601.69451.27.
12. McKee AC, et al. Chronic traumatic encephalopathy in athletes: progressive tauopathy after repetitive head injury. J Neuropathol Exp Neurol. 2009;68(7):709–35. https://doi.org/10.1097/NEN.0b013e3181a9d503.
13. Braak H, Braak E. Demonstration of amyloid deposits and neurofibrillary changes in whole brain sections. Brain Pathol. 1991;1(3):213–6. https://doi.org/10.1111/j.1750-3639.1991.tb00661.x.
14. Schmidt ML, et al. Tau isoform profile and phosphorylation state in dementia pugilistica recapitulate Alzheimer's disease. Acta Neuropathol. 2001;101(5):518–24. https://doi.org/10.1007/s004010000330.

15. Kanaan NM, et al. Characterization of early pathological tau conformations and phosphorylation in chronic traumatic encephalopathy. J Neuropathol Exp Neurol. 2016;75(1):19–34. https://doi.org/10.1093/jnen/nlv001.
16. Falcon B, et al. Novel tau filament fold in chronic traumatic encephalopathy encloses hydrophobic molecules. Nature. 2019;568(7752):420–3. https://doi.org/10.1038/s41586-019-1026-5.
17. Kovacs GG, et al. Aging-related tau astrogliopathy (ARTAG): harmonized evaluation strategy. Acta Neuropathol. 2016;131(1) https://doi.org/10.1007/s00401-015-1509-x.
18. Kovacs GG, Lee VM, Trojanowski JQ. Protein astrogliopathies in human neurodegenerative diseases and aging. Brain Pathol. 2017;27(5):675–90. https://doi.org/10.1111/bpa.12536.
19. Braak H, Braak E. Neuropathological stageing of Alzheimer-related changes. Acta Neuropathol. 1991;82(4):239–59. https://doi.org/10.1007/BF00308809.
20. Hsu ET, et al. Astrocytic degeneration in chronic traumatic encephalopathy. Acta Neuropathol. 2018;136(6):955–72. https://doi.org/10.1007/s00401-018-1902-3.
21. Mufson EJ, et al. Progression of tau pathology within cholinergic nucleus basalis neurons in chronic traumatic encephalopathy: a chronic effects of neurotrauma consortium study. Brain Inj. 2016;30(12):1399–413. https://doi.org/10.1080/02699052.2016.1219058.
22. Kelley CM, Perez SE, Mufson EJ. Tau pathology in the medial temporal lobe of athletes with chronic traumatic encephalopathy: a chronic effects of neurotrauma consortium study. Acta Neuropathol Commun. 2019;7(1):207. https://doi.org/10.1186/s40478-019-0861-9.
23. Holleran L, et al. Axonal disruption in white matter underlying cortical sulcus tau pathology in chronic traumatic encephalopathy. Acta Neuropathol. 2017;133(3):367–80. https://doi.org/10.1007/s00401-017-1686-x.
24. McKee AC, et al. TDP-43 proteinopathy and motor neuron disease in chronic traumatic encephalopathy. J Neuropathol Exp Neurol. 2010;69(9):918–29. https://doi.org/10.1097/NEN.0b013e3181ee7d85.
25. Mackenzie IR, Neumann M. Reappraisal of TDP-43 pathology in FTLD-U subtypes. Acta Neuropathol. 2017;134(1):79–96. https://doi.org/10.1007/s00401-017-1716-8.
26. Stein TD, et al. Beta-amyloid deposition in chronic traumatic encephalopathy. Acta Neuropathol. 2015;130(1):21–34. https://doi.org/10.1007/s00401-015-1435-y.
27. McKee AC, et al. The spectrum of disease in chronic traumatic encephalopathy. Brain. Oxford University Press. 2013;136(1):43–64. https://doi.org/10.1093/brain/aws307.
28. McKee AC, et al. The first NINDS/NIBIB consensus meeting to define neuropathological criteria for the diagnosis of chronic traumatic encephalopathy. Acta Neuropathol. 2016;131(1):75–86. https://doi.org/10.1007/s00401-015-1515-z.
29. Stern RA, et al. Clinical presentation of chronic traumatic encephalopathy. Neurology. 2013;81(13):1122–9. https://doi.org/10.1212/WNL.0b013e3182a55f7f.
30. Mez J, et al. Clinicopathological evaluation of chronic traumatic encephalopathy in players of American football. JAMA. 2017;318(4):360–70. https://doi.org/10.1001/jama.2017.8334.
31. Koga S, Dickson DW, Bieniek KF. Chronic traumatic encephalopathy pathology in multiple system atrophy. J Neuropathol Exp Neurol. 2016;75(10):963–70. https://doi.org/10.1093/jnen/nlw073.
32. Small GW, et al. PET scanning of brain tau in retired national football league players: preliminary findings. Am J Geriatr Psychiatr. 2013;21(2):138–44. https://doi.org/10.1016/j.jagp.2012.11.019.
33. Barrio JR, et al. In vivo characterization of chronic traumatic encephalopathy using [F-18]FDDNP PET brain imaging. Proc Natl Acad Sci U S A. National Academy of Sciences. 2015;112(16):E2039–47. https://doi.org/10.1073/pnas.1409952112.
34. Bieniek KF, et al. Chronic traumatic encephalopathy pathology in a neurodegenerative disorders brain bank. Acta Neuropathol. 2015;130(6):877–89. https://doi.org/10.1007/s00401-015-1502-4.
35. Hay JR, et al. Blood-brain barrier disruption is an early event that may persist for many years after traumatic brain injury in humans. J Neuropathol Exp Neurol. 2015;74(12):1147–57. https://doi.org/10.1097/NEN.0000000000000261.

36. Marchi N, et al. Consequences of repeated blood-brain barrier disruption in football players. PLoS One. 2013;8(3):e56805. https://doi.org/10.1371/journal.pone.0056805.
37. Cherry JD, et al. Microglial neuroinflammation contributes to tau accumulation in chronic traumatic encephalopathy. Acta Neuropathol Commun. 2016;4(1):112. https://doi.org/10.1186/s40478-016-0382-8.
38. Cherry JD, et al. CCL11 is increased in the CNS in chronic traumatic encephalopathy but not in Alzheimer's disease. PLoS One. 2017;12(9):e0185541. https://doi.org/10.1371/journal.pone.0185541.
39. Lee P-C, et al. Traumatic brain injury, paraquat exposure, and their relationship to Parkinson disease. Neurology. 2012;79(20):2061–6. https://doi.org/10.1212/WNL.0b013e3182749f28.
40. Lee PC, et al. Head injury, α-synuclein genetic variability and Parkinson's disease. Eur J Neurol. 2015;22(5):874–8. https://doi.org/10.1111/ene.12585.
41. Adams JW, et al. Lewy body pathology and chronic traumatic encephalopathy associated with contact sports. J Neuropathol Exp Neurol. 2018;77(9):757–68. https://doi.org/10.1093/jnen/nly065.
42. Kenney K, et al. Dementia after moderate-severe traumatic brain injury: coexistence of multiple proteinopathies. J Neuropathol Exp Neurol. 2018;77(1):50–63. https://doi.org/10.1093/jnen/nlx101.
43. Tagge CA, et al. Concussion, microvascular injury, and early tauopathy in young athletes after impact head injury and an impact concussion mouse model. Brain J Neurol. 2018;141(2):422–58. https://doi.org/10.1093/brain/awx350.
44. Goldstein LE, et al. Chronic traumatic encephalopathy in blast-exposed military veterans and a blast neurotrauma mouse model. Sci Transl Med. 2012;4(134):134ra60. https://doi.org/10.1126/scitranslmed.3003716.

Index

T. A. Schweizer, A. J. Baker (eds.), *Tackling the Concussion Epidemic*,
https://doi.org/10.1007/978-3-030-93813-0

M

N

P